PROGRESS TOWARD RESTORING THE EVERGLADES

The Second Biennial Review - 2008

Committee on Independent Scientific Review of Everglades Restoration Progress

Water Science and Technology Board

Board on Environmental Studies and Toxicology

Division on Earth and Life Studies

NATIONAL RESEARCH COUNCIL
OF THE NATIONAL ACADEMIES

THE NATIONAL ACADEMIES PRESS
Washington, D.C.
www.nap.edu

THE NATIONAL ACADEMIES PRESS 500 Fifth Street, N.W. Washington, D.C. 20001

NOTICE: The project that is the subject of this report was approved by the Governing Board of the National Research Council, whose members are drawn from the councils of the National Academy of Sciences, the National Academy of Engineering, and the Institute of Medicine. The members of the panel responsible for the report were chosen for their special competences and with regard for appropriate balance.

This report was produced under assistance of Cooperative Agreement No. W912EP-04-2-0001 with the Department of the Army. Support for this project was provided by the U.S. Department of the Interior and the South Florida Water Management District. Any opinions, findings, conclusions, or recommendations expressed in this publication are those of the author(s) and do not necessarily reflect the views of the organizations or agencies that provided support for the project.

International Standard Book Number-13: 978-0-309-12574-1
International Standard Book Number-10: 0-309-12574-X

Additional copies of this report are available from the National Academies Press, 500 Fifth Street, N.W., Lockbox 285, Washington, D.C. 20055; (800) 624-6242 or (202) 334-3313 (in the Washington metropolitan area); Internet, http://www.nap.edu.

Printed in the United States of America

THE NATIONAL ACADEMIES

Advisers to the Nation on Science, Engineering, and Medicine

The **National Academy of Sciences** is a private, nonprofit, self-perpetuating society of distinguished scholars engaged in scientific and engineering research, dedicated to the furtherance of science and technology and to their use for the general welfare. Upon the authority of the charter granted to it by the Congress in 1863, the Academy has a mandate that requires it to advise the federal government on scientific and technical matters. Dr. Ralph J. Cicerone is president of the National Academy of Sciences.

The **National Academy of Engineering** was established in 1964, under the charter of the National Academy of Sciences, as a parallel organization of outstanding engineers. It is autonomous in its administration and in the selection of its members, sharing with the National Academy of Sciences the responsibility for advising the federal government. The National Academy of Engineering also sponsors engineering programs aimed at meeting national needs, encourages education and research, and recognizes the superior achievements of engineers. Dr. Charles M. Vest is president of the National Academy of Engineering.

The **Institute of Medicine** was established in 1970 by the National Academy of Sciences to secure the services of eminent members of appropriate professions in the examination of policy matters pertaining to the health of the public. The Institute acts under the responsibility given to the National Academy of Sciences by its congressional charter to be an adviser to the federal government and, upon its own initiative, to identify issues of medical care, research, and education. Dr. Harvey V. Fineberg is president of the Institute of Medicine.

The **National Research Council** was organized by the National Academy of Sciences in 1916 to associate the broad community of science and technology with the Academy's purposes of furthering knowledge and advising the federal government. Functioning in accordance with general policies determined by the Academy, the Council has become the principal operating agency of both the National Academy of Sciences and the National Academy of Engineering in providing services to the government, the public, and the scientific and engineering communities. The Council is administered jointly by both Academies and the Institute of Medicine. Dr. Ralph J. Cicerone and Dr. Charles M. Vest are chair and vice chair, respectively, of the National Research Council.

www.national-academies.org

COMMITTEE ON INDEPENDENT SCIENTIFIC REVIEW OF EVERGLADES RESTORATION PROGRESS[1]

WILLIAM L. GRAF, *Chair*, University of South Carolina, Columbia
STEVEN R. BEISSINGER, University of California, Berkeley
LINDA K. BLUM, University of Virginia, Charlottesville
DONALD F. BOESCH, University of Maryland, Center for Environmental Science, Cambridge
FRANK W. DAVIS, University of California, Santa Barbara
CHARLES T. DRISCOLL, Syracuse University, New York
JOAN G. EHRENFELD, Rutgers University, New Brunswick, New Jersey
CHRIS T. HENDRICKSON, Carnegie Mellon University, Pittsburgh, Pennsylvania
WILLIAM P. HORN, Birch, Horton, Bittner, and Cherot, Washington, DC
WAYNE C. HUBER, Oregon State University, Corvallis
DAVID H. MOREAU, University of North Carolina, Chapel Hill
JEAN-YVES PARLANGE, Cornell University, Ithaca, New York
K. RAMESH REDDY, University of Florida, Gainesville

NRC Staff

STEPHANIE E. JOHNSON, Study Director, Water Science and Technology Board
DAVID J. POLICANSKY, Scholar, Board on Environmental Studies and Toxicology
DOROTHY K. WEIR, Research Associate, Water Science and Technology Board

[1]The activities of this committee were overseen and supported by the National Research Council's Water Science and Technology Board and Board on Environmental Studies and Toxicology (see Appendix I for listing). Biographical information on committee members and staff is contained in Appendix J. Note: William G. Boggess, Oregon State University, served on the committee until October 2007, when he resigned for personal reasons.

Acknowledgments

Many individuals assisted the committee and the National Research Council staff in their task to create this report. We would like to express our appreciation to the following people who have provided presentations to the committee and served as guides during the field trips:

Kenneth Ammon, South Florida Water Management District
Stuart Appelbaum, U.S. Army Corps of Engineers
Nicholas Aumen, National Park Service
Laura Brandt, U.S. Fish and Wildlife Service
Edwin Brown, U.S. Army Corps of Engineers
Eric Bush, U.S. Army Corps of Engineers
Wanda Caffie-Simpson, South Florida Water Management District
Doug Chaltry, U.S. Fish and Wildlife Service
Major Dominic Ciaramitaro, U.S. Army Corps of Engineers
Carlos Coronado-Molina, South Florida Water Management District
Orlando Diaz, University of Florida
Peter Doering, South Florida Water Management District
Dennis Duke, U.S. Army Corps of Engineers (formerly)
Vic Engel, National Park Service
Sharon Ewe, Florida International University
Donald Fox, Florida Fish and Wildlife Conservation Commission
Jack Gentile, Harwell Gentile & Associates
Lawrence Gerry, South Florida Water Management District
Andrew Gottlieb, South Florida Water Management District
Chris Graham, U.S. Army Corps of Engineers
Paul Gray, Audubon
Susan Gray, South Florida Water Management District
David Hallac, National Park Service
Gary Hardesty, U.S. Army Corps of Engineers

Matthew Harwell, U.S. Fish and Wildlife Service
James Heaney, University of Florida
Lorraine Heisler, U.S. Fish and Wildlife Service
Eliza Hines, Everglades Partners Joint Venture
Todd Hopkins, U.S. Fish and Wildlife Service
Jim Jackson, South Florida Water Management District
Thomas James, South Florida Water Management District
Robert Johnson, National Park Service
Ray Judah, Lee County Commission
Wiley Kitchens, University of Florida
Greg Knecht, Florida Department of Environmental Protection
Kevin Kotun, National Park Service
Beth Lewis, South Florida Water Management District
Julie Lockwood, Rutgers University
Joette Lorion, Miccosukee Tribe of Indians
Linda McCarthy, Florida Department of Agriculture and Consumer Services
Carol Mitchell, National Park Service
Matthew Morrison, South Florida Water Management District
Jana Newman, South Florida Water Management District
Jayantha Obeysekera, South Florida Water Management District
John Ogden, Audubon, former South Florida Water Management District
Jose Otero, South Florida Water Management District
Pam Repp, U.S. Fish and Wildlife Service
Terry Rice, Miccosukee Tribe of Indians
Terrence "Rock" Salt, U.S. Department of the Interior
Leonard Shabman, Resources for the Future
Bruce Sharfstein, South Florida Water Management District
Patti Sime, South Florida Water Management District
Fred Sklar, South Florida Water Management District
Paul Souza, U.S. Fish and Wildlife Service
Hilary Swain, Archbold Biological Station
Susan Sylvester, South Florida Water Management District
Kimberley Taplin, U.S. Army Corps of Engineers
Thomas Teets, South Florida Water Management District
David Tipple, U.S. Army Corps of Engineers
Paul Trimble, South Florida Water Management District
Shelley Trulock, U.S. Army Corps of Engineers
Tom Van Lent, Everglades Foundation
Elmar von Kurzbach, U.S. Army Corps of Engineers

Greta von Unruh, Economic Development Research Institute of Palm Beach County
Paul Warner, South Florida Water Management District
Rebecca Weiss, U.S. Army Corps of Engineers
Tori White, U.S. Army Corps of Engineers
Eric Wood, Princeton University
Dewey Worth, South Florida Water Management District

Preface

In 1881 Hamilton Disston, a Philadelphia investor, began a grand project in the Everglades wilderness of Florida to drain the wetlands and convert them to an agricultural cornucopia. The Everglades once encompassed about 3 million acres, with its "River of Grass" extending southward from the area north of Lake Okeechobee to a sweeping confluence with Florida Bay at the southern end of the Florida peninsula. Disston's project in the northern reaches of the Everglades eventually failed, but "reclamation" efforts continued. When Napoleon Bonaparte Broward became governor of Florida in 1904, he initiated a massive investment and development plan that began the wholesale modification of the Everglades for agriculture with water supply and flood control for the growing cities along the coastal margins. During this early period, environmental protectionists like Frank M. Chapman of the American Museum of Natural History worked tirelessly to protect endangered birds and their habitats. By the end of the 20th century, more than half the Everglades had disappeared, and the remainder was an ecosystem in rapid decline. In 1999, the federal and state governments combined their efforts in the Comprehensive Everglades Restoration Plan (CERP) to save the remaining Everglades along with their iconic wildlife, while at the same time providing water and flood protection for the region's rapidly increasing human population.

The CERP is a complex, multibillion-dollar project managed by the U.S. Army Corps of Engineers (USACE) and the South Florida Water Management District (SFWMD) that was projected to require 40 years for completion. With 68 separate subprojects requiring sophisticated scientific knowledge of the ecosystem and creation of new technologies for water management, CERP represents a research, planning, implementation, and construction challenge unlike any other. In authorizing the CERP, the U.S. Congress mandated periodic independent reviews of progress restoring the natural system in the Everglades. In compliance with this requirement, the USAC, in coordination with the SFWMD and the Department of the Interior, arranged with the National Research Council

(NRC) of the National Academies the establishment of the Committee on Independent Scientific Review of Everglades Restoration Progress (CISRERP), which submits formal reports to Congress on a biennial basis.

The NRC has previously reviewed (for the South Florida Ecosystem Restoration Task Force) such specific aspects of the Everglades restoration as the management of science for decision making, general science and engineering perspectives on water storage, and the management of science for particular parts of the ecosystem such as Florida Bay. The CISRERP reviews for Congress, however, are more all encompassing, and they provide a broad picture of both science and engineering and the contributions of these endeavors to restoration. These more general reviews cannot touch upon every aspect of the overall project, so exploration of some representative examples supplements the general statements in the reports. The committee provided its first biennial report in 2006, examining the initiation of the CERP with its emphasis on planning, identifying embryonic progress in projects related to the CERP, specifying that there were no scientific impediments that should stand in the way of restoration progress, and offering a philosophic approach to managing science and restoration.

This second biennial report continues the NRC review of Everglades restoration progress. During this exacting process, I have been privileged to work with committee members who are among the nation's leading experts in their respective fields. The committee members served without compensation (except for expenses), and they have generously contributed their time and talents as their donations in service to the state and the nation. The committee includes experts in biological, hydrologic, and geographic sciences, hydrologic and systems engineering, project administration, law, and policy. The committee met seven times over the course of 18 months, with five meetings in Florida that permitted the committee to hear testimony from researchers, planners, and decision makers associated with the USACE and SFWMD, as well as from representatives of interest groups and private citizens. The report generated by this diverse committee is a consensus document.

In late June 2008, after the committee had completed its deliberations and was about to send its report for external review, the state of Florida announced its intention to enter into negotiations to acquire almost 300 square miles of the Everglades Agricultural Area from U.S. Sugar Corporation. Given the timing of the announcement late in the committee's reporting cycle, the committee was unable to assess the implications of the land purchase for the CERP in any detail in this report. The purchase of these lands could have some important implications for water quality and possibly water storage for the Everglades, and the committee does draw attention to these in appropriate places in the report, but these issues will undoubtedly be analyzed in greater detail in future biennial reviews.

The committee could not have accomplished its task without the help of the numerous NRC staff members associated with this review, including Stephen Parker (Director of the Water Science and Technology Board). His broad vision and effective management style have been keys to our success. Three staff members in particular were our partners in this effort: Stephanie Johnson, David Policansky, and Dorothy Weir. Stephanie Johnson is a true Everglades expert whose outstanding knowledge and understanding of the science, engineering, and administrative aspects of the CERP suffuse this report. Her encyclopedic capabilities to find information, absorb its essence, analyze its implications, and write about its consequences have been a key to the committee's success. David Policansky has long been a partner of committees engaged in Everglades oversight and review, applying his extensive biological knowledge and sound scientific sense. His service with this review committee and his contributions to the reporting process exemplify his fine ability to tease out the nuances in what is one of the most complicated ecosystems and restorations that any of us has ever seen. Dorothy Weir made it possible for the committee to do its job, adroitly managing every meeting: from the preliminary planning, through the management of minute procedural details, to the concluding summary processes. Her assistance in creating the final report has been, simply, indispensable.

This report has been reviewed in draft form by individuals chosen for their diverse perspectives and technical expertise in accordance with procedures approved by the NRC's Report Review Committee. The purpose of this independent review is to provide candid and critical comments that will assist the institution in making its published report as sound as possible and to ensure that the report meets institutional standards for objectivity, evidence, and responsiveness to the study charge. The review comments and draft manuscript remain confidential to protect the integrity of the deliberative process. We wish to thank the following individuals for their review of this report: Jean M. Bahr, University of Wisconsin-Madison; Patrick L. Brezonik, University of Minnesota; Elvin R. "Vald" Heiberg III, independent consultant; Judith L. Meyer, University of Georgia; Leonard Shabman, Resources for the Future; Alan D. Steinman, Annis Water Resources Institute; Myron F. Uman, former Associate Executive Officer, National Research Council; Thomas Van Lent, The Everglades Foundation; and Jeffrey R. Walters, Virginia Polytechnic Institute and State University.

Although the reviewers listed above have provided many constructive comments and suggestions, they were not asked to endorse the conclusions or recommendations nor did they see the final draft of the report before its release. The review of this report was overseen by Frank H. Stillinger, Princeton University, and Kenneth W. Potter, University of Wisconsin, Madison. Appointed by the NRC, they were responsible for making certain that an independent examination of this report was carried out in accordance with institutional procedures and that all review comments were carefully considered. Responsibility for the final content of this report rests entirely with the authoring committee and the institution.

Hamilton Disston, Napoleon Bonaparte Broward, and Frank M. Chapman would not recognize today's Florida. Nevertheless, many of those developers' dreams have been realized in hydrologic control systems of canals, ditches, levees, control structures, and pumps, and they would have approved of the productive agriculture and bustling cities of the region. The preservationists have succeeded in establishing sprawling refuges and a national park. Disston, Broward, and Chapman likely would be amazed that the state and the nation have committed themselves to restoring and maintaining substantial parts of the natural system while at the same time providing ecosystem services for the human population. But the three were big thinkers, and in adapting to the present-day goals of combined environmental quality and economic development, they would probably approve of the CERP: bold, challenging, and complex, but with great potential for public good. We offer this report as our contribution to the realization of that lofty goal for the Everglades.

William L. Graf
Chair

Contents

Summary

The Florida Everglades, uniquely shaped by the slow flow of water, is one of the world's treasured ecosystems. However, an extensive water control infrastructure, designed to increase regional economic productivity through improved flood control, urban water supply, and agricultural production, has changed the landscape of South Florida. The vast area of sawgrass plains, ridges, sloughs, and tree islands once supported a high diversity of plant and animal life, but remnants of the original Everglades now compete for vital water with urban and agricultural interests, and contaminated runoff from these two activities impairs their waters. The Comprehensive Everglades Restoration Plan (CERP), a joint effort led by the state and the federal government launched in 2000, seeks to reverse the general decline of the ecosystem in the midst of a changing human and environmental context. This unprecedented project envisioned the expenditure of billions of dollars in a multi-decadal effort to achieve ecological restoration by restoring the hydrologic characteristics of the Everglades, where feasible, and to create a water system that simultaneously serves the needs of the natural and the human systems of South Florida.

The Committee on Independent Scientific Review of Everglades Restoration Progress was established in 2004 in response to a request from the U.S. Army Corps of Engineers (USACE), with support from the South Florida Water Management District (SFWMD) and the U.S. Department of the Interior (DOI), based on Congress's mandate in the Water Resources Development Act of 2000 (WRDA 2000). The committee is charged to submit biennial reports that review the CERP's progress in restoring the natural system (see Chapter 1). This is the committee's second report in a series of biennial evaluations.

The committee concludes that the CERP is bogged down in budgeting, planning, and procedural matters and is making only scant progress toward achieving restoration goals. Meanwhile, the ecosystems that the CERP is intended to save are in peril, construction costs are escalating, and population growth and associated development increasingly make accomplishing the goals of the CERP more difficult. Lack of timely restoration progress by the CERP, to date, has been largely due to the complex federal planning process and the need to resolve conflicts among agencies and stakeholders. However, future restoration progress is likely to be limited by the availability of funding and the current authorization and funding mechanisms. In periods of restricted funding and limited capability to move forward on many fronts, the ability to set priorities and implement them is critical. Much good science has been developed to support the restoration efforts, and the foundations of adaptive management have been established to support the CERP. To avert further declines, CERP planners should address major project planning and authorization hurdles and move forward expeditiously with projects that have the most potential for contributing to natural system restoration progress in the South Florida ecosystem.

SOUTH FLORIDA ECOSYSTEM RESTORATION

Several South Florida restoration programs, including the CERP—the largest of the initiatives—are now under way. The CERP, led by the USACE and the SFWMD, consists primarily of projects to increase storage capacity (e.g., conventional surface-water reservoirs, aquifer storage and recovery, in-ground reservoirs), improve water quality (e.g., stormwater treatment areas [STAs]), reduce loss of water from the system (e.g., seepage management, water reuse, conservation), and reestablish pre-drainage hydrologic patterns wherever possible (e.g., removing barriers to sheet flow, rainfall-driven water management). The largest portion of the budget is devoted to water storage and conservation and to acquiring the lands needed for those projects.

The CERP builds upon other activities of the state and the federal government aimed at restoration (hereafter, non-CERP activities), many of which are essential to the success of the CERP in achieving its restoration goals. These include Modified Water Deliveries to Everglades National Park (Mod Waters) and the Kissimmee River Restoration—projects that will alter hydrologic patterns to more closely resemble pre-drainage conditions. Several non-CERP projects address water quality issues, including the Everglades Construction Project (construction of over 44,000 acres of STAs) and restoration of Lake Okeechobee. In addition, research on and management of invasive species is important to the overall restoration program. Finally, the state of Florida's Acceler8 initiative is a mix of expedited projects that were identified in the CERP and some non-CERP projects. In Chapter 2 of the report, the committee analyzes the broader context for the South Florida ecosystem restoration efforts and presents the following conclusions and recommendations:

Population growth and associated development will make restoration more difficult. Increasing water demands from an expanding population in Florida could create competition with ecosystem restoration when supplies are limited. Agriculture faces an uncertain future in South Florida, particularly in the Everglades Agricultural Area, which intervenes directly in the flow of water between Lake Okeechobee and Everglades National Park and influences the movement of water, sediment, and nutrients for the rest of the system. The use of "smart growth" principles that integrate the needs of environmental restoration with human demographic changes can lessen the negative impacts of population growth if cities, counties, the state, and CERP planners are all involved.

Human-induced climate change is likely to impact the effectiveness of CERP projects, and CERP planners should assess and factor into planning and implementation the most recent projections of the impacts of climate change in South Florida. Precipitation, evapotranspiration, and the intensity of rainfall events in South Florida are all expected to change during the current century. Impending climate change should not be an excuse for delay or inaction in the restoration but instead provides further motivation to restore the resilience of the ecosystem. The CERP Guidance Memorandum on climate change recommends consideration of sea-level rise and changes in precipitation quantity, distribution, and evapotranspiration in all CERP planning, but new analysis of impacts based on assumptions about higher sea-level rise are needed. Among those possible changes that should be assessed and factored into planning and implementation are: changes in the water budget, including increasing human demands for water; changes in the return frequency and intensity of hurricanes; the effects of climate change on the distribution of biota in the Everglades ecosystem; and impacts of projected sea-level rise on the hydro-geomorphology of the estuaries and the mangrove zone.

Ongoing delay in South Florida ecosystem restoration not only has postponed improvements to the hydrologic condition but also has allowed ecological decline to continue. Recent water management strategies have not produced conditions that are conducive to restoring the Cape Sable seaside sparrow and appear to be negatively impacting the snail kite. Tree islands have undergone a multi-decadal decline in both number and surface area—a trend that appears likely to continue until significant CERP and non-CERP restoration progress has been made. In the past decade, Lake Okeechobee has experienced continued water quality and habitat degradation. Meanwhile, the number and area of influence of invasive species are increasing and represent very real challenges to restoration efforts.

In the face of these numerous challenges, Everglades restoration efforts are even more essential to improve the condition of the South Florida eco-

system and strengthen its resiliency as it faces additional stresses in the future. If ecological resilience is not restored, the possibility exists that environmental changes could precipitate rapid and deleterious state changes that might be very difficult or impossible to reverse. Unless near-term progress is achieved on major restoration initiatives, including CERP and non-CERP efforts, opportunities for restoration may close with further loss of species numbers and habitat deterioration, and the Everglades ecosystem may experience irreversible losses to its character and functioning.

PROGRESS IN PROGRAM IMPLEMENTATION: BUILDING THE FOUNDATION FOR ADAPTIVE MANAGEMENT

The initial National Research Council (NRC) biennial review of restoration progress noted that in the first 6 years after the WRDA 2000 was authorized, actual construction progress was limited, and most of the CERP accomplishments were programmatic. In 2008, most CERP accomplishments remain programmatic, including the monitoring and assessment plan, development of modeling tools, and other ways in which the foundations of adaptive management are being built in support of the restoration. Congress mandated an adaptive management approach for the CERP to facilitate restoration progress despite some scientific and engineering uncertainty, and as of 2008, nearly all of the elements needed to implement a decision-making framework using adaptive management have been produced (see Chapter 6). These elements include:

- Documents describing the adaptive management process and all aspects of performance assessment,
- Conceptual ecological models to support monitoring and assessment, and
- An information and data management system and the Interagency Modeling Center to support assessment and planning aspects of decision making.

These are significant accomplishments, and their importance should not be underestimated; however, the CERP adaptive management scheme could be improved by addressing several major issues, which are summarized in the text that follows.

For monitoring and assessment information to adequately support CERP adaptive management, a robust program of ecological monitoring should remain a priority. While monitoring in and of itself does not ensure restoration progress, without monitoring to understand ecosystem response to project implementation from local to whole ecosystem scales, uninformed management decisions will be made with potentially undesirable ecosystem consequences. A well-justified and documented set of performance measures has been developed, and a scientifically robust process for updating, refining, and adding to the set of performance measures is in place. The periodic review of performance measures should consider ways to make sure that the total number of variables monitored is appropriate to their purposes for informing decisions and to the funding available for monitoring efforts. It also is important to match the frequency of monitoring with the speed of change of the variables that are monitored and to increase reliance on remotely sensed data-collection methods. Revisions of the monitoring and assessment system should be firmly grounded in the use of the data for planning and management decision making.

The 2007 System Status Report achieved its stated objectives to test the monitoring and assessment plan and to establish as long a baseline as possible to capture the natural variance of CERP performance measures. The first System Status Report serves as the reference that will be used to gauge system response as CERP projects are implemented, and it is extremely valuable. Insights learned during the production of the report should be incorporated into the revision of the Monitoring and Assessment Plan (MAP) and the conceptual ecological models, as needed, and for re-prioritization of the performance measures. To maximize the usefulness of System Status Reports for adaptive management, the interagency body called Restoration, Coordination, and Verification (RECOVER) should develop succinct summaries in future reports that clearly address whether the interim and longer-term goals are being met; if not, why; and what CERP operations or design changes are most likely to move ecosystem response closer to the interim goals.

Integrated hydrologic, ecological, and water quality modeling tools are needed for science to have a fully developed role in CERP decision making and ecosystem management. CERP planning and assessment of performance indicators are dependent on modeling tools; as model development and implementation lag, so does access to more accurate and functional tools. Models are needed for each ecological indicator (performance measures) to compare predicted and monitored indicator responses for effective adaptive management decision making. This will occur only when

- ecological modeling and data management activities are fully incorporated and funded in the CERP's Interagency Modeling Center;
- water quality and sediment transport models become routinely available and integrated with the new Regional Simulation Model; and
- these physical-chemical models can be readily linked to ecological models.

Shrinking CERP resources means that the trade-off between using staff for model development versus using them for model production runs for CERP planning favors the latter. This committee recognizes that resources are limited but notes that model development is a long-term proposition and should continue with as much support as possible so the tools required to restore and manage the ecosystem are available in the future.

CERP PROJECT PLANNING AND IMPLEMENTATION

The attempt to restore an ecosystem as large and complex as the Everglades is an unprecedented challenge. Despite programmatic accomplishments and the beginning of construction for some projects identified in the CERP, natural system restoration has been delayed. The South Florida ecosystem continues to suffer as a result of a complex and sometimes contentious planning process, funding uncertainties, lack of clear restoration priorities that are central to restoration, and statutory and regulatory impediments. In Chapter 3, the committee analyzes progress in CERP planning and implementation and makes the following conclusions and recommendations:

It is too early to evaluate the response of the ecosystem to CERP projects because none have been completed. Construction completion for the first CERP components has not been achieved through mid-2008, and key foundational pre-CERP projects, such as Mod Waters, remain far behind schedule. If limited natural system restoration progress continues, frustration will further increase among stakeholders and agency staff, and public support for restoration is likely to diminish. Actual construction and implementation of key non-CERP and CERP projects are the only means to arrest the degradation and to assure that natural system restoration begins. State efforts to construct projects in spite of funding limitations and other serious obstacles to progress are commendable. Some partial benefits have been produced from phased construction in the Picayune Strand Restoration (wetland restoration) and Acme Basin B (stormwater treatment) projects. Additionally, several non-CERP activities are positive harbingers of future CERP programs and indicate that when project implementation does occur, bona fide ecological restoration benefits will be demonstrated. For example, the success of the Kissimmee River Restoration effort continues to be the most important piece of evidence that restoration of a natural system is possible in the Everglades region.

The state of Florida should continue its active land acquisition efforts, accompanied by monitoring of and regular reporting on land conversion patterns in the South Florida ecosystem. Land management for a successful CERP depends on purchasing particular sites within the project area and protecting more general areas within the South Florida ecosystem that could help meet the broad restoration goals. The committee commends the state of Florida for its aggressive and effective financial support for acquiring important parcels, including the impressive recent announcement that the state will enter into negotiations for the potential purchase of 187,000 acres of land in the EAA from U.S. Sugar Corporation for $1.75 billion. The acquisition of this large amount of land has the potential to alter basic CERP plans, but because of uncertainty in the timing and structure of the purchase and the possibility of numerous land exchanges made after the purchase, direct effects of the deal are impossible to predict and may not be seen for a decade or more.

The complex project planning and approval process has been a major cause of delays for CERP projects to date. The greatest challenge in the project planning process has been developing technically sound project plans that are acceptable to the many agencies and stakeholders involved. The process of resolving disagreements among agencies and stakeholders has led to lengthy delays in the development of some project implementation reports that can be submitted to Congress for authorization. The infrequent and unpredictable federal authorization mechanism for CERP projects has caused some additional problems and attendant delays. The committee judges that the lack of federal funding in the first eight years of the CERP is not the most serious cause of the CERP delays. Instead, the slow pace of federal funding has largely been a symptom of the problems caused by the complex and lengthy CERP planning and authorization process for each project. However, now that three CERP projects have approval for their project implementation reports and congressional authorization, funding limitations will certainly create additional constraints to CERP progress in the years ahead. Non-CERP and CERP projects will increasingly compete for limited state and federal funding, while project costs increase due to inflationary pressures and scope changes. Both state and federal partners are facing budget constraints, and dramatic state budget cuts in FY09 threaten to affect the speed of restoration progress.

Deficiencies in CERP system-wide planning are affecting the delivery of natural system restoration benefits. The CERP lacks a systematic approach to analyze the costs and benefits across multiple projects in support of project planning. Fundamentally, the CERP is designed as a system of related projects (i.e., components) that work together in the aggregate to produce overall restoration benefits. Without a system-wide planning process, it is not clear how system benefits can be optimized for any one project without any systematic consideration of other projects. The next added increment is a benefits evaluation method that considers benefits only from the proposed and previously authorized projects and, as currently implemented in the Everglades, it undermines system-wide restoration planning and sequencing. The current planning process also appears to reward the least contentious projects, regardless of their potential contribution to ecosystem restoration. Without clear priorities for project planning and funding, projects with large potential restoration benefits may see lengthy restoration delays while other, less-contentious projects that address only isolated portions of the ecosystem may tie up available funding. During the 5-year review of the Programmatic Regulations, the USACE should address deficiencies and impediments in the CERP planning process that are affecting restoration progress. CERP planners should also develop mechanisms to improve system-wide planning and decision making for the CERP.

Developing a realistic schedule and sound project sequence is a critical need for the restoration effort. In this time of increasing fiscal pressures, it is critical that CERP planners find a means to prioritize and properly sequence restoration projects so that public funds are allocated by the degree to which the projects are essential to restoration of the South Florida ecosystem, rather than by local stakeholder support or the order of authorization. Public Web-based reporting on project progress, delays, and anticipated completion dates should be more transparent than it is currently.

The executive and legislative branches of the federal government should consider departing from traditional project-by-project review, authorization, and yearly funding to benefit both the CERP and other multi-component ecosystem restoration projects across the nation. It may be far more efficacious—scientifically, managerially, and economically—to design a different approach for comprehensive restoration programs that provides assured funding over a multiple-year period.

The incremental adaptive restoration (IAR) concept proposed in the initial NRC biennial review has stimulated creative restoration approaches to Everglades restoration but has not yet been fully applied. The prior committee's recommendation to apply IAR has been widely embraced by implementing agencies at all levels of organization as well as by various stakeholders, but an effort to apply IAR to an integrated group of Southern Everglades restoration projects was discontinued. CERP planners, however, are using the IAR concept in planning the Biscayne Bay Coastal Wetlands and C-111 Spreader Canal projects. The most effective applications of the IAR concept will probably be in the incremental execution of project components that produce significant outcomes but are of a scope and scale that can be feasibly implemented and assessed. Because most of the desired ecological changes are likely to take years or decades to respond to IAR actions, agencies should emphasize assessing variables that are leading indicators of likely long-term ecological responses as they develop IAR strategies.

To reduce restoration delays, CERP planners should develop a stronger conceptual basis for multi-species recovery planning and management. Although implementation of the Endangered Species Act (ESA) has increasingly become focused on single species management, the statute does provide various mechanisms that can reduce the threat of legitimate litigation and facilitate the recovery and management of multiple-listed species. However, effective multi-species management under the ESA requires a high level of integration of scientific knowledge about individual species and species interactions to understand risks and trade-offs during construction and under alternative water management regimes. It also requires strong federal leadership and a high level of trust and cooperation among the regulatory and management agencies and other stakeholders to allow for learning, compromise, and decision making under uncertainty. In addition, jeopardy determinations for endangered species and associated litigation are a significant, unresolved challenge for adaptive management and IAR. Currently, there is no scientifically credible operational plan for managing multiple species at risk in South Florida. To expedite multi-species restoration under the ESA, the DOI should immediately initiate and lead the development of a South Florida multi-species adaptive management strategy, including both science and policy dimensions, to accompany the existing South Florida Multi-species Recovery Plan.

CASE STUDY ANALYSES OF RESTORATION PROGRESS

The committee evaluated two restoration efforts in detail—Mod Waters and Lake Okeechobee—to better understand the progress and challenges in the restoration of the South Florida ecosystem.

Modified Water Deliveries to Everglades National Park

The history of the Mod Waters project is one of the most discouraging stories in Everglades restoration (see Chapter 4). The project, which would provide crucial first steps toward ecological restoration within Everglades National Park, has been plagued by changes in direction and scope, parochial interests, debilitating litigation, enormous cost escalation due both to inflation and to plan modifications, unanticipated engineering constraints (e.g., Tamiami Trail integrity), and lack of coordinated leadership from the responsible agencies. How the project will be funded (i.e., involving the National Park Service, USACE, Florida Department of Transportation) is a further complicating factor. While some events may have been unavoidable, the overall outcome has been the loss of support from Congress—the ultimate source of funding for the project—and the loss of enthusiasm—or even understanding—from the public. Worst of all, the history of delay further damages Everglades National Park. Completion of Mod Waters is crucial to the success of Everglades restoration and the CERP projects that follow. If this relatively modest restoration project cannot proceed and provide some restoration benefits, the outlook for the CERP is dismal.

Without completion of Mod Waters, central components of the CERP cannot proceed, and ecological conditions in the Everglades ecosystem will continue to deteriorate. Nineteen years have passed since the Mod Waters project was authorized, and the restoration of water flows has not occurred, even though it is a critical foundation project for the CERP. Political leadership and the timely provision of funding are essential if progress on Mod Waters and the associated delivery of restoration benefits to Everglades National Park are going to occur.

Strong leadership, focused on building and maintaining support among stakeholders and overcoming conflicts, is essential for Everglades restoration projects to achieve their restoration goals. If there is insufficient political leadership to align research, planning, funding, and management with restoration goals agreed upon by the stakeholders, the CERP will be likely to result in an abbreviated series of disconnected projects that ultimately fail to meet the restoration goals. Other lessons for the CERP that can be learned from the struggles faced during the planning and implementation of the Mod Waters project include benefits of early agreement on project scope and objectives, the need for a clear project management structure, and the need to anticipate adapting project plans over time.

The reduced scope of Mod Waters attainable with the 2008 recommended plan for modifying Tamiami Trail (alternative 3.2.2.a) provides some environmental benefits but shifts increased responsibility (and cost) to the CERP to achieve authorized Mod Waters goals. The 2008 recommended plan represents a substantially smaller step toward restoration than was originally envisioned for Mod Waters. The recommended alternative also is less cost-effective than other alternatives when benefits are considered as habitat units per dollar spent. Although it is critical to move ahead and implement it quickly, the recommended alternative should be viewed only as a first step toward restoration. Moreover, it should be recognized that moving forward with the 2008 recommended plan increases the urgency to proceed more quickly to implement the additional necessary Tamiami Trail modifications through the CERP or some other mechanism, so that the restoration benefits for Everglades National Park outlined in the WRDA 2007 conference report can be achieved as soon as possible.

Lake Okeechobee

Lake Okeechobee is a critical linchpin of the South Florida ecosystem. However, both high and, more recently, very low water levels, as well as poor water quality, presently plague the lake. The challenges of water quality and water quantity in the lake have two critical ramifications for the entire ecosystem: the lake supports important elements of the region's biota, and the lake has the potential to serve as a major source of water storage and water supply for downstream ecosystems, a potential that will become more critical if other planned and proposed sources of water storage do not become available. Based on an analysis of Lake Okeechobee's condition and current restoration plans (see Chapter 5), the committee presents the following conclusions and recommendations:

An integrated, system-wide view of water quality management is essential to the achievement of restoration goals for the South Florida ecosystem. Good data are available to understand the local dynamics of phosphorus and other contaminants, but a system-wide accounting is lacking for water and phosphorus as well as other important contaminants, such as sulfur, mercury, and nitrogen. A system-wide accounting is needed to determine the mechanisms of contaminant transport, to assess the implications of upstream changes on downstream habitats, to determine appropriate management actions, and to evaluate system-wide progress to improve water quality. It also is crucial to determine to what degree the current status of the lake represents a changed condition that will resist restoration.

Recent water quality restoration initiatives in the Northern Everglades are not likely to achieve the stated water quality goals (40 ppb total phosphorus in the lake and 140 metric tons per year phosphorus input load) by the year 2015, and it might take decades for these goals to be met using current strategies. Using the "no-action alternative" to manage internal phosphorus loads in the lake is likely to delay achieving in-lake concentration goals by several decades, as concluded by the SFWMD. Also, although the Northern Everglades initiative's technical plan identifies management measures to reduce phosphorus loads, the strategies probably are not adequate to reduce external phosphorus loads sufficiently. More significant remediation strategies in the lake and its watershed will probably be needed to reduce the legacy phosphorus in the system and meet the stated goal.

Although the Northern Everglades plan represents a sizable effort, it will not be easy or inexpensive to reverse the lake's decline in water quality. The lake's importance in the ecosystem, however, justifies the devotion of considerable resources to the lake.

In the near term, restoration planners should consider the consequences of the likely failure to achieve the phosphorus goals on the South Florida ecosystem restoration and develop alternative approaches. Alternatives may involve significant reallocation of priorities among restoration projects and/or significant changes to water quality criteria downstream. Restoration planners should consider the needs for additional STAs and should investigate methods to improve the long-term ability of STAs to remove phosphorus. In-lake treatment of phosphorus may also be needed to expedite the rehabilitation of Lake Okeechobee as external loads are reduced.

Given concerns about the financial and technical feasibility of aquifer storage and recovery (ASR) at the large scale proposed in the CERP, additional opportunities for water storage should be investigated, and Lake Okeechobee may be an important component of those alternatives. Several important water storage projects are under development through the CERP and Acceler8, and opportunities for upstream water storage are being considered within the Northern Everglades initiative. Nevertheless, alternative storage options should be considered as possible contingencies to ASR—the primary source of new water storage for the CERP, but for which there are concerns about financial and technical feasibility—including synergistic opportunities related to modifications of the Herbert Hoover Dike. This committee encourages CERP planners to consider a wide array of water storage alternatives and their costs and benefits.

Short-term and long-term trade-offs will be necessary in the rehabilitation of Lake Okeechobee and northern estuaries. Given the current altered state of the whole system, goals for the lake, the northern estuaries, and other downstream interests might not be mutually compatible in all respects. As a result, trade-offs will have to be made. Modeling and adequate, reliable data will be needed to evaluate these trade-offs.

OVERALL EVALUATION OF PROGRESS AND CHALLENGES

If the sweeping vision of environmental restoration of the Everglades is to be realized, demonstrable progress needs to come soon. Even though the science and engineering that support the restoration program have been of high quality, to date, the CERP has not been effective in halting the decline of the South Florida ecosystem. Instead, the CERP is currently mired in a complex federal planning and approval process, while project costs continue to rise and development threatens to foreclose some restoration options, and funding limitations are likely to add further delays in the years ahead. To do nothing is to do harm. If the CERP continues on its present course, at its current pace, the system will continue to lose some of its vital parts, and more importantly, the restoration effort will lose the support of the public at large. Clear funding priorities, modifications to the project planning, authorization, and funding process, and strong political leadership are needed to support system-wide restoration and to begin to reverse the decades of decline.

1

Introduction

The Florida Everglades is one of the world's treasured ecosystems (Davis and Ogden, 1994). Uniquely shaped by the slow flow of water, its vast terrain of sawgrass plains, ridges, sloughs, and tree islands used to support a high diversity of plant and animal life. This natural landscape also served as a sanctuary for Native Americans and subsequently as a vast storehouse of natural resources that fueled the region's development. Large-scale economic development and urbanization, however, diminished the natural resources, and by the mid- to late-20th century, many of the area's defining natural characteristics had been lost. The remnants of the original Everglades (see Figure 1-1 and Box 1-1) now compete for vital water with urban and agricultural interests, and contaminated runoff from these two activities impairs their waters.

Recognition of past declines in environmental quality, combined with continuing threats to the natural character of the remaining Everglades, led to initiation of the Comprehensive Everglades Restoration Plan (CERP) in the late 1990s. This unprecedented project envisioned the expenditure of billions of dollars in a multi-decadal effort to achieve ecological restoration by restoring the hydrologic characteristics of the Everglades, where feasible, and to create a water system that simultaneously serves the needs of both the natural and the human systems of South Florida. Within this 21st-century social, economic, and political latticework, the restoration of the South Florida ecosystem is now under way, representing one of the most ambitious ecosystem renewal projects ever conceived. This report represents the second independent assessment of the progress of the CERP by the Committee on Independent Scientific Review of Everglades Restoration Progress (CISRERP) of the National Research Council (NRC).

THE NATIONAL RESEARCH COUNCIL AND EVERGLADES RESTORATION

The NRC has been providing scientific and technical advice related to the Everglades restoration since 1999. The NRC's Committee on the Restora-

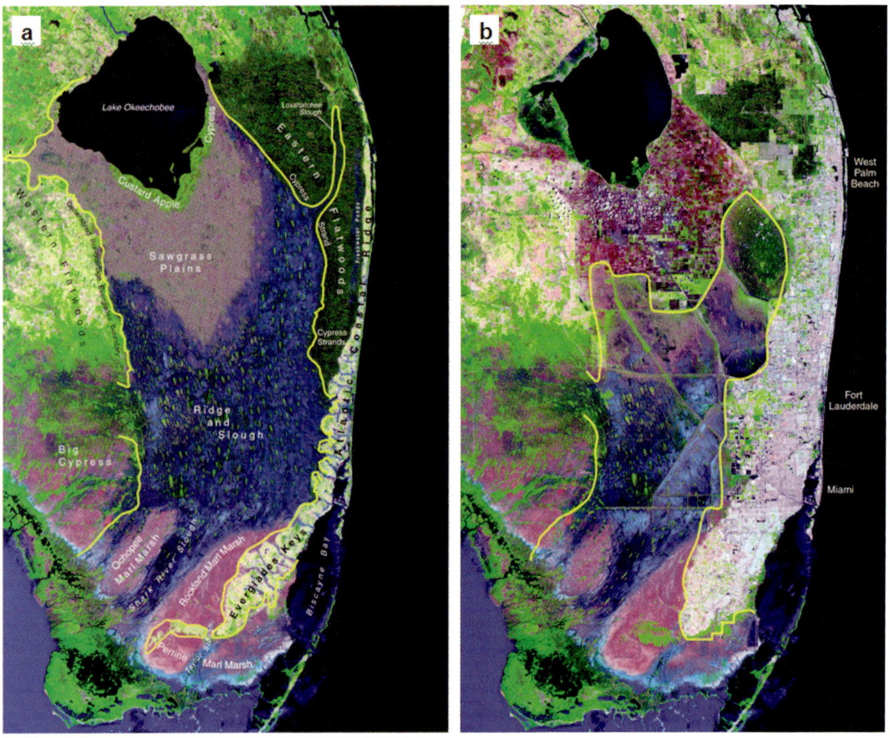

FIGURE 1-1 Reconstructed (a) pre-drainage (circa 1850) and (b) current (1994) satellite images of the Everglades ecosystem.

NOTE: The yellow line in (a) outlines the historical Everglades ecosystem, and the yellow line in (b) outlines the remnant Everglades ecosystem as of 1994.

SOURCE: Courtesy of C. McVoy, J. Obeysekera, and W. Said, South Florida Water Management District.

tion of the Greater Everglades Ecosystem (CROGEE), which operated from 1999 until 2004, was formed at the request of the South Florida Ecosystem Restoration Task Force (hereafter, simply Task Force), and the committee produced six reports (NRC, 2001, 2002a, 2002b, 2003a, 2003b, 2005). The NRC's Panel to Review the Critical Ecosystem Studies Initiative produced an additional report in 2003 (NRC, 2003c; see Appendix A). The Water Resources Development Act of 2000 (WRDA 2000) mandated that the U.S. Department of the Army, the Department of the Interior, and the state of Florida, in

BOX 1-1
Geographic Terms

To minimize confusion, this box defines some key geographic terms used throughout this report.

- The **Everglades,** the **Everglades ecosystem,** or the **remnant Everglades ecosystem** refers to the present areas of sawgrass, marl prairie, and other wetlands south of Lake Okeechobee (Figure 1-1b).
- The **original, historical, or pre-drainage Everglades** refers to the areas of sawgrass, marl prairie, and other wetlands south of Lake Okeechobee that existed prior to the construction of drainage canals beginning in the late 1800s (Figure 1-1a).
- The **Everglades watershed** is the drainage that encompasses the Everglades ecosystem but also includes the Kissimmee River watershed and other smaller watersheds north of Lake Okeechobee that ultimately supply water to the Everglades ecosystem.
- The **South Florida ecosystem** (also known as the Greater Everglades Ecosystem; see Figure 1-2) extends from the headwaters of the Kissimmee River near Orlando through Lake Okeechobee and the Everglades into Florida Bay and ultimately the Florida Keys. The boundaries of the South Florida ecosystem are determined by the boundaries of the South Florida Water Management District, the southernmost of the state's five water management districts, although they approximately delineate the boundaries of the South Florida watershed. This designation is important and is helpful to the restoration effort because, as many publications have made clear, taking a watershed approach to ecosystem restoration is likely to improve the results, especially when the ecosystem under consideration is as water dependent as the Everglades (NRC, 1999, 2004).

The following represent legally defined geographic terms used in this report:

- The **Everglades Protection Area** is defined in the Everglades Forever Act as comprising Water Conservation Areas (WCAs) 1 (the Arthur R. Marshall Loxahatchee National Wildlife Refuge), 2A, 2B, 3A, and 3B and Everglades National Park.
- The **natural system** is legally defined in the Water Resources Development Act of 2000 (WRDA 2000) as all land and water managed by the federal government or the state within the South Florida ecosystem (see Figure 1-3). "The term 'natural system' includes (i) water conservation areas; (ii) sovereign submerged land; (iii) Everglades National Park; (iv) Biscayne National Park; (v) Big Cypress National Preserve; (vi) other Federal or State (including a political subdivision of a State) land that is designated and managed for conservation purposes; and (vii) any tribal land that is designated and managed for conservation purposes, as approved by the tribe" (WRDA 2000).

Many maps in this report include shorthand designations that use letters and numbers for man-made additions to the South Florida ecosystem. For example, canals are labeled C-#; levees and associated borrow canals as L-#; and structures, such as culverts, locks, pumps, spillways, control gates, and weirs, as S-#.

consultation with the Task Force, establish an independent scientific review panel to evaluate progress toward achieving the natural system restoration goals of the CERP. The NRC's CISRERP was therefore established in 2004 under contract with the U.S. Army Corps of Engineers. After publication of the first biennial review (NRC, 2007; see Appendix B for the report summary), some members rotated off the committee and some new members were added.

The committee is charged to submit biennial reports that address the following items:

1. An assessment of progress in restoring the natural system, which is defined by section 601(a) of WRDA 2000 as all of the land and water managed by the federal government and state within the South Florida ecosystem (see Figure 1-3 and Box 1-1);
2. A discussion of significant accomplishments of the restoration;
3. A discussion and evaluation of specific scientific and engineering issues that may impact progress in achieving the natural system restoration goals of the plan; and
4. An independent review of monitoring and assessment protocols to be used for evaluation of CERP progress (e.g., CERP performance measures, annual assessment reports, assessment strategies, etc.).

To help it evaluate restoration progress, the committee met seven times over the course of this review; received briefings at its public meetings from agencies, organizations, and individuals involved in the restoration, as well as from the public; and took several field trips to sites with restoration activities (see Acknowledgments). In addition to information received at the meetings, the committee based its assessment of progress on information in relevant CERP and non-CERP restoration documents. The committee's conclusions and recommendations also were informed by a review of relevant scientific literature and the experience and knowledge of the committee members in their fields of expertise.

The committee was unable to consider in any detail new materials received after May 1, 2008. In late June 2008, after the committee had completed its deliberations and was about to send its report for external review, the state of Florida announced its potential purchase of 187,000 acres (almost 300 square miles) of the Everglades Agricultural Area from U.S. Sugar Corporation (see Figure 1-3). Given the timing of the announcement late in the committee's reporting cycle—coupled with the considerable uncertainty surrounding the fate of lands involved in the purchase and associated land trades, which are not yet defined (Cave, 2008)—the committee was unable to assess the implications of the land

FIGURE 1-2 The South Florida ecosystem. © International Mapping Associates

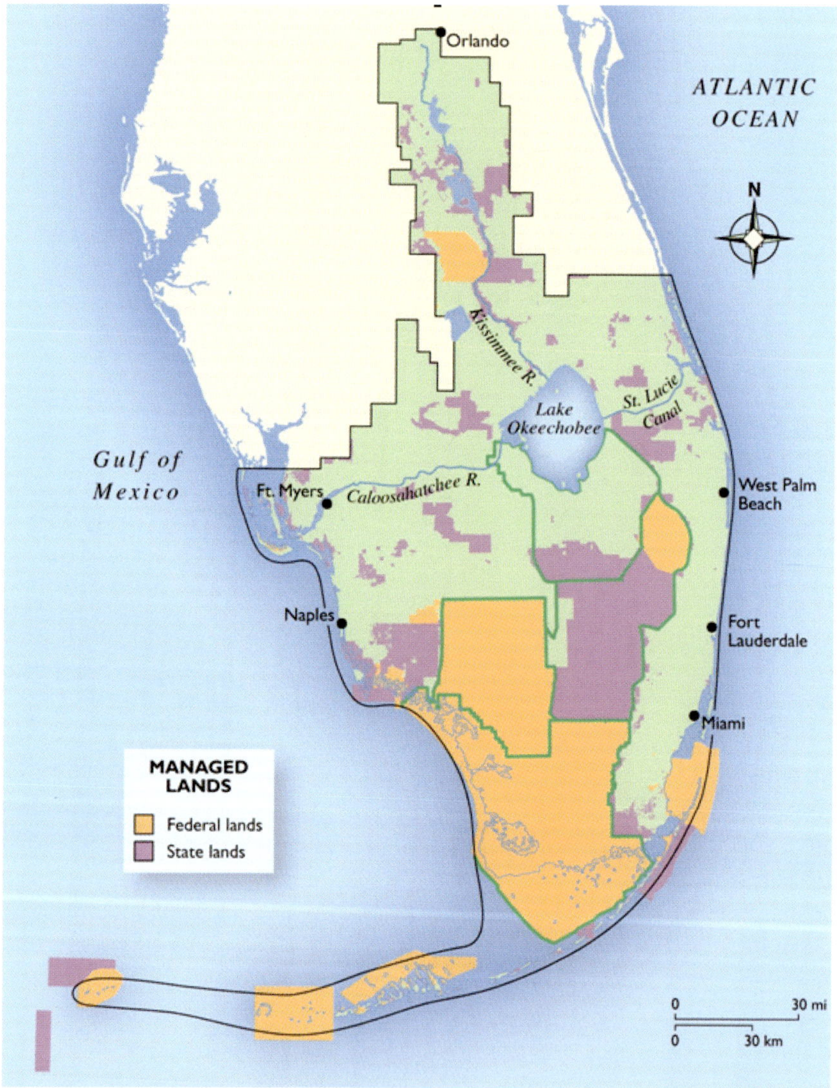

FIGURE 1-3 Land and waters managed by the state of Florida and the federal government as of December 2005 for conservation purposes within the South Florida ecosystem.

SOURCE: Based on data compiled by Florida State University's Florida Natural Areas Inventory (*http://www.fnai.org/gisdata.cfm*).

purchase for the CERP in any detail in this report. The purchase of these lands could have some important implications for water quality and possibly water storage for the Everglades, and the committee does draw attention to these in appropriate places in the report, but these issues will undoubtedly be analyzed in greater detail in future biennial reviews.

REPORT ORGANIZATION

Chapter 2 is an overview of the CERP in the context of other ongoing restoration activities and discusses the restoration goals that guide the overall effort. Restoration challenges and their implications for achievement of restoration goals by the CERP are discussed by analyzing recent changes to the natural system and larger-scale changes associated with population growth and climate change, which provide additional challenges for the CERP but make its implementation more rather than less urgent.

Progress on program implementation for the CERP is discussed in Chapter 3, including progress in implementing key CERP and non-CERP activities and issues encountered during project implementation (Tasks 1, 2, and 3). Project management, sequencing, and finances (addressing Tasks 2 and 3) are also discussed. The chapter includes an analysis of CERP implementation delays and recommendations for improving CERP planning and funding processes.

In Chapter 4 the committee analyzes the progress of the Modified Water Deliveries to Everglades National Park project (also known as Mod Waters), an essential foundation project for the CERP that was authorized in 1989. The project affects the timing and implementation of many other parts of the restoration, and it reflects the difficulties that might face restoration planners in other aspects of the CERP.

Lake Okeechobee is a critical component of the South Florida ecosystem. Reducing phosphorus loads to and in Lake Okeechobee and increasing water storage are essential for rehabilitation of the lake and the northern estuaries and for using the lake's water for restoring the southern part of the ecosystem. Thus, the challenges and progress in the rehabilitation of Lake Okeechobee are described in Chapter 5.

Adaptive management for the CERP requires the support of effective monitoring and assessment protocols and adequate hydrologic and ecological models. Therefore, recent developments from the monitoring and assessment program and modeling issues are discussed in Chapter 6 as the foundations of adaptive management (Tasks 2 and 4).

A synthesis of the report's key messages is provided in Chapter 7.

2

The Restoration in Context

This chapter sets the stage for the second of this committee's biennial assessments of restoration progress in the South Florida ecosystem. Background is provided for understanding the restoration project by defining the ecosystem decline, restoration goals, the needs of a restored ecosystem, and the specific activities of the restoration project. The Everglades and its restoration are also discussed in the larger context of human activities in South Florida and climate change. Finally, the chapter provides a view of important recent changes in the ecosystem, including tree islands, invasive species, and endangered bird populations.

BACKGROUND

The Everglades once encompassed about 3 million acres of slow-moving water and associated biota that stretched from Lake Okeechobee in the north to Florida Bay in the south (Figures 1-1a and 2-1a). The conversion of the uninhabited Everglades wilderness into an area of high agricultural productivity and cities was a dream of 19th-century investors, and projects begun between 1881 and 1894 affected the flow of water in the watershed north of Lake Okeechobee. By the late 1800s, more than 50,000 acres north and west of the lake had been drained and cleared for agriculture (Grunwald, 2006). These early projects included straightening the channel of the Kissimmee River and connecting Lake Okeechobee to the Caloosahatchee River and, ultimately, the Gulf of Mexico. In 1907 Governor Napoleon Bonaparte Broward created the Everglades Drainage District to construct a vast array of ditches, canals, dikes, and "improved" channels. By the 1930s, Lake Okeechobee had a second outlet, through the St. Lucie Canal, leading to the Atlantic Ocean, and 440 miles of other canals altered the hydrology of the Everglades (Blake, 1980). After hurricanes in 1926 and 1928 resulted in disastrous flooding from Lake Okeechobee, the U.S. Army Corps of Engineers (USACE) replaced the small berm that bordered

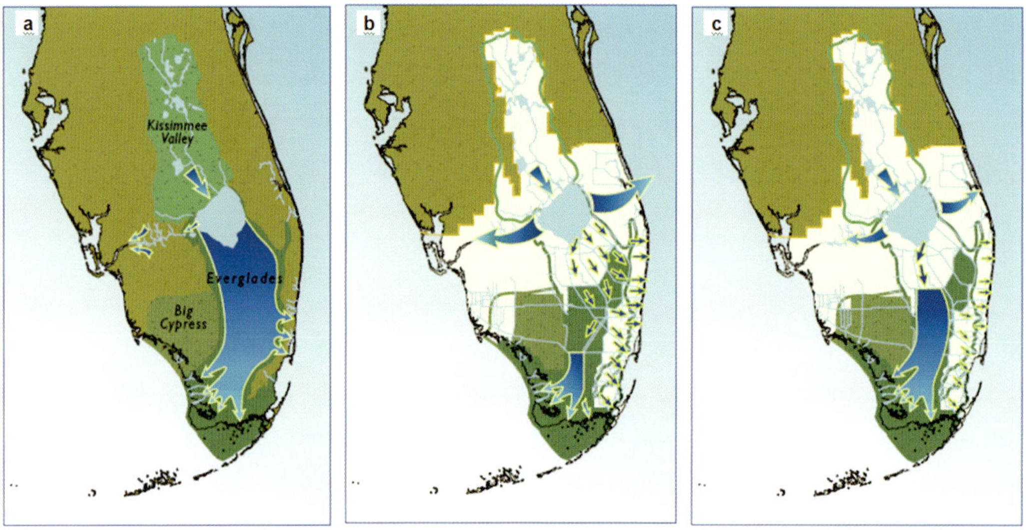

FIGURE 2-1 Water flow in the Everglades under (a) historical conditions, (b) current conditions, and (c) conditions envisioned upon completion of the Comprehensive Everglades Restoration Plan (CERP).

SOURCE: Graphics provided by USACE, Jacksonville District.

the southern edge of the lake with the massive Herbert Hoover Dike that now encircles the lake. The hydrologic end product of these drainage activities was the drastic reduction of water storage within the system and an increased susceptibility to drought and desiccation in the southern reaches of the Everglades (NRC, 2005).

After further flooding in 1947 and increasing demands for improved agricultural production and flood control for the expanding population centers on the southeast Florida coast, the U.S. Congress authorized the Central and Southern Florida (C&SF) Project. This USACE project provided flood control with the construction of a levee along the eastern boundary of the Everglades to prevent flows into the southeastern urban areas, established the 700,000-acre Everglades Agricultural Area (EAA) south of Lake Okeechobee (see Box 2-1), and created a series of water conservation areas (WCAs) in the remaining space between the lake and Everglades National Park (Light and Dineen, 1994). The eastern levee isolated about 100,000 acres of Everglades ecosys-

BOX 2-1
The Everglades Agricultural Area

Making the land in the Everglades Agricultural Area (EAA) (see Figure 1-3) suitable for agriculture was one of the original primary objectives of the C&SF Project (Lodge, 2005). Preliminary assessments in the late 1940s identified the peat soils just south of the southern rim of Lake Okeechobee as ideal for agriculture (Jones, 1948). Between 1950 and 1973, the USACE constructed a major dike on the east side of the agricultural area, established water delivery and drainage canals, and added pumps and control gates to manage water for agriculture. They also created the water conservation areas (WCAs) as temporary holding ponds that could accept surplus water during wet periods and provide additional water for agriculture during dry periods. Lake Okeechobee could also be managed to supply water in dry periods and accept excess water in wet periods. All of the EAA was designed for agricultural production, except for two fairly small wildlife management areas (WMAs): Rotenberger WMA and Holey Land WMA (Lodge, 2005). When the EAA was complete in the early 1970s, it subsumed 27 percent of the pre-drainage Everglades; for comparison, the WCAs occupy 37 percent, and Everglades National Park covers about 20 percent (Lodge, 2005; Secretary of Interior, 1994). By the 1990s, the EAA agricultural landscape had evolved into its present general form: about 85 percent of its area is devoted to sugar production, and 4 percent or less each is used for sod, vegetables, pasture, and non-specific general agriculture cultivation.[a]

The peat soils (Histosols) of the EAA accumulated under marshy conditions, but drainage by the C&SF Project caused the soils to shrink, while oxidation further reduced their volume. These processes continue today, and the surface of the EAA subsides about 1 inch (2.5 cm) per year. Peak agricultural production in the EAA probably occurred in the 1980s, before subsidence of the soils began to take its toll on productivity (Snyder and Davidson, 1994). Sugar, the most important crop in the EAA, requires soil depths that are at least 3 feet (Jurenas, 1992). Preliminary soil surveys in the EAA showed that most of the soils were at least 5 feet deep in 1912, but by 2003 most soil depths had declined to depths less than 3 feet. In some cases, sugar cane was being grown on soils as thin as 1 foot, an unsustainable practice (Snyder, 2004).

---

[a]*http://www.florida-agriculture.com/agfacts.htm.*

tem, making it available for development (Lord, 1993). Urban and agricultural development has reduced the Everglades to about one-half its pre-drainage size (Davis and Ogden, 1994; Figure 1-1b) and has contaminated its waters with phosphorus, nitrogen, mercury, and pesticides. Associated drainage and flood-control structures, including the C&SF Project, have diverted large quantities of water to the coastal areas, thereby reducing the freshwater inflows and natural water storage that defined the ecosystem (see Box 2-2; Figure 2-1b).

The profound hydrologic alterations were accompanied by many changes in the biotic communities in the ecosystem, including reductions and changes

BOX 2-2
Everglades Time Line:
Significant Events in South Florida Ecosystem Management

1934 Everglades National Park is authorized.

1948 Congress authorized the Central and Southern Florida Flood Control Project to control the water flow in the Everglades. From 1949 to 1969, USACE and the Central and Southern Florida Flood Control District built and operated the project works.

1968 Biscayne National Park is established as a national monument; expanded to a national park in 1980.

1972 Florida Water Resources Act establishes fundamental water policy for Florida, attempting to meet human needs and sustain natural systems; puts in place a comprehensive strategic program to preserve and restore the Everglades Ecosystem.

1974 Big Cypress National Preserve is created.

1983 Florida Governor's Save Our Everglades Program outlines a six-point plan for restoring and protecting the South Florida Ecosystem so that it functions more like it did in the early 1900s.

1987 Florida Surface Water Improvement and Management Act requires the five Florida water management districts to develop plans to clean up and preserve Florida lakes, bays, estuaries, and rivers.

1989 Modified Water Deliveries to Everglades National Park Project was authorized.

1990 Florida Preservation 2000 Act establishes a coordinated land acquisition program at $300 million per year for 10 years to protect the integrity of ecological systems and to provide multiple benefits, including the preservation of fish and wildlife habitat, recreation space, and water recharge areas.

1992 Federal and state parties enter into a consent decree on Everglades water quality issues in federal court. Under the agreement, all parties committed themselves to achieving both the water quality and quantity necessary to protect and restore the unique ecological characteristics of the Arthur R. Marshall Loxahatchee National Wildlife Refuge and Everglades National Park.

Water Resources Development Act (WRDA) of 1992 authorizes the Kissimmee River Restoration Project and the C&SF Project Restudy, a comprehensive review study for restoring the hydrology of South Florida.

1994 Florida Everglades Forever Act enacted into state law the settlement provisions of federal-state water quality litigation and provided a financing mechanism for the state to advance water quality improvements in the Everglades by constructing over 44,000 acres of stormwater treatment areas (STAs) for water entering the Everglades Protection Area. The act also requires the South Florida Water Management District to ensure that best management practices (BMPs) are being used to reduce phosphorus in waters discharged into the STAs from the EAA and other areas. The rulemaking process by which the numeric total phosphorus criterion of 10 parts per billion (ppb) was proposed for the Everglades Protection Area also was established by this act.

1996 WRDA 1996 formally establishes the intergovernmental South Florida Ecosystem Restoration Task Force to coordinate the restoration effort among the state, federal, tribal, and local agencies. It authorizes the USACE to implement the critical restoration projects (see Box 2-3).

Section 390 of the Farm Bill grants $200 million to conduct restoration activities in the South Florida Ecosystem.

1999 WRDA 1999 extends Critical Restoration Project authority until 2003; authorizes two pilot infrastructure projects proposed in the Comprehensive Everglades Restoration Plan (CERP).

Florida Forever Act improves and continues the coordinated land acquisition program initiated by the Florida Preservation 2000 Act of 1990; commits $300 million per year for 10 years.

2000 WRDA 2000 authorized the CERP as a framework for modifying the Central and Southern Florida Project to increase future water supplies, with the appropriate timing and distribution, for environmental purposes so as to achieve a restored Everglades ecosystem, while at the same time meeting other water-related needs of the ecosystem. WRDA 2000 includes $1.4 billion in authorizations for 10 initial Everglades infrastructure projects, four pilot projects, and an adaptive management and monitoring program; also grants programmatic authority for projects with immediate and substantial restoration benefits at a total cost of $206 million; establishes a 50 percent federal cost share for implementation of the CERP and for operation and maintenance.

Florida legislature passes the Lake Okeechobee Protection Act, a phased, comprehensive program designed to restore and protect the lake.

2003 Programmatic Regulations are issued which establish a procedural framework and set specific requirements that guide implementation of the CERP to ensure that the goals and purposes of the CERP are achieved.

2004 State of Florida unveils plan to expedite restoration of America's Everglades (Acceler8).

2005 State of Florida announces the Lake Okeechobee Estuary Recovery Plan to help restore the ecological health of Lake Okeechobee and the St. Lucie and Caloosahatchee estuaries.

2007 The Florida state legislature authorized the Northern Everglades and Estuaries Protection Program which expanded the Lake Okeechobee Protection Act to strengthen protection for the Northern Everglades by restoring and preserving the Lake Okeechobee, Caloosahatchee, and St. Lucie watersheds, including the estuaries.

WRDA 2007 authorizes three projects under the CERP: the Indian River Lagoon-South Project, Picayune Strand Restoration, and the Site 1 Impoundment Project. WRDA 2007 also increases funding limits for WRDA 1996 critical projects and for three WRDA 1999 authorized pilot projects.

2008 State of Florida announces that it will begin negotiations to acquire 187,000 acres of farmland in the EAA from U.S. Sugar Corporation for $1.75 billion for the purpose of restoration.

SOURCES: SFERTF (2006); *http://everglades.fiu.edu/reclaim/timeline/index.htm; http://www.washingtonpost.com/wp-dyn/content/article/2008/06/24/AR2008062401140.html.*

in the composition, distribution, and abundance of the populations of wading birds. Today, the federal government has listed 67 plant and animal species in South Florida as threatened or endangered, with many more included on state lists. Some distinctive Everglades habitats, such as custard-apple forests and peripheral wet prairie, have disappeared altogether, while other habitats are severely reduced in area (Davis and Ogden, 1994; Marshall et al., 2004). Approximately 1 million acres are contaminated with mercury (McPherson and Halley, 1996). Phosphorus from agricultural runoff has impaired water quality in large portions of the Everglades and has been particularly problematic in Lake Okeechobee (Flaig and Reddy, 1995). The Caloosahatchee and St. Lucie estuaries, including parts of the Indian River Lagoon, have been greatly altered by high and extremely variable freshwater discharges that bring nutrients and contaminants (Doering, 1996; Doering and Chamberlain, 1999).

At least as early as the 1920s, private citizens were calling attention to the degradation of the Florida Everglades (Blake, 1980). However, by the time Marjory Stoneman Douglas's classic book *The Everglades: River of Grass* was published in 1947 (the same year that Everglades National Park was dedicated), the South Florida ecosystem had already been altered extensively. Prompted by concerns about deteriorating conditions in Everglades National Park and other parts of the South Florida ecosystem, the public, as well as the federal and state governments, directed increasing attention to the adverse ecological effects of the flood-control and irrigation projects beginning in the 1970s (Kiker et al., 2001; Perry, 2004). By the late 1980s it was clear that various minor corrective measures undertaken to remedy the situation were insufficient. As a result, a powerful political consensus developed among federal agencies, state agencies and commissions, Native American tribes, county governments, and conservation organizations that a large restoration effort was needed in the Everglades (Kiker et al., 2001). This recognition culminated in the CERP, which builds on other ongoing restoration activities of the state and federal governments to create one of the most ambitious and extensive restoration efforts in the nation's history.

Ecosystem Restoration Goals for the Everglades

Several goals have been articulated for the restoration of the South Florida ecosystem, reflecting the various restoration programs. The South Florida Ecosystem Restoration Task Force (hereafter, simply the Task Force), an intergovernmental body established to facilitate coordination in the restoration effort, has three broad strategic goals: (1) "get the water right," (2) "restore, preserve, and protect natural habitats and species," and (3) "foster compatibility of the built and natural systems" (SFERTF, 2000). These goals encompass, but are not limited to, the CERP. The Task Force works to coordinate and build consensus among the many non-CERP restoration initiatives that support these broad goals.

The goal of the CERP, as stated in the Water Resources Development Act of 2000 (WRDA 2000), is "restoration, preservation, and protection of the South Florida Ecosystem while providing for other water-related needs of the region, including water supply and flood protection." The Programmatic Regulations (33 CFR 385.3; see Box 2-2) that guide implementation of the CERP further clarify this goal by defining restoration as "the recovery and protection of the South Florida ecosystem so that it once again achieves and sustains the essential hydrological and biological characteristics that defined the undisturbed South Florida ecosystem." These defining characteristics include a large-areal extent of interconnected wetlands, extremely low concentrations of nutrients in freshwater wetlands, sheet flow, healthy and productive estuaries, resilient plant communities, and an abundance of native wetland animals (DOI and USACE, 2005). Although development has permanently reduced the areal extent of the Everglades ecosystem, the CERP hopes to recover many of the Everglades' original characteristics and natural ecosystem processes. At the same time, the CERP is charged to maintain current levels of flood protection (as of 2000) and provide for other water-related needs, including water supply, for a rapidly growing human population in South Florida (DOI and USACE, 2005).

Although the CERP contributes to each of the Task Force's three goals, it focuses primarily on restoring the hydrologic features of the undeveloped wetlands remaining in the South Florida ecosystem, on the assumption that improvements in ecological conditions will follow. Originally, "getting the water right" had four components—quality, quantity, timing, and distribution. However, the hydrologic properties of flow, encompassing the concepts of direction, velocity, and discharge, have been recognized as an important component of getting the water right that had previously been overlooked (NRC, 2003c; SCT, 2003). Numerous studies have supported the general approach to restoration of getting the water right (Davis and Ogden, 1994; NRC, 2005; SSG, 1993), although it is widely recognized that recovery of the native habitats and species in South Florida may require restoration efforts in addition to getting the water right, such as controlling exotic species and reversing the decline in the spatial extent and compartmentalization of the natural landscape (SFERTF, 2000; SSG, 1993).

The goal of ecosystem restoration can seldom be the exact re-creation of some historical or preexisting state because physical conditions, driving forces, and boundary conditions usually have changed and are not fully recoverable. Rather, restoration is better viewed as the process of assisting the recovery of a degraded ecosystem to the point where it contains sufficient biotic and abiotic resources to continue its functions without further assistance in the form of energy or other resources from humans (NRC, 1996; Society for Ecological Restoration International Science & Policy Working Group, 2004). In addition, in this report the committee sometimes uses the term *ecosystem rehabilitation* when the objective is to improve conditions in a part of the South Florida ecosystem to at least some minimally acceptable level to allow the restoration of the larger ecosystem to advance. This is particularly a focus in Chapter 5, "Lake Okeechobee and Its Place in the Restoration of the South Florida Ecosystem." Implicit in the understanding of ecosystem restoration is the recognition that natural systems are self-designing and dynamic and, therefore, it is not possible to know in advance exactly what can or will be achieved. Thus, ecosystem restoration is an enterprise with some scientific uncertainty in methods or outcomes that requires continual testing of assumptions and monitoring of progress. Additional challenges in defining and implementing restoration goals are discussed in the initial National Research Council (NRC) biennial review (NRC, 2007).

What Natural System Restoration Requires

Restoring the South Florida ecosystem to a desired ecological landscape requires reestablishment of the critical processes that sustained its historical functions. Although getting the water right is the oft-stated and immediate goal, the restoration will be recognized as successful if it restores the distinctive characteristics of the historical ecosystem to the remnant Everglades (DOI and USACE, 2005). Getting the water right is a means to an end, not the end in itself. The hydrologic and ecological characteristics of the historical Everglades serve as restoration goals for a functional (albeit reduced in size) Everglades ecosystem. The first Committee on Independent Scientific Review of Everglades Restoration Progress (CISRERP) review identified five critical components of Everglades restoration:

1. Enough water storage capacity combined with operations that allow for appropriate volumes of water to support healthy estuaries and the return of sheet flow through the Everglades ecosystem while meeting other demands for water;
2. Mechanisms for delivering and distributing the water to the natural system in a way that resembles historical flow patterns, affecting volume, depth, velocity, direction, distribution, and timing of flows;
3. Barriers to eastward seepage of water so that higher water levels can be maintained in parts of the Everglades ecosystem without compromising the current levels of flood protection of developed areas as required by the CERP;

4. Methods for securing water quality conditions compatible with restoration goals for a natural system that was inherently extremely nutrient poor, particularly with respect to phosphorus; and

5. Retention, improvement, and expansion of the full range of habitats by preventing further losses of critical wetland and estuarine habitats and by protecting lands that could usefully be part of the restored ecosystem.

If these five critical components of restoration are achieved and the difficult problem of invasive species can be managed, then the basic physical, chemical, and biological processes that created the historical Everglades can once again work to create a functional mosaic of biotic communities that resemble what was distinctive about the historical Everglades. Even if the restored system does not exactly replicate the historical system, or reach all of the biological, chemical, and physical targets, the reestablishment of natural processes and dynamics should result in a viable and valuable Everglades ecosystem. The central principle of ecosystem management is to provide for the natural processes that historically shaped an ecosystem, because ecosystems are characterized by the processes that regulate them. If the conditions necessary for those processes to operate are met, recovery of species and communities is far more likely than if humans attempt to specify every constituent and element of the ecological system (NRC, 2007).

Restoration Activities

Several restoration programs, including the largest of the initiatives, the CERP, are now ongoing. The CERP often builds upon non-CERP activities (also called "foundation projects"), many of which are essential to the effectiveness of the CERP. The following section provides a brief overview of the CERP and some of the major non-CERP activities. Details of the progress in implementing the CERP projects are described in Chapter 3, and a few projects are discussed in more detail in Chapters 4–5.

Comprehensive Everglades Restoration Plan

WRDA 2000 authorized the CERP as the framework for modifying the C&SF Project. Considered a blueprint for the restoration of the South Florida ecosystem, the CERP is led by two organizations with considerable expertise managing the water resources of South Florida—the USACE, which built most of the canals and levees throughout the region, and the South Florida Water Management District (SFWMD), the state agency with primary responsibility for operating and maintaining this complicated water collection and distribution system.

The CERP conceptual plan (USACE and SFWMD, 1999; also called the Yellow Book) proposes major alterations to the C&SF Project in an effort to reverse decades of ecosystem decline. The Yellow Book includes roughly 50 major projects consisting of 68 project components to be constructed at a cost of approximately $10.9 billion (estimated in 2004 dollars; DOI and USACE, 2005; Figure 2-2). Major components of the restoration plan focus on restoring the quantity, quality, timing, and distribution of water for the natural system. These major CERP components include the following:

- **Conventional surface-water storage reservoirs**, which will be located north of Lake Okeechobee, in the St. Lucie and Caloosahatchee basins, in the EAA, and in Palm Beach, Broward, and Miami-Dade counties, will provide storage of approximately 1.5 million acre-feet.
- **Aquifer storage and recovery** is a highly engineered approach that proposes to use a large number of wells built around Lake Okeechobee, in Palm Beach County, and in the Caloosahatchee basin to store water approximately 1,000 feet below ground; the approach has not yet been tested at the scale proposed.
- **In-ground reservoirs** will store water in quarries created by rock mining.
- **Stormwater treatment areas (STAs)** are constructed wetlands that will treat agricultural and urban runoff water before it enters natural wetlands.[1]
- **Seepage management** approaches will prevent unwanted loss of water from the natural system through levees and groundwater flow; the approaches include adding impermeable barriers to the levees, installing pumps near levees to redirect lost water back into the Everglades, and holding water levels higher in undeveloped areas between the Everglades and the developed lands to the east.

[1]Although some STAs are included among CERP projects, the USACE has recently clarified its policy on federal cost sharing for water quality features. A memo from the Assistant Secretary of the Army (Civil Works) (USACE, 2007d) states: "Before there can be a Federal interest to cost share a WQ [water quality] improvement feature, the State must be in compliance with WQ standards for the current use of the water to be affected and the work proposed must be deemed essential to the Everglades restoration effort...This determination must be based on some finding other than the project is a part of CERP and generally will aid the restoration effort." The memo goes on to state, "the Yellow Book specifically envisioned that the State would be responsible for meeting water quality standards." Therefore, it appears that until the water flowing into the project features meets existing water quality requirements or unless a special exemption is granted for projects deemed "essential to Everglades restoration," the state is responsible for 100 percent of the costs of CERP water quality project features.

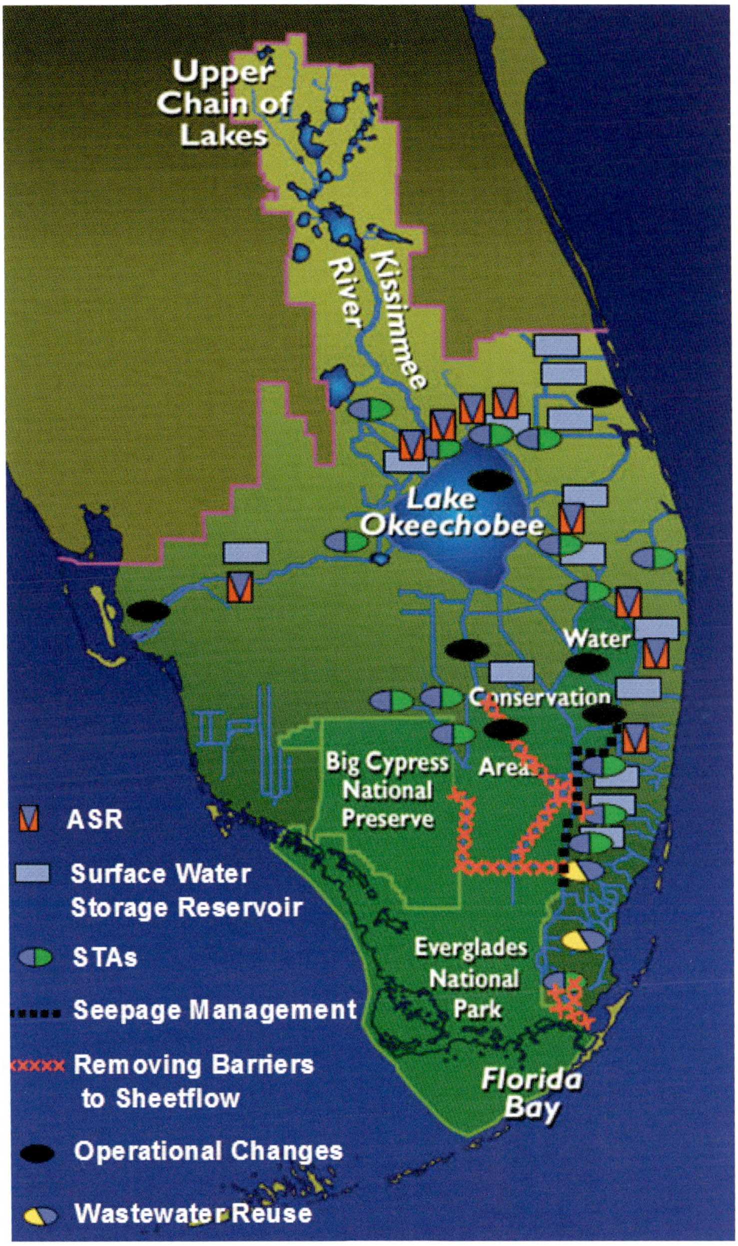

FIGURE 2-2 Major project components of the CERP.

SOURCE: Courtesy of Laura Mahoney, USACE.

- **Removing barriers to sheet flow,** including 240 miles of levees and canals, will reestablish shallow sheet flow of water through the Everglades ecosystem.
- **Rainfall-driven water management** will be created through operational changes in the water delivery schedules to the WCAs and Everglades National Park to mimic more natural patterns of water delivery and flow through the system.
- **Water reuse and conservation** strategies will build additional water supply in the region; two advanced wastewater treatment plants are proposed for Miami-Dade County in order to clean wastewater to a standard which would allow it to be discharged to wetlands along Biscayne Bay or to recharge the Biscayne aquifer.

The largest portion of the budget is devoted to storage and water conservation projects and to acquiring the lands needed for them (see NRC, 2005).

The modifications to the C&SF Project embodied in the CERP are expected to take more than three decades to complete, and to be effective, they require a clear strategy for managing and coordinating restoration efforts. The Everglades Programmatic Regulations specifically require coordination with other agencies at all levels of government, although final responsibility ultimately rests with the USACE and SFWMD. WRDA 2000 endorses the use of an adaptive management framework for the restoration process, and the Programmatic Regulations formally establish an adaptive management program that will "assess responses of the South Florida ecosystem to implementation of the Plan; . . . [and] seek continuous improvement of the Plan based upon new information resulting from changed or unforeseen circumstances, new scientific and technical information, new or updated modeling; information developed through the assessment principles contained in the Plan; and future authorized changes to the Plan." An interagency body called Restoration, Coordination, and Verification (RECOVER) has been established to ensure that sound science is used in the restoration. The RECOVER leadership group oversees the monitoring and assessment program that will evaluate the progress of the CERP toward restoring the natural system and will assess the need for changes to the plan through the adaptive management process. Progress in developing these essential programmatic aspects of the CERP is discussed in Chapter 6.

In 2004, Florida launched Acceler8, a plan to hasten the pace of project implementation, and committed $1.5 billion of its portion of the state-federal cost share for the CERP by 2011 for this initiative. The objectives of Acceler8 are to provide immediate environmental and water supply benefits and to serve as a foundation for subsequent restoration efforts. Florida's Acceler8 comprises

11 project components identified in the CERP and some non-CERP components (see Table 3-2 in Chapter 3 for a listing of state-accelerated projects; for further discussion of Acceler8, see NRC, 2007).

Non-CERP Restoration Activities

When Congress authorized the CERP in WRDA 2000, the SFWMD, the USACE, the National Park Service (NPS), and the U.S. Fish and Wildlife Service were already implementing several activities intended to restore key aspects of the Everglades ecosystem. These non-CERP initiatives are critical to the overall restoration progress. In fact, the effectiveness of the CERP was predicated upon the completion of many of these projects. These projects include Modified Water Deliveries to Everglades National Park (Mod Waters), C-111 (South Dade), and the Critical Projects (see Box 2-3). Several additional projects are also either under way or in planning stages to meet the broad restoration goals for the South Florida ecosystem and associated legislative mandates. They include extensive water quality initiatives, such as the Everglades Construction Project, and programs to establish best management practices to reduce nutrient loading.

BOX 2-3
Non-CERP Restoration Activities in South Florida

The following represent the major non-CERP initiatives currently under way in support of the South Florida ecosystem restoration (Figure 2-3). Progress on these non-CERP projects is discussed in Appendix C.

Kissimmee River Restoration Project

This project, authorized by Congress in 1992, aims to reestablish the historical river-floodplain system at the headwaters of the Everglades watershed and, thereby, restore biological diversity and functionality. The project plans to backfill 22 miles of the 56-mile C-38 canal and restore 43 miles of meandering river channel in the Kissimmee River. The project includes a comprehensive evaluation program to track ecological responses to restoration. Completion is expected by 2012 (SFWMD and FDEP, 2005).

Everglades Construction Project

The Everglades Forever Act (see Box 2-2) required the state of Florida to construct 45,000 acres of STAs to reduce the loading of phosphorus into the Arthur R. Marshall Loxahatchee National Wildlife Refuge, the WCAs, and Everglades National Park. These STAs are part of the state's long-term plan for achieving water quality goals, including the total phosphorus criterion of 10 parts per billion (ppb).[a]

continued

BOX 2-3 Continued

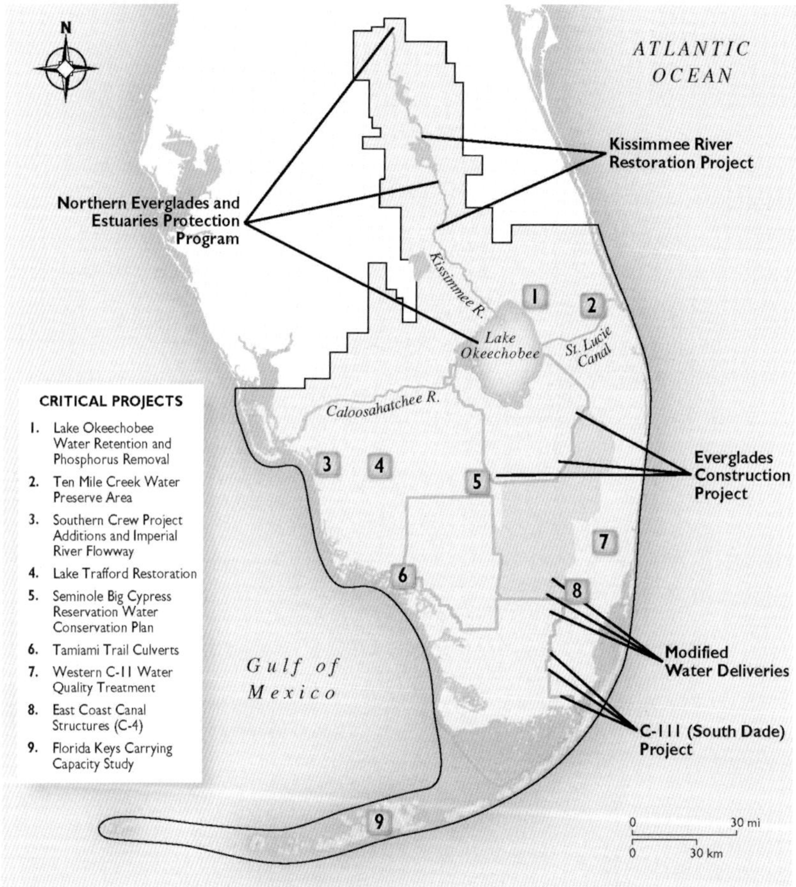

FIGURE 2-3 Locations of major non-CERP initiatives. © International Mapping Associates

Modifications to the C&SF: C-111 (South Dade) Project

This project is designed to improve hydrologic conditions in Taylor Slough and the Rocky Glades of the eastern panhandle of Everglades National Park and increase freshwater flows to northeast Florida Bay, while maintaining flood protection for urban and agricultural development in south Miami-Dade County. The project plan includes a tieback levee with pumps to capture groundwater seepage to the east, detention areas to increase groundwater levels and thereby enhance flow into Everglades National

Park, and backfilling or plugging several canals in the area. A Combined Structural and Operational Plan that will integrate the goals of the Mod Waters and C-111 (South Dade) projects and protect the quality of water entering Everglades National Park (DOI and USACE, 2005).

Modified Water Deliveries to Everglades National Park Project (Mod Waters)

This federally funded project, authorized in 1989, is designed to restore more natural hydrologic conditions in Everglades National Park (see also Chapter 4). The project includes levee modifications and installation of a seepage control pump to increase water flow into WCA 3B and northeastern portions of Everglades National Park. It also includes providing flood mitigation to about 60 percent of the 8.5-square-mile area (a low-lying but partially developed area on the northeast corner of Everglades National Park) and raising portions of Tamiami Trail. Mod Waters is a prerequisite for the first phase of "decompartmentalization" (i.e., removing some barriers to sheet flow), which is part of the CERP[b] (DOI and USACE, 2005).

Northern Everglades and Estuaries Protection Program

In 2007, the Florida legislature expanded the Lake Okeechobee Protection Act (LOPA) to include protection and restoration of the Lake Okeechobee watershed and the Caloosahatchee and St. Lucie estuaries. The legislation, being implemented as the Northern Everglades and Estuaries Protection Program, will focus resources on restoration efforts for Lake Okeechobee and the Caloosahatchee and St. Lucie estuaries. The new laws include $54 million for Lake Okeechobee and an additional $40 million for the Caloosahatchee and St. Lucie rivers. The Lake Okeechobee Watershed Construction Project Phase II Technical Plan, issued in February 2008 in accordance with LOPA, consolidated the numerous initiatives already underway through Florida's Lake Okeechobee Protection Plan and Lake Okeechobee and Estuary Recovery Plan.

Critical Projects

Congress gave programmatic authority for the Everglades and South Florida Ecosystem Restoration Critical Projects in WRDA 1996, with modification in WRDA 1999 and WRDA 2007. These were small projects that could be quickly implemented to provide immediate and substantial restoration benefits such as improved quality of water discharged into WCA 3A and Lake Okeechobee and more natural water flows to estuaries. Examples of the Critical Projects include the Florida Keys Carrying Capacity Study, Lake Okeechobee Water Retention and Phosphorus Removal, Seminole Big Cypress Reservation Water Conservation Plan, Tamiami Trail Culverts, Ten Mile Creek Water Preserve Area, and the Lake Trafford Restoration (DOI and USACE, 2005)[c]. See also Appendix C.

---

[a] *http://www.sfwmd.gov/org/erd/longtermplan/index.shtml.*

[b] See *http://www.saj.usace.army.mil/dp/mwdenp-c111/index.htm* for more information on Mod Waters and the C-111 (South Dade) project.

[c] See *http://www.saj.usace.army.mil/projects* for more information on and the status of the Critical Projects.

LARGE-SCALE INFLUENCES ON THE CERP

The South Florida ecosystem restoration efforts take place within a multidimensional context that includes the influence of large-scale human and environmental processes. From an ecological perspective, many of these processes appear as threats to the integrity of the South Florida ecosystem. These threatening processes, reviewed in the following sections, include expansion of the human population accompanied by land use changes, climate change, and sea-level rise that will broadly affect the South Florida ecosystem. The section ends with a discussion of how CERP planners are attempting to address these issues.

Human Population Growth, Land-use, and Water Demand

A primary objective of the CERP is the restoration and maintenance of an Everglades ecosystem that functions substantially more like the pre-drainage system than the present disrupted system. This restoration effort takes place in a changing human context as Florida's population continues to grow. The implications to restoration of expanded urban landscapes, increased demands for water supply, and higher land values that accompany population growth are discussed.

Population Growth and Land Use Changes

One of the primary drivers of ecosystem change in its largest sense for South Florida is the growth of its human population. During the 19th century, population increases in Florida were small, and it was not until the late 1960s that Floridians numbered more than 5 million (Figure 2-4). The present (2007) population is 18 million; over the past 10 years, on average, there has been a net daily gain of nearly 1,000 people.[2] Estimates of the overall rate of growth in the state mask local rates that may temporarily be much higher. Three of the 10 fastest-growing cities in the country (with populations greater than 100,000) are in Florida: Miramar in Broward County, growing at 39.5 percent per year; Port St. Lucie in St. Lucie County, growing at 33.4 percent; and Cape Coral in Lee County, growing at 25.1 percent. All three cities are in areas adjacent to remnants of the pre-drainage Everglades. This population growth, along with other factors, has had a strong effect on property values: between 2000 and 2006, housing prices doubled (Durrenburger et al., 2007). In addition to permanent residents, each year Florida hosts more than 80 million temporary tourists and

[2]See *http://edr.state.fl.us/conferences/population/FDEC0807_pop_change.pdf.*

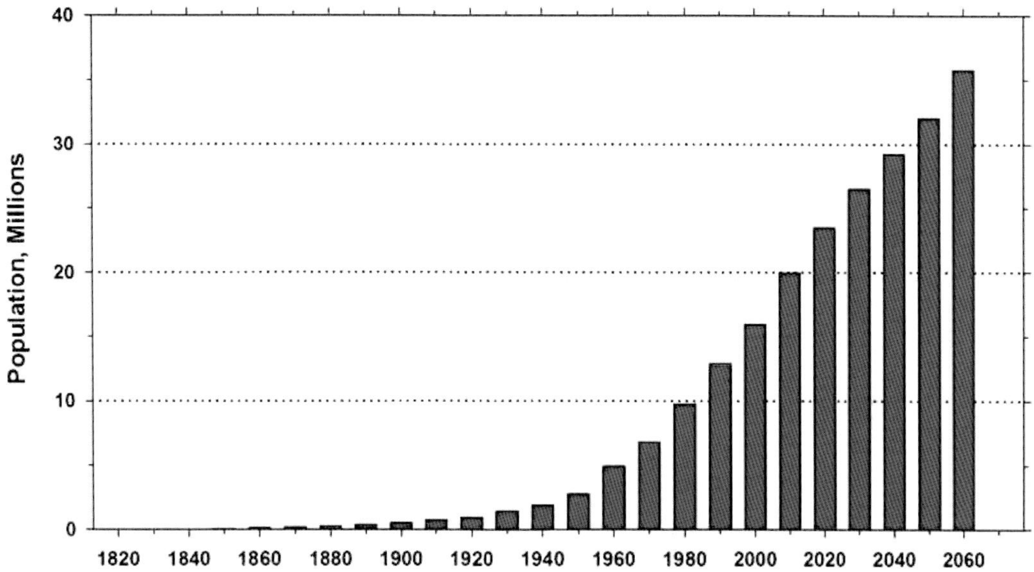

FIGURE 2-4 Florida population by decade, actual data 1830–2000; projection 2010–2060.

SOURCES: 1830–1970, U.S. Census Bureau (1975); 1980–2000, Florida Office of Economic and Demographic Research (2007) online at *http://Edr.state.fl.us/population/FLPopChange.pdf*; 2010–2030, U.S. Census Bureau (2005); 2040 and 2060, Zwick and Carr (2006); 2050, NPG (2002). There is some minor variability among various census counts and projections, but the overall trends are the same regardless of data source.

1 million "snowbirds" who take up temporary winter residence (Gurnett, 2001; NPG, 2002).

By 2040, the approximate original ending date for construction of CERP projects, Zwick and Carr (2006) have projected that the state's population will be more than 29 million, and by 2060, nearly 36 million—twice the present population. These projections are based on long-term population growth patterns, and they represent educated best guesses about what the future will be like. Projections of population growth are difficult because the many influences on migration, birth, and death rates cannot be fully foreseen. However, the trend of substantial growth of the state's population is well established and has been relatively consistent at the decade scale of analysis for more than 170 years. The growth has slowed at times and accelerated at others, but the overall trend has remained in spite of numerous boom and bust cycles and real estate episodes. The population growth in South Florida over the past decade exceeded

SFWMD's estimates determined during the development of the CERP (L. Gerry, SFWMD, personal communication, 2008), and data released by the state of Florida in November 2007 indicate continued population growth despite recent economic reversals, albeit below the decadal average (Florida Association of Realtors, 2007).

Land Use and Development Density Changes. Population growth in Florida is usually associated with increasing urbanization of the landscape. Presently, development converts 860 acres (1.34 sq mi, or 3.5 sq km) per day from undeveloped forest, wetland, or agricultural uses to urban landscapes (NPG, 2002). If urbanization continues at this pace, all the undeveloped land in the state would be developed in just 60 years.[3] By some general measures, half the original Everglades has disappeared through conversion to agriculture and urban uses (Davis and Ogden, 1994). Projections of the geographic distribution of the urbanized area of Florida shows that South Florida in particular is likely to be transformed by 2060 (Figures 2-5 and 2-6). Significant components of this urbanization have been projected to be in areas closely associated with the southward flows of water that will be needed to nourish the remnant Everglades ecosystem (Figure 2-7). In particular, potential land use changes in the EAA (see Box 2-1) have direct bearing on the prospects for restoration of the Everglades because the EAA influences the movement of water, sediment, and nutrients for the rest of the remnant Everglades ecosystem.

Land use change appears to be inevitable for the EAA. As subsidence makes the land less productive for sugar cane (see Box 2-1), other types of agricultural and nonagricultural land uses are being considered, including suburban development, rock mining, and ecosystem restoration. Water management in the EAA has a substantial effect on the overall water budget and water quality of South Florida (which includes stormwater runoff), thus land use changes in the area will have far-reaching effects (Alvarez et al., 1994). Development could alter water flows, introduce new sources of pollution, and alter the landscape of South-Central Florida to create an ecosystem even less like the predrainage conditions than with agriculture in place. Although considerable attention has been paid to agricultural sources of water pollution, urban areas are likely to become increasingly important nutrient sources in the coming decades because of population growth and sprawl.

---

[3]Estimated time to total urbanization of Florida at the present rate is the result of a conversion rate of 860 ac per day = 1.34 sq mi per day = 490.47 sq mi per year. Florida includes 65,755 sq mi, 59,772 sq mi of which are land surface; at present 48.8 percent of the land surface is in crops, range, or wild lands such as forest, a total of 29,169 sq mi. That area divided by 490.47 sq mi per year yields an estimate of 59.5 years.

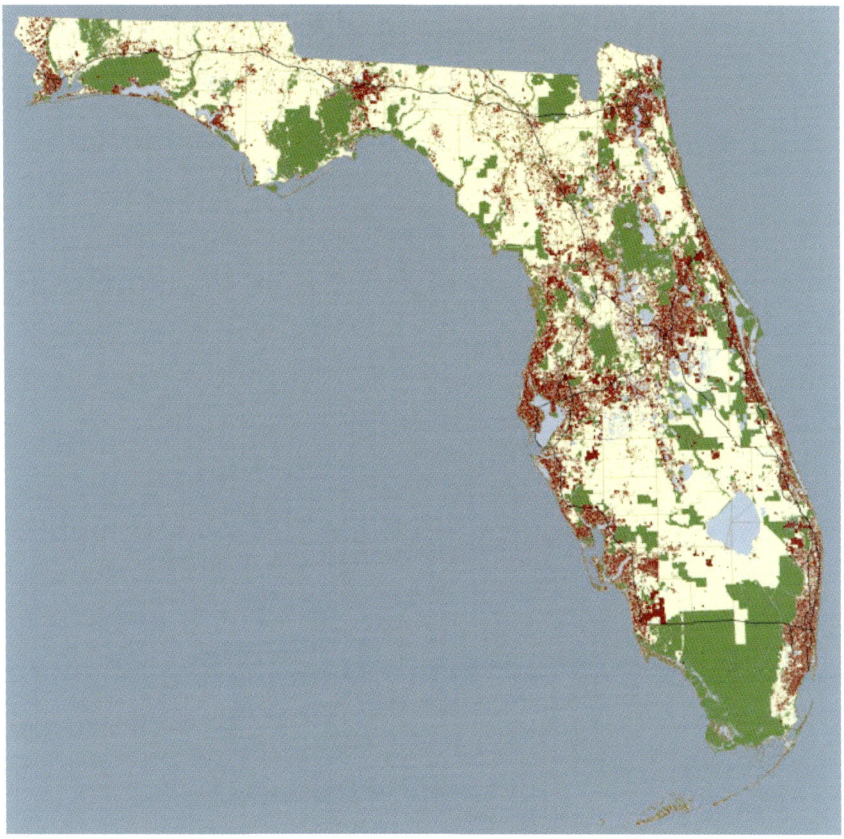

FIGURE 2-5 Distribution of urbanized areas of Florida in 2006.

SOURCE: Courtesy of 1000 Friends of Florida (2006).

A sweeping change in land ownership in the EAA may be in the offing. On June 24, 2008, the state of Florida announced that it will begin negotiations with U.S. Sugar Corporation to purchase 187,000 acres (292 square miles) of land in the EAA for $1.75 billion (Achenbach, 2008). Although the company will retain use of the land for 6 years following the purchase, eventual public ownership of this substantial area would open many unforeseen possibilities for Everglades restoration. The acquisition might protect large areas from urban and commercial development, and if certain exchanges were made, a development-free corridor might connect Lake Okeechobee with valuable undeveloped lands to the south.

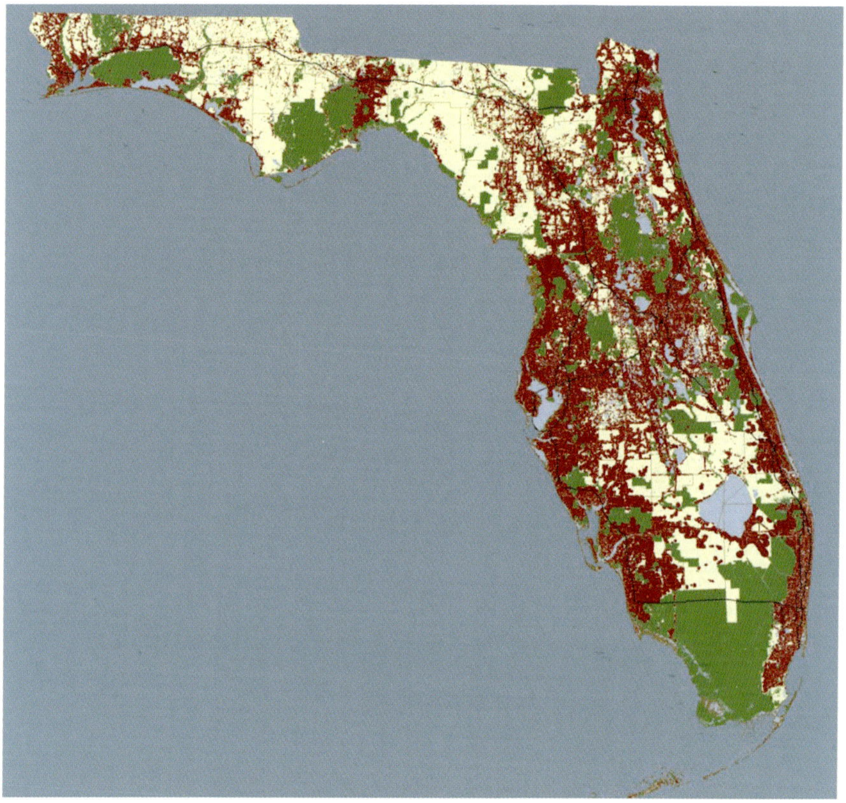

FIGURE 2-6 Projected distribution of urbanized areas of Florida in 2060, a time period after the anticipated completion of CERP.

SOURCE: Courtesy of 1000 Friends of Florida (2006).

The acquisition might also allow for enlarged STAs to reduce phosphorus loads into the Everglades ecosystem and to expand the overall treatment and/or water storage capacity. The size and location of the U.S Sugar land, combined with the potential to swap the land for other strategically located parcels, may lead to some rethinking of the strategies in the original CERP plan (Stokstad, 2008). Given that the state will not take ownership for six years and that many land trades may be required after that time, the effects of the purchase may not be seen for a decade or more.

The urban projections in Figures 2-6 and 2-7 assume that urban densities

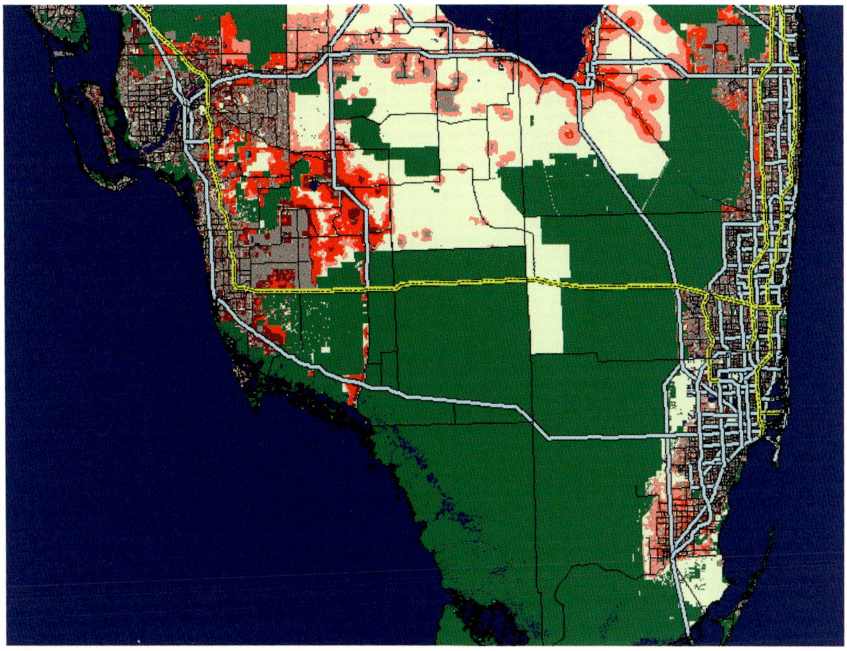

FIGURE 2-7 Projections of urban growth (in red) expected between 2006 and 2060 in areas likely to directly affect CERP projects.

SOURCE: Zwick and Carr (2006). Courtesy of 1000 Friends of Florida.

(people/urban acre) will remain constant through 2060, whereas actual patterns could be quite different depending on how counties respond to the threat of continuing urban sprawl. Many innovative approaches to managing sprawl and smart development strategies can work to protect the Everglades restoration efforts from negative outside influences (see Regional and CERP Planning in Response to Large Scale Influences, later in this chapter). For this reason, the committee endorses efforts by CERP planners to estimate and account for increasing population and attendant land use changes and urban sprawl, and the committee recommends that these efforts be done in close consultation with state and county planning agencies.

Population Growth and Water Demand

Projections by the SFWMD show that urban expansion, even if accompanied by conservation measures, will result in increased demands for water from the regional hydrologic system (Gulf Engineers and Consultants and Taylor Engineering, 2003). In the year 2000, users in the SFWMD service area[4] received slightly more than 1,070 million gallons per day (MGD). Over the course of a year, this amount is about 1.2 million acre-feet, an amount 22 percent greater than the amount of the surface water presently flowing into Everglades National Park (based on the 1995 base, USACE and SFWMD, 1999).[5] The SFWMD's projections indicate that under the conditions of the most likely scenario for population growth and water conservation, by 2050 water use in the district will top 1,600 MGD, an amount that is 60 percent more than at present, and almost twice the amount of the present surface water flow into Everglades National Park (Gulf Engineers and Consultants and Taylor Engineering, 2003). The water protected by the "Savings Clause"[6] of the original CERP agreements will not be adequate for the needs of a growing human population, and whether the limited available fresh water supplies will be sufficient for both developed and conserved lands remains an unanswered question. The SFWMD and the state of Florida are committed to addressing potential water shortage problems by implementing increasingly strict conservation rules and management, reuse of water where possible, and promulgating consumption rules. The state also has in place a permitting system to manage applications of new users for existing water supplies (see Regional and CERP Planning in Response to Large Scale Influences, later in this chapter). Large cities may increasingly need to turn to conservation, water reuse, or desalination to expand their water supplies (NRC, 2008).

Additional Implications of Population Growth

If Florida's population growth trends of the past 170 years continue, as many planners and researchers anticipate, large population increases will change the context within which the CERP will evolve during the anticipated construction

[4]Includes Palm Beach, Broward, and Miami-Dade Counties; the Florida Keys portion of Monroe County; and portions of Martin, Okeechobee, Glades, Hendry, and Lee Counties.

[5]The calculations comparing water demand with surface flows into Everglades National Park are based on the following data: 1,070 MGD = 1,070,000,000 = 3,284 ac ft/d, or 1,198,550 ac ft/yr. Surface water inflows to Everglades National Park, according to the 1995 Base Primary Water Budget Components, are 915,000 ac ft, including flood-control discharges and environmental water supply flows from the WCAs and environmental water supply flows from the lower east coast.

[6]Enabling legislation for CERP mandated a "Savings Clause" that stipulated restoration must be conducted in such a way that water supply to urban and agricultural users being used at the time of the authorization (1999) would not be reduced.

time for the project. Urban growth, or sprawl, reduces the area of more natural surfaces and often consumes buffer areas around conserved lands. With continued urban sprawl, the Everglades is in danger of becoming an undeveloped area surrounded on all its land sides by high-density urban areas constructed directly up to the boundaries of conserved lands, a phenomenon already occurring in some areas. The replacement of ecotones, zones of gradual change from one ecosystem to another, by these sharp boundaries would adversely affect management of the remaining natural areas and could further endanger listed species. Thus, even if the CERP were to be completed as planned, the outcomes may be different from anticipated.

Increasingly large urban areas would also affect South Florida's natural areas indirectly because of increasing demands for power and transportation infrastructure. Population growth drives the construction of new generating facilities. Coal burning power plants are a primary source of mercury in the atmosphere over the Everglades, and atmospheric deposition is a major source of mercury contamination of the landscape and waters of the conserved areas and associated terrestrial and aquatic biota. The search for new clean energy sources will therefore have direct implications on water quality in the Everglades. Energy use, the generation of greenhouse gases, and climate change will also have important implications for the CERP. The movement and storage of water in South Florida demands considerable energy for pumping, and if some aspects of the CERP are constructed as envisioned, the demand for energy to be used in water management will increase, which will expand the state's carbon footprint unless specific methods are designed to reduce power demands. The state of Florida has dedicated itself to an ambitious plan of reducing greenhouse gas emissions to 1990 levels by 2025 and to 80 percent below 1990 levels by 2050; to achieve this goal, the governor has pledged wide-ranging plans to increase energy efficiency and pursue more renewable and alternative energy sources (State of Florida, 2007). Consequently, energy conservation measures will likely be an increasingly important part of the design of CERP projects.

Aggressive road and highway construction accompanies population growth and urban sprawl, and such construction is highly likely to affect Everglades restoration in an adverse way. The demand for building materials, particularly road aggregate and cement, is already acute, driving up construction costs for CERP projects. Continued and increasing pressure on finite construction materials would mean that funds appropriated for the CERP projects would not provide as much construction in the future as they have in the past.

The implications of an increasing human population for the completion of the CERP include the need for a continuing community commitment to the project. As more people move into the state, in many cases for a perceived quality of life that includes conserved natural areas, it is not clear whether they will be willing to pay for the restoration and preservation of such lands or willing to undertake other sacrifices, such as more expensive water, extensive water conservation measures, and geographical limits to sprawl. The influx of new Floridians will introduce new dynamics into the state's economy and politics, testing the durability of agreements hammered out by past decision makers and political leaders. In the end, the fate of the CERP is likely to rest in the hands of people who are not even in Florida yet, and who have no affinity for either the project or the ecosystem it is designed to restore and preserve.

Many of the implications of the increasing human population in South Florida result in increasing difficulties in accomplishing the goals of the CERP, but these population processes are not completely out of society's control. Communities can seek to find ways to manage the growth in a manner that is responsible for both the natural and the human environment. Use of "smart growth" principles and similar innovative approaches can lessen the negative impacts of population growth. Well-managed planning that integrates the needs of environmental restoration and human demographic changes can be effective if it engages cities, counties, the state, and CERP planners. Regardless of the effectiveness of such integrated planning efforts, the natural values of the Everglades ecosystem are much likely to fare better with a completed CERP in place.

Climate Change

Both the Everglades ecosystem restoration and the growing human population of Florida, with its demands for increasing amounts of water, will take place in an uncertain hydro-climate. In the following section, the implications for the CERP of short-term variability and long-term climate change and sea-level rise are considered.

Natural Climate Variability

Variations in rainfall occur from year to year and decade to decade. The experience of the 20th century clearly demonstrates the high variability in rainfall that creates alternating wet and dry periods (Figure 2-8), and such short-term variability (over a few years) can be expected to continue. Because of the variability, CERP planning does not depend on a constant set of moisture conditions; rather, it considers a 36-year precipitation record with extended periods of drought and is working to incorporate an even longer precipitation record to accommodate more natural hydrologic variability in the planning process.

Some of this variability is due to random events, regional-scale climatology,

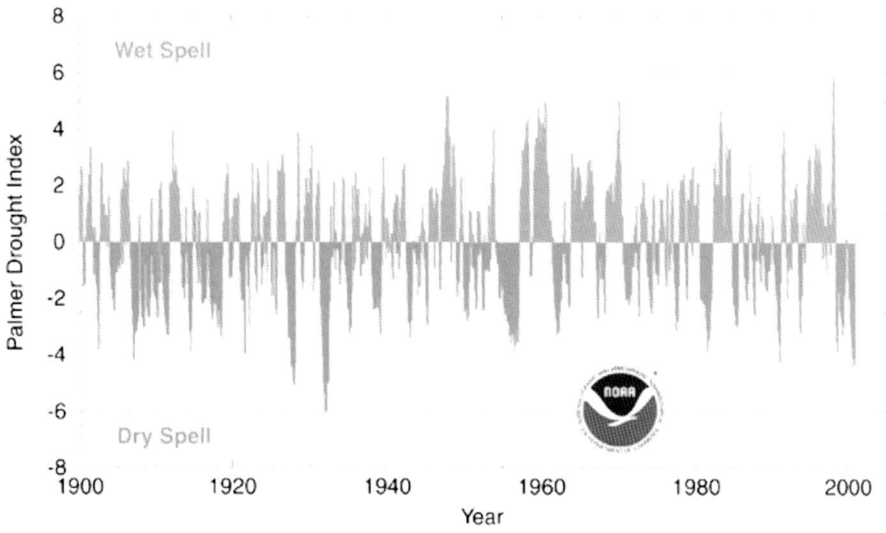

FIGURE 2-8 Fluctuations of the Palmer Drought Severity Index for Florida for the 20th century show that variability is common, with extreme variation. The Palmer Drought Severity Index measures meteorological conditions over a relatively short period (in this case, years) and compares them with long-term (in this case a century) averages. Thus, the index compares any given year with its long-term context. The graph shows the variability of moisture available in the Florida ecosystem; droughts are common on a century-long time scale.

SOURCE: *http://www.ncdc.noaa.gov/img/climate/research/2000/dec/fl0000pdi_pg.gif.*

or hurricanes passing over or near Florida, but an understanding of the importance of climate variations occurring on frequencies of several years or decades and caused by basin-scale ocean-atmosphere dynamics is slowly emerging. The most familiar cyclic event is the El Niño/Southern Oscillation, which has a relatively short cycle of 3 to 7 years. During its La Niña phase, it brings warmer and drier fall and winter seasons to Florida, often resulting in an increase in the number of forest and glades fires, such as in 1999, 2000, 2006, late 2007, and 2008. Because they operate over longer cycles, the Atlantic Multidecadal Oscillation (AMO) and Pacific Decadal Oscillation exert interacting effects that are pervasive over North America (McCabe et al. 2004) but are difficult to resolve in the instrumental record. During the AMO "warm phase," for example, annual rainfall in South Florida may be greater but more variable than at other times (RECOVER, 2007c).

In addition to these short-term and cyclic variations in climate that affect the timing and amounts of precipitation and evapotranspiration, and thus the water budget, water resource planners and managers will have to take into account that major changes in Earth's climate systems are already taking place as a result of global warming (IPCC, 2007). These changes have led a prominent group of hydrologists to proclaim that stationarity, the notion that natural systems fluctuate within an unchanging envelope of variability that can be defined by past observations, is "dead" (Milly et al., 2008). Climate change is undermining a basic assumption that historically has been the basis of planning for water supplies, demands, and risks and must henceforth be taken into account.

Human-induced Climate Change

The growing evidence of human-induced climate change, brought to a head in the more definitive fourth assessment of the Intergovernmental Panel on Climate Change (IPCC) in 2007, has shaped public perception of the immediacy and inevitability of global warming and the urgency of action to mitigate the accumulation of greenhouse gases. Many states, including Florida, have taken steps to mitigate emissions and develop strategies to adapt to the changing climate (State of Florida, 2007). In short order, taking into account the effects of 21st-century climate change in environmental management and ecosystem restoration has gone from an exercise that was too uncertain or too politically sensitive to an expectation for credible planning.

The IPCC reports present projections for climate change on global and continental scales, and the outputs of the supporting model runs from multiple modeling centers are available in archives. These model results are available as maps that have been statistically "downscaled" to a finer spatial resolution[7] that are being used in regional and U.S. climate impact assessments (e.g., Union of Concerned Scientists, 2006). The downscaled models project air temperatures in South Florida warming by about 2°F throughout the year by mid-century and by 3 to 5°F, depending on the trajectory of greenhouse gas emissions by the end of the century. Although the climate models produce much more variable projections for precipitation, model averages indicate decreases in precipitation in all seasons except the fall. The projected decreases are modest (generally 10 percent or less) but are greater toward the tip of the peninsula, approaching 30 percent toward the end of the century under higher-emissions scenarios.

---

[7]See *http://gdo-dcp.ucllnl.org/downscaled_cmip3_projections/#Datapercent3Apercent20Complete percent20Archives.*

Although projections of precipitation have much more uncertainty than those for temperature, the fact that most models project decreases in precipitation, consistent with the general expectation for drier subtropics, and the more certain increase in evapotranspiration suggest that stationarity—assuming that past observations define future probabilities—is not a good assumption for water resource planning in South Florida.

Hurricane frequency and intensity are important to Everglades hydrology and water supply because hurricanes deliver large quantities of water over short time periods, affecting water levels in the Everglades and Lake Okeechobee and increasing the risk of urban flooding. Although there is much interannual variability in hurricane frequency related to a variety of complex factors, there has been an increase in the number of hurricanes since the 1980s that may be attributable to favorable atmospheric circulation patterns related to the AMO (Figure 2-9). The relationship of hurricanes to global warming, however, has been hotly debated in the scientific community (Mooney, 2007). A recent synthesis report of the U.S. Climate Change Science Program produced a consensus that the destructive potential of Atlantic tropical storms and hurricanes increased since 1970 in association with warming of sea surface temperatures (Figure 2-10), but a similar relationship with the frequency of hurricanes could not be drawn (Karl et al., 2008). The consensus concluded that it is likely that hurricane winds and rainfall will increase in response to the expected continued warming of sea surface temperatures, but changes in hurricane frequency cannot be predicted with any confidence. Even more recent modeling studies have projected a reduction in hurricane and tropical storm frequencies under warming scenarios but, at the same time, increased storm intensity (Emanuel et al., 2008) or near-storm rainfall rates (Knutson et al., 2008). It is clear that episodic incursions of hurricanes will continue to be a feature of Everglades hydrology and likely that hurricanes will intensify even if they are not more frequent. Thus, CERP components will need to be resilient enough to accommodate such radical short-term changes in water quantity in the system.

Sea-Level Rise

Changes in sea level will also have significant effects on restoration options and requirements for the Everglades. The entire watershed of the Everglades has a land surface of very low relief: the highest elevation in the basin is only 65 feet above mean sea level, and elevations in the area south of Lake Okeechobee are 12 feet or less. Gradients for water flows are as little as an inch per mile. Under these topographic conditions, even small changes in mean sea level are likely to have far-reaching effects that will alter the general character of the physical

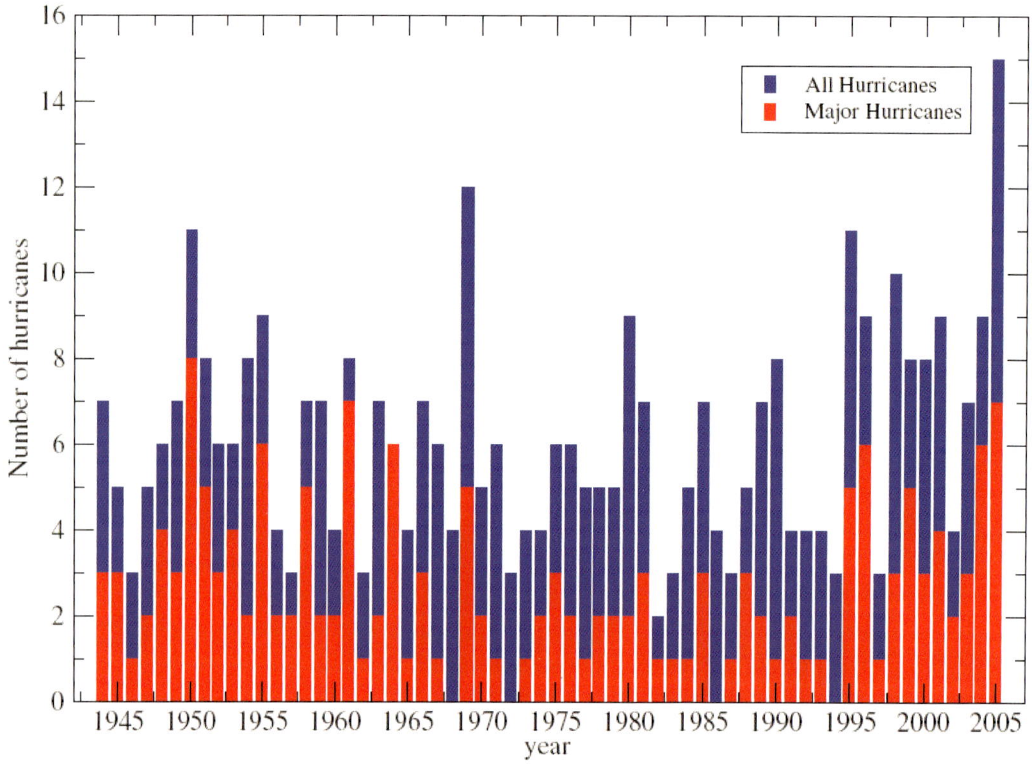

FIGURE 2-9 Number of hurricanes in the Atlantic Basin, 1945–2005, showing a general decline from about 1950 to about 1994, but somewhat larger numbers since 1994.

SOURCE: NOAA (2006).

environmental context of the Everglades and even the general shape of the lower Florida peninsula. In fact, the specter of three feet or more of sea-level rise has prompted some to question the wisdom of restoring the Everglades at all if it is consigned to be inundated in the near future (Dean, 2008).

Over most of the last 2,500 years of this interglacial period, relative sea-level rise in South Florida averaged about 1.6 inch/century (Wanless et al., 1994). However, relative sea-level rise of approximately 9 inches/century was observed during the 20th century based on tide gauges at Key West and Miami Beach.[8] The IPCC's fourth assessment (IPCC, 2007) projected a rise in global sea level of

[8]See *http://tidesandcurrents.noaa.gov/sltrends/sltrends.shtml.*

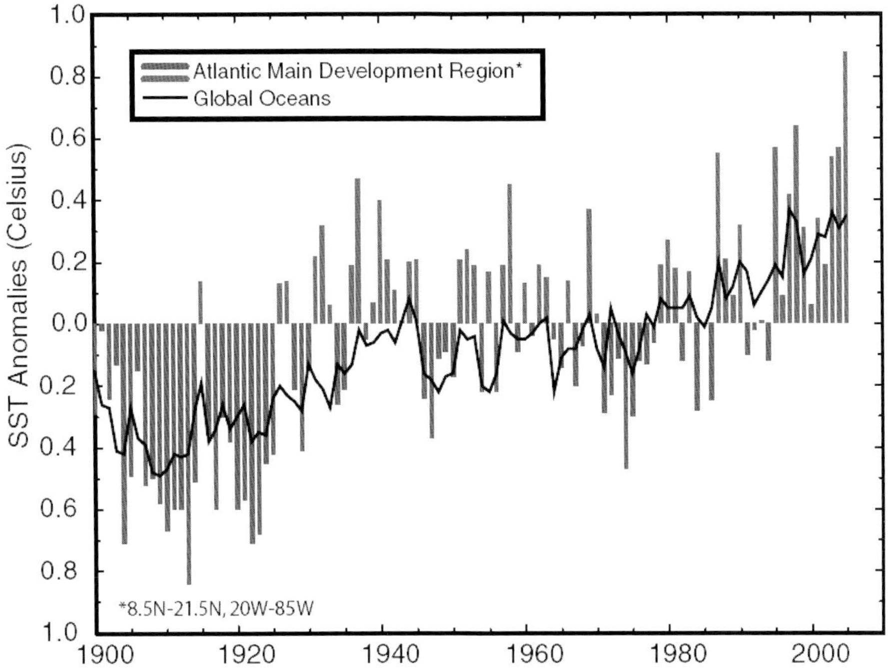

FIGURE 2-10 Sea surface temperature anomalies in global oceans and in the part of the Atlantic Ocean where hurricanes originate, which may influence hurricane climatology and, ultimately, Everglades hydrology.

SOURCE: NOAA (2006).

7 to 15 inches over the present century under its lowest greenhouse gas emissions scenario (B1) and 9 to 20 inches under a high-emissions scenario (A2). When adjusted for differences between South Florida gauge estimates and global mean sea-level rise during the 20th century (reflecting vertical land movement and other local factors), the IPCC projections suggest a 4- to 9-inch rise in South Florida by mid-century and a 9- to 17- (lower emissions) or 11- to 22-inch (higher emissions) rise toward the end of this century. However, the IPCC projections specifically excluded estimation of additional sea-level rise that might be due to further acceleration in the melting of glaciers, ice caps, and polar ice sheets. Extrapolation of the recently observed acceleration of loss of ice volume (Meier et al., 2007), as well as statistical extrapolation based on the relationship of sea surface temperature and sea level (Rahmstorf, 2007), both suggest that under a high-emissions scenario (continued growth in greenhouse gas emissions throughout the

century), sea level could rise by another 14 inches or so beyond the IPCC projections. Considering these high-side risk projections collectively, sea level in South Florida could possibly rise as much as 14 inches by mid-century and 36 inches (3 feet) by the end of the century, if the growth of emissions is not reversed.

CERP Guidance Memorandum 016.00 (USACE and SFWMD, 2004a) provides probability distributions for sea-level rise in South Florida to be used in project planning. It indicates a most-probable scenario of sea-level rise of 0.8 feet (about 10 inches) for 2050 and 1.7 feet (20 inches) by 2100. The 14- and 36-inch projections developed represent an approximation of how high sea level could conceivably rise with accelerated melting based on current scientific understanding. Of course, the uncertainty in these projections increases with time into the future. The CERP Guidance Memorandum projected sea-level rise with the probability of 10 percent exceedance at 14 and 32 inches, for 2050 and 2100, respectively (Table 2-1), and these are very similar to the reasonable upper-end projections. However, to plan based only on the most-probable (mean) sea-level rise of 0.8 feet in 2050, as the Guidance Memorandum suggests, disregards the skewed nature of the probability distribution and the risks of greater acceleration of sea-level rise.

The Science and Technology Committee of the Miami-Dade County Climate Change Task Force (hereafter, simply the Science and Technology Committee) (2007) suggested that sea-level rise of up to 5 feet could occur by the end of the century. While there are considerable uncertainties about the rate of melting of polar ice sheets (the reason that the IPCC declined to project this contribution), the most-recent results of glaciologists' research suggest that sea-level rise of much more than 3 feet this century is not very likely (Meier et al., 2007). Nonetheless, the Science and Technology Committee (2007) correctly pointed out the substantial impacts of even a 2- to 3-foot rise in sea level in low-lying and geologically porous Miami-Dade County, particularly when spring high tides and storm surges are added to the changes in mean sea level. Moreover, if relative sea level does rise by 3 feet during this century, it will very likely reflect accelerating and unstoppable melting of polar ice that portends even higher rates of sea-level rise during the next century.

Sea-level rise has significant consequences for Everglades ecosystem restoration. Salt-tolerant mangroves will expand at the expense of freshwater wetlands (Figure 2-11). The degree to which wetlands will survive inundation depends on the rate at which sediments and soils are accumulated. Higher sea levels also affect the flow gradients in the lower Everglades and hydraulic head differences that affect seepage and operations of control structures. These were evaluated during the development of the CERP by simulating the effects of a sea-level rise of 0.5 feet (15 cm) (Trimble et al., 1998). Few effects were seen on the interior

TABLE 2-1 Probability Distribution of Sea-Level Rise for Miami Beach for Years 2025, 2050, 2075, and 2100

Percent chance exceedence	2025		2050		2075		2100	
	cm	ft	cm	ft	cm	ft	cm	ft
90	7	0.2	13	0.4	20	0.7	27	0.9
80	9	0.3	17	0.6	26	0.9	36	1.2
70	11	0.4	20	0.7	30	1.0	42	1.4
60	12	0.4	22	0.7	34	1.1	46	1.5
50	13	0.4	24	0.8	37	1.2	51	1.7
40	14	0.5	27	0.9	41	1.4	56	1.8
30	16	0.5	29	1.0	44	1.5	62	2.0
20	17	0.6	32	1.1	49	1.6	70	2.3
10	20	0.7	37	1.2	57	1.9	81	2.7
5	22	0.7	41	1.4	63	2.1	92	3.0
1	27	0.9	49	1.6	77	2.5	118	3.9
Mean	13	0.4	25	0.8	38	1.3	53	1.7

SOURCE: USACE and SFWMD (2004a).

hydrology of South Florida, but lower east coast water supply cutbacks and peak-stage flooding in some areas increased significantly. The above considerations require new analysis of impacts based on higher sea-level rise assumptions, and the CERP Guidance Memorandum should be amended accordingly.

Adaptation to Climate Change

The range of possibilities of climate and sea level change during the 21st century do not indicate that Everglades restoration is either infeasible or futile, only that the changing conditions will have to be taken into account and adapted to. Rising sea level is likely to change the character of the lower Everglades. Plant communities will have different distributions from those at present, and water flows are likely to change, but dynamic aquatic and terrestrial habitats are likely to continue to be part of the remaining undeveloped South Florida ecosystem. Moreover, impending climate change should not be an excuse for delay or inaction but, rather, as motivation to avoid irreversible losses and restore the resilience of the ecosystem.

The impacts of the long-term climatic fluctuations and changes in temperature, precipitation, and sea level, however, are significant to the CERP in many ways. Among those possible changes that should be assessed and factored into planning and implementation are the following:

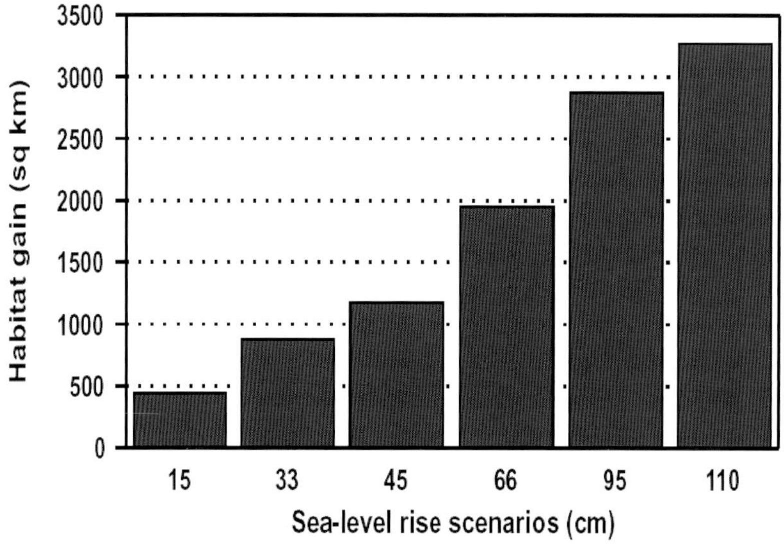

FIGURE 2-11 Projections of the expansion of mangrove habitat in the lower Everglades under various sea-level scenarios based on elevation gradients and plant succession models.

SOURCE: Doyle (2003).

• Changes in the water budget and its variability, including the amount of precipitation and its temporal distribution (changing seasonality and frequency of intense precipitation events and droughts) and the effects of increased evapotranspiration under the warmer conditions expected. CERP managers are beginning to build into their hydrologic models the capability to accommodate climate changes on time scales of about 50 years, and particularly to account for AMO influences (RECOVER, 2007c); the inclusion of longer-scale adjustments is possible. Because of increasing demands for water for a growing urbanized population, even small changes in the amount of available water may pose management challenges. Potential changes in water availability should be factored into targets for water levels and flows and ecosystem restoration, considerations of the greater frequency of fires, and plans for preservation of endangered species at particular risk, such as the snail kite. The effects of climate change on human demands for water should also be considered, as should setting time limits of 20–25 years on water allocation permits so that the next generation can reevaluate apportionment of the potentially changing water supplies.

• Changes in the return frequency and intensity of hurricanes and tropical storms. While it is not possible at this time to project the frequency of land-falling hurricanes in Florida, the evolving research in this area could better inform risk analyses.

• Effects of temperature changes on the distribution of plants and animals in the Everglades ecosystem, including implications for invasive species.

• Consequences of increasing concentrations of carbon dioxide on plant growth, biodeposition of carbonate sediments, and soil building processes. Management approaches to enhance sediment accretion in the lower Everglades to keep pace with sea-level rise should be evaluated.

• Impacts of projected sea-level rise on the hydro-geomorphology of northern and southern estuaries, saltwater intrusion, and transgressions of the mangrove zone.

Regional and CERP Planning in Response to Large-scale Influences

CERP planners are cognizant of major large-scale influences on the restoration, including population growth, water demand, land use change, short- and long-term climate variability, and sea-level rise. Local, state, and federal officials at the planning stage are attempting to prepare for the consequences of these influences.

The SFWMD consistently updates its population growth projections and factors them into their water plans. Their objective is to identify future needs now so that adequate supplies are in place when they are needed (Balbin, 2008). In anticipation of higher demands for water, South Florida governments have initiated efforts to protect water for the environment while making the most-efficient use of existing water supplies. The SFWMD's Regional Water Supply Availability Rule limits consumptive use withdrawal of water from Everglades water bodies and requires consumptive users to develop alternative water supplies for increased allocation. Southeastern coastal areas may opt to develop deeper aquifers, for example, and most areas will need to institute stringent conservation measures and water reuse strategies. The SFWMD established an alternative water supply grant program that provides funding to local municipalities developing desalination or water reuse facilities.

Land use management is largely in the hands of counties in Florida. Counties create plans for acceptable use and try to accommodate the urban sprawl that results from population growth (e.g., Miami-Dade County, 2008). The SFWMD reviews the plans of each county from the standpoint of water supply and assists counties in defining potential needs for water. Each county in the Everglades watershed anticipates some growth management, but because the authority for land use planning is fragmented among the counties, there is no central clearinghouse for coordination. However, the state of Florida established the Rural Land Stewardship Area (RLSA) program in 2001 to provide a mechanism for counties to designate such areas to prevent urban sprawl, protect natural resources, and promote rural economic activity (Florida Department of Community Affairs, 2007). Under the RLSA program, some counties such as Collier and St. Lucie Counties have tried applying growth management tools such as Transfer of Development Rights to steer development away from agriculturally important or environmentally sensitive areas and concentrate it in or near existing urban areas. If implemented appropriately, such efforts could significantly reduce South Florida's urban footprint and its environmental impact. It appears that current and proposed RLSAs, which already total over 900,000 acres in the CERP region, are not being designed or implemented in consultation with restoration efforts, although the RLSA process would seem to be a good forum for aligning county growth management planning with restoration objectives and activities.

As described in the previous section on climate change, the USACE and SFWMD are working to accommodate short- and long-term climate change, and a guidance memorandum (USACE and SFWMD, 2004a) was issued to provide advice about climate change for CERP project planning. CERP assessments, evaluations, and management recommendations all are taking short- and long-term climate change into account. RECOVER has stated that the physical characteristics of CERP facilities will need flexibility to accommodate anticipated changes, and their operational plans will include a built-in resilience to deal with climate changes and sea-level rise (RECOVER, 2006f). SFWMD modelers responsible for predicting the behavior of a restored Everglades hydrologic system are analyzing the implications of intrusions of salt water and fluctuations in water supply that result from short-term climate changes. They are also working to downscale global circulation models to anticipate the long-term climate changes at regional scales that are useful for Central and South Florida (L. Gerry, SFWMD, personal communication, 2008). State and local officials are also anticipating the possible effects of sea-level rise, particularly along Florida's southeast coast (e.g., Science and Technology Committee, 2007).

Many of these efforts, however, are in their infancy, and it is too early to evaluate how effective these planning strategies will be to mitigate these large-scale influences on the CERP.

RECENT CHANGES IN THE SOUTH FLORIDA ECOSYSTEM

NRC (2007) described some recent trends in the ecology and hydrology of the South Florida ecosystem, demonstrating that the natural system will continue to move away from conditions that support natural ecosystem processes until greater progress is made in implementing CERP and non-CERP projects. More recent trends suggest that the ecosystem is at risk and that some important components are showing continued declines. In this section, the status and trends for tree islands, several bird species, Lake Okeechobee, and exotic and invasive species are reviewed as examples of recent changes to the ecosystem that compromise its resiliency. These examples are critical components of the South Florida ecosystem, and because each involves numerous aspects of water quantity, quality, flow, and biology, they serve as indicators of the status of functional components of the system. Finally, the concept of regime shift is discussed, based on the committee's concerns that continued declines may lead to ecosystem conditions that may be very difficult to restore.

Tree Islands

Tree islands are visually striking and ecologically critical habitats in the Everglades landscape (Figure 2-12). These small and relatively dry patches of

FIGURE 2-12 Aerial view of tree island landscape in Shark River Slough.

SOURCE: Courtesy of Ross and Jones (2004).

trees and woody shrubs set amidst the grasses, sedges, and aquatic plants of the large expanses of flooded land provide unique and vital resources for wildlife foraging and nesting. They are found over a large area, from the Loxahatchee Wildlife Refuge through the WCAs and in the Shark River Slough of Everglades National Park (Ross and Jones, 2004). They are refugia during high waters, and they are biodiversity "hotspots" for both plants and animals (Armentano et al., 2002; Gawlik and Rocque, 1998; van der Valk and Sklar, 2002).

Over the past decade, several studies have documented decreases in the extent of tree island habitats. Hofmockel (1999) reported that between 1953 and 1995, WCA-2A lost 87 percent of its tree islands. The most recent analysis of tree island change (Sklar and van der Valk, 2002; Van der Valk and Sklar, 2002) used aerial images to document changes in WCA-3 from the 1940s to 1995 (Figure 2-13). The primary period of tree island area loss occurred between 1950–1970, with lower rates of loss before and after. This analysis suggested a decrease in total tree island area of 67 percent and a decline of 54 percent in the number of islands in WCA-3. This loss is generally attributed to changes in both water levels and fire frequencies (Brandt et al., 2002). Sklar (2007a) predicted that tree island numbers and areal extent will continue to decline due to muck

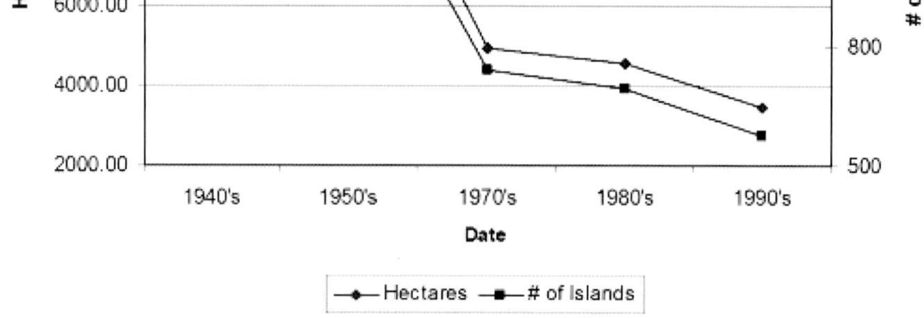

FIGURE 2-13 Area and number of tree islands in WCA-3A between the 1940s and the 1990s.

SOURCE: Courtesy of Sklar (2007a).

fires if restoration is further delayed. With further delays, resilience to hydrologic variability may also decline, creating lethal flooding stress when historic water levels are ultimately restored.

Endangered Everglades Birds

Population trends over the past 5 to 10 years are quite variable among the Everglades' most-high-profile and threatened bird species, but several downward population trends are quite clear and appear to be related to water levels and their management within the southern Everglades project area.

Snail Kites

The snail kite (*Rostrahamus sociabilis*), a specialized hawk that feeds almost solely on freshwater snails of the genus *Pomacea,* has been listed under the Endangered Species Act since 1967 (Beissinger, 1990; Snyder and Snyder, 1969; Stieglitz and Thompson, 1967; Sykes et al., 1995). This high degree of diet specialization makes the snail kite dependent on flooded wetlands to feed and nest and vulnerable to population declines if it is unable to find snails, such as during regional droughts (Beissinger, 1995). Destruction of Everglades wetlands and the drying of marshes caused a population decline to approximately 50–75 individuals in the late 1960s and early 1970s (Stieglitz and Thompson, 1967; Sykes, 1979), but kite populations in Florida made a remarkable recovery to over 3,500 individuals in 1999 (Martin et al., 2007a) following a decade of relatively high water levels. Snail kite numbers over the past 5 years, however, have not recovered from the major drought of 2000–2001 (Martin et al., 2007a; Figure 2-14). In 2007, when water levels were very low in many areas in Florida, kites declined by 27 percent to about 1,200 individuals (J. Martin, University of Florida, personal communication, 2008). Kite reproduction in WCA-3A, the largest and most consistently used area of snail kite critical habitat, has declined precipitously during the past 5 years. No young were known to fledge from WCA-3A in 2005 and 2007, and only 9 of 81 (11 percent) nests successfully produced young in 2006 (Martin et al., 2007b).

Declines of the kite in WCA-3A may be partially attributable to the manner in which water is managed. According to the current Interim Operational Plan (IOP), water is held behind the S-12 structures from November to March (see Figure 2-15). When water is shifted rapidly to the south into Everglades National Park from April through June, it results in rapid recession rates that can leave kite nests vulnerable to terrestrial predators, further reducing the rate of survival of juveniles after they fledge (Martin et al., 2007b). Water management and climate

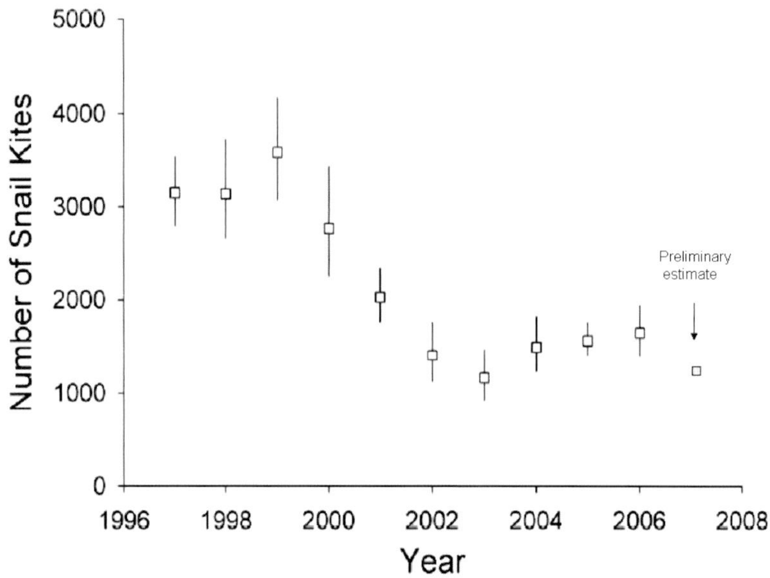

FIGURE 2-14 Estimates of the mean and 95 percent confidence interval for snail kite population size in Florida between 1997 and 2007.

SOURCE: USFWS (2007) based on data supplied by W. Kitchens, University of Florida.

factors (e.g., the AMO) have also led to deeper levels of inundation during the wet season in WCA-3A over the past decade compared to that which occurred between the mid-1960s and the early 1990s.[9] These deeper water depths are also having negative effects on snail populations and are changing wetland plant species composition to less-favorable communities for kites and snails (Darby et al. 2008; Karunaratne et al., 2006). Apple snail (*Pomacea paludosa*) abundance appears to have declined substantially within WCA-3A (Darby et al., 2005).

Cape Sable Seaside Sparrow

The Cape Sable seaside sparrow (*Ammodramus maritimus mirabilis*; hereafter, simply CSSS), which was listed as an endangered species in 1968, is a morphologically, genetically, and ecologically unique subspecies of seaside

[9]For more information, see water depth data from 1966 to the present at *http://www.fgcu.edu/bcw/wca3a/wca3a.htm* (sites 62, 64, and 65).

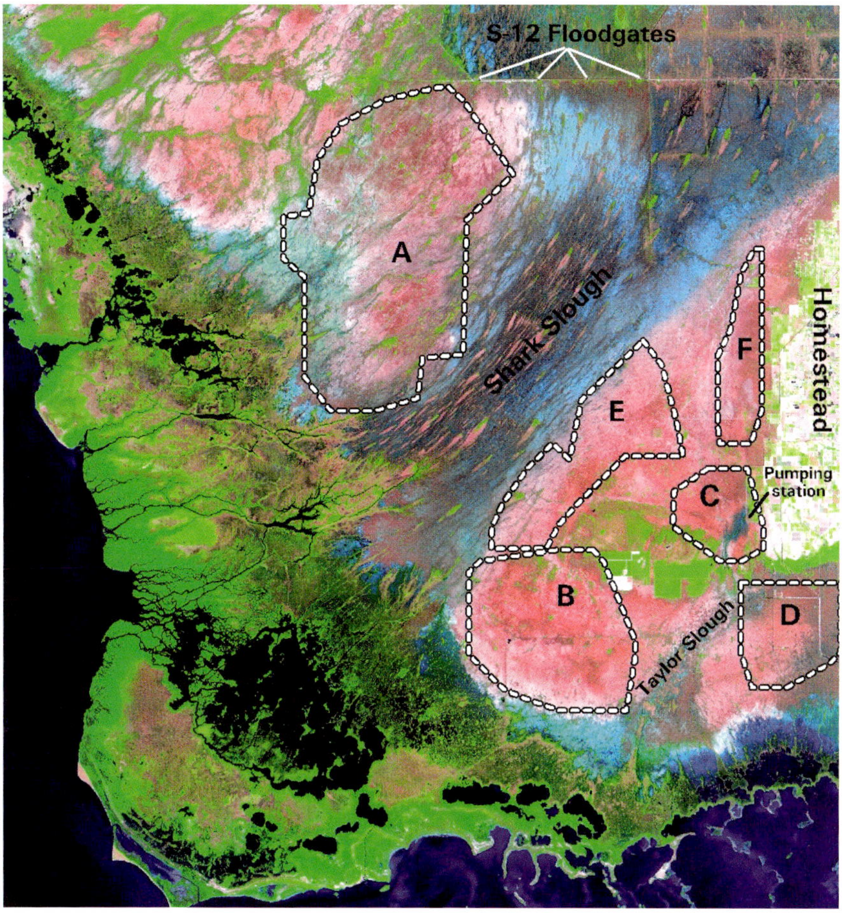

FIGURE 2-15 Locations of Cape Sable seaside sparrow subpopulations. Cape Sable is the landmass on the lower left (southwest) edge of the Florida peninsula.

SOURCE: Courtesy of Pimm et al. (2002) .

sparrow that is restricted to the Everglades ecosystem (Kushlan et al., 1982; McDonald, 1988). The CSSS is now distributed in what has been termed 6 subpopulations within the marl prairies (Figure 2-15), although presently only two of these areas support populations with more than 100 individuals (Hallac et al., 2007; Pimm et al., 2002; Walters et al., 2000). Large declines in the proportion of area occupied by CSSSs within its range have been demonstrated across all the subpopulations between 1981 and 1992 (Cassey et al., 2007).

During the past 5 to 10 years, the total number of CSSSs appears to have remained relatively stable (Figure 2-16), numbering around 3,000 individuals, but the number of subpopulations has declined from 6 to 4. Most individuals are in 2 subpopulations (B and E) that support 80–90 percent of the remaining individuals (SEI, 2007). Subpopulations B and E appear to have remained stable in recent years, but other subpopulations have declined. Most notable has been the precipitous decline of subpopulation A on the northwest side of Shark River Slough, which was estimated to have supported several thousand individuals in 1992 (Walters et al., 2000), 128 birds in 2003, and only 64 sparrows in 2007 (D. Hallac, NPS, personal communication, 2008). This decline occurred despite the attempts to reduce water flows across the S-12 structures from November

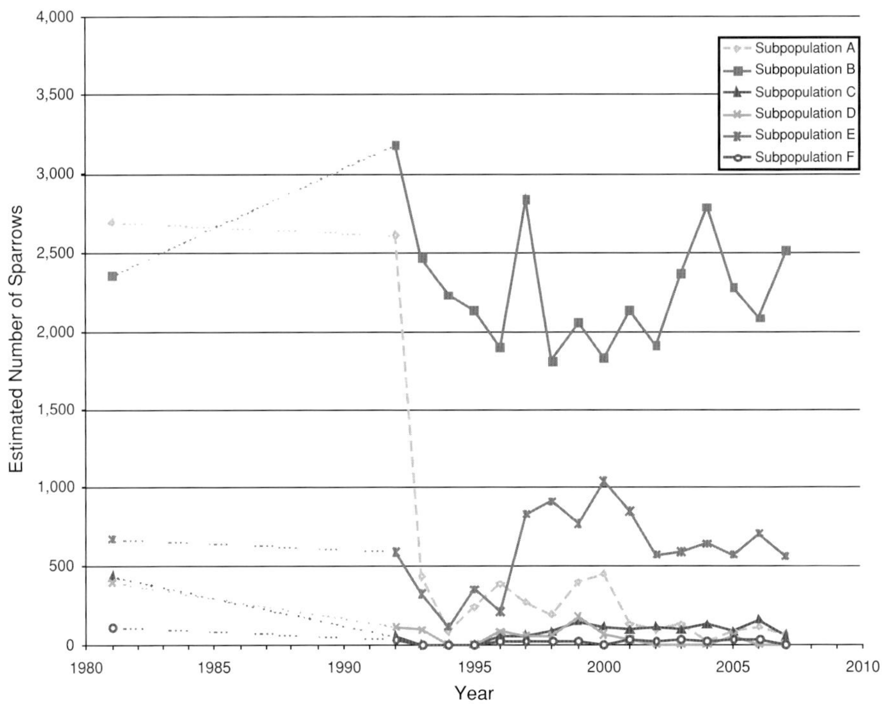

FIGURE 2-16 Estimated number of Cape Sable seaside sparrows by subpopulation.

SOURCE: Based on data received from D. Hallac, NPS, personal communication (2008).

through April to increase nesting success of sparrows in subpopulation A. Several of the other subpopulations have also exhibited recent population declines, and subpopulations D and F had no birds detected in 2007 (Figure 2-16).

Water management is integrally linked to the survival of the CSSS (Nott et al., 1998). Nests of these sparrows are susceptible to inundation if water levels rise quickly (Nott et al., 1998), and recent work suggests that nest predation may also be linked to water levels, as rates appear to increase under both high and low conditions (Baiser and Lockwood, 2006; Lockwood et al., 2006). Moreover, water level conditions that permit multiple brooding appear to be very important if CSSS populations are to increase (Walters et al., 2000). A change in habitat suitability as a result of shifts in hydrologic characteristics may have resulted in the drastic decline of subpopulations A and D (Pimm et al., 2002; SEI, 2007).

Wood Storks and Other Wading Birds

Although most wading birds are not listed under the Endangered Species Act, with the exception of the wood stork (*Mycteria americana*), they are considered important indicators of Everglades ecosystem functioning. The total number of wading birds using Everglades National Park and WCA-3A has increased over the past 5 years from about 330,000 to 500,000, although their distribution has changed. Four of seven species of wading birds (great egrets, great blue herons, wood storks, and white ibis) have shown an annual increase over the past 5 years, and a larger number of wood storks were identified in 2006 and 2007 than had been documented in the previous 40 years in the United States (Cook and Herring, 2007). However, large colonies of wading birds are now found in more northern areas, particularly northeastern WCA-3A, rather than in the southern Everglades where they historically occurred (Crozier and Cook, 2004). Only one species—the great white heron, which primarily resides in Florida Bay—has shown a decline (Alvarado and Bass, 2007).

Restoration Delays Further Endanger Everglades Birds

While wading birds responded favorably to environmental conditions in the Everglades over the past 5 years, endangered Everglades birds have not recovered. Recovering endangered species requires identifying and ameliorating the causes of population decline. Declines in endangered Everglades bird populations often are not gradual; instead, they occur after some catastrophic event (e.g., drought or flooding, hurricanes, fires). All populations are subject to these effects, but when populations decline to a few hundred or a few thousand individuals, their resiliency to recover from natural perturbations is greatly reduced, further increasing their risk of extinction due to natural climate and environmental variations.

A panel of independent experts (SEI, 2007) who reviewed the current situation of endangered birds in the Everglades recently concluded that the status quo of flow conditions and water schedules in the WCAs and Everglades National Park "is not an option if the goal is to restore the ecosystem and prevent the extinction of critically endangered species. Incomplete implementation of emergency measures and failure to complete more major plans in a timely way increases the risks to endangered species. Moreover it makes it more difficult and more expensive to recover them."

Ongoing delays in the Mod Waters project (see Chapter 4) not only have postponed improvements to the hydrologic conditions but has allowed ecological decline to continue. As discussed previously, recent water management strategies (i.e., the IOP) have not produced conditions that are conducive to restoring the sparrow and appear to be negatively impacting the kite (SEI, 2007). The recent Sustainable Ecosystems Institute's (SEI's) avian ecology panel (SEI, 2007) stated that "continuing degradation of the ecosystem has reached the point that there is immediate concern about both of these species (CSSS and Snail Kite) rather than just the former." That same SEI panel also stated that completing Mod Waters is critical to maintaining healthy avian populations in the Everglades.

Fortunately, none of the endangered bird species exists in such small numbers that they are in immediate danger of extinction. There is still time to rectify the situation through restoration activities, even if some populations are impacted as a necessary consequence of the early transitions of restoration. Nevertheless, positive ecosystem changes may take years or decades to occur once restoration activities are enacted. Further loss of numbers and habitat deterioration due to delays in completing Mod Waters reduces the opportunity for adaptive management, may preclude allowing some incidental take, and increases the chance that an extreme weather event (e.g., hurricane or drought) could imperil the existing populations.

Lake Okeechobee

Lake Okeechobee has been profoundly altered by the combination of diking and connection to the coastal estuaries (St. Lucie to the east and Caloosahatchee to the west). The changes have affected not only the amounts and flows of water but also water quality, especially the overload of phosphorus. As a result, the biotic communities in the lake and in the estuaries also have undergone significant changes (SFWMD and FDEP, 2005, 2008a) (see Chapter 5 for a more detailed discussion). For example, the goal for shoreline water clarity in Lake

Okeechobee (100 percent visibility to the lake bed from May through September) was met less than 10 percent of the time during the past 5 years (SFWMD and FDEP, 2008a). Submerged aquatic vegetation declined substantially, and other components of the vegetation experienced large changes (SFWMD and FDEP, 2008a).

The zooplankton and fish also experienced changes in recent years, particularly in response to four hurricanes in 2004 and 2005. In general, piscivorous fishes declined, while omnivores and planktivores increased, with marked declines in the fish species of greatest recreational and commercial interest, particularly largemouth bass and various species of sunfish (Johnson et al., 2007). The changes in water quantity and quality in Lake Okeechobee have been accompanied by declines in most of the water bird populations, although the cause-effect mechanisms are not well understood for all the species (e.g., Beissinger and Snyder, 2002; Rodgers, 2007).

Exotic and Invasive Species

Invasive, nonindigenous (nonnative) species are a large and expanding threat to the South Florida ecosystem. There are at least 32 invasive nonindigenous plant species and over 150 nonindigenous animals found in the South Florida ecosystem. Although many invasive species are widespread and occupy large areas (e.g., Brazilian pepper [*Schinus terebinthifolius*]), some of the most damaging invasive plants have been brought under control through a combination of vigorous control efforts and introduction of biocontrol agents (Ferriter et al., 2008). The most striking success is for *Melaleuca*; only 7,000 acres are currently heavily infested—down from nearly 500,000 acres in 1993 (TAME Melaleuca, 2004)—although approximately 270,000 acres remain under maintenance control (Ferriter et al., 2008; Morgan and Allen, 2007).

Despite effective control mechanisms for some invasive, nonindigenous plant species and $21 million per year spent on exotics control, both the number and the abundance of invasive species continue to increase (e.g., Figure 2-17). However, few quantitative data are available to adequately track changes in the number, abundance, and distribution of invasive species, especially invasive animals. None of the eight physiographic regions within the South Florida ecosystem is currently free of exotic species invasion, and three are predicted to decline in quality over the next 1–2 years due to the expansion of invasive nonindigenous plant species (Table 2-2; Ferriter et al., 2008). Recent proposals to develop biofuel plantations growing giant reed (*Arundo donax*), a known invasive species near the Everglades, have sparked concern that a new source of problematic species is developing in the region (Rosenthal, 2008). Even if Ever-

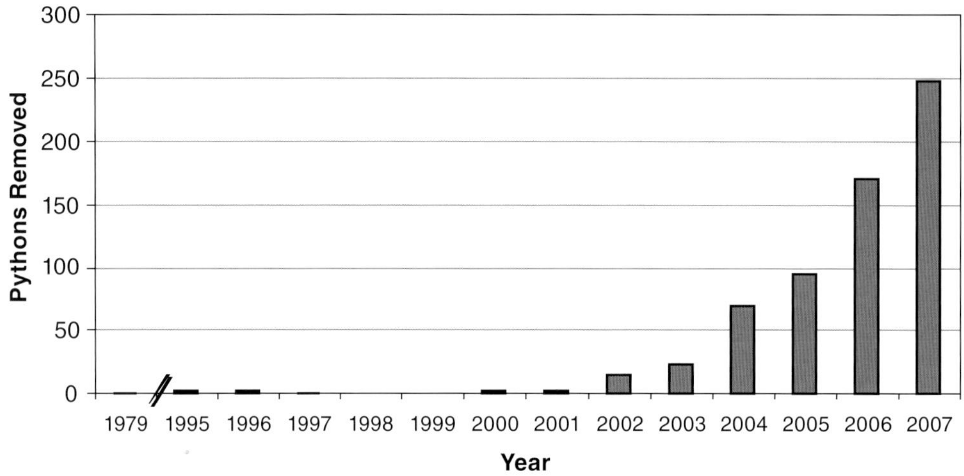

FIGURE 2-17 Number of Burmese pythons removed from the Greater Everglades region between 1979 and 2007. Increase largely reflects the result of reproduction, dispersal, and establishment in the park and on adjacent lands, with some increase in reporting effort.

SOURCE: S. Snow, NPS, personal communication (2008).

glades hydrology is effectively restored, the continuing introduction and spread of damaging species could threaten the restoration of ecological integrity.

Current Trends and Regime Shifts

The observed patterns of species and habitat decline and increasing threats from invasive nonindigenous species need to be understood within the context of regime shifts (also called alternate stable states). The ability of an ecosystem to resist change in its configuration, given a disturbance or change in environmental conditions, is referred to as ecological resilience (Gunderson, 2000; van Nes and Scheffer, 2004). However, given sufficient environmental changes, the ecosystem can undergo a relatively rapid transition to a new configuration of species and processes, which in turn remains stable over a wide range of environmental conditions (Beisner et al., 2003; Carpenter and Gunderson, 2001; Gunderson, 2000; Mayer and Rietkerk, 2004; Scheffer and Carpenter, 2003; Scheffer et al., 2001). Both model analysis and observational studies show that as the threshold condition for regime shift is approached, it may require only a small additional change to precipitate a large change in the system configuration, as it assumes

TABLE 2-2 Invasive Nonindigenous Plant Indicator Status for Components of the South Florida Ecosystem

	Florida Keys	Southern Estuaries	Greater Everglades	Big Cypress	Northern Estuaries West	Northern Estuaries East	Lake Okee-chobee	Kissimmee River
2006 Overall Status	Y	Y	Y	R	R Y	Y	Y	Y R
2007 Overall Status	Y	Y	Y	R Y	R Y	Y	Y	Y R
1-2 Year Prognosis	Y G	Y R	Y	Y	Y R	Y R	Y	Y R
Number of Plant Species Rated "Yellow-Red" Or "Red"	3	3	3	5	12	6	2	7
Number of Serious Animal Species	5	3	8	5	4	3	7	8

NOTE: A stoplight indicator system (see below) integrates these components of invasive species impact, considering: (1) number of exotic plant species present; (2) number, abundance, and frequency of new exotic species; (3) abundance of established invasive species in new locations; (4) location and density of invasive exotic plants; (5) rate of spread; and (6) effectiveness of control actions/programs, measured as reduction in spatial extent.

Red (R)= Severe negative condition, or one is expected in near future.

Red/Yellow (R | Y) = Currently a negative condition but there are reasonable control efforts under way. However, without continued efforts this species may revert to a severe situation.

Yellow/Red (Y | R) = Problem was previously localized or not too severe but appears to be progressing. Condition generally due to inaction. Without attention and resources, the situation may develop or become red.

Yellow (Y) = Situation is improving and either is stable or moving toward stabilizing, or the species is still very localized.

Yellow/Green (Y | G) = Significant progress is being made and situation is moving toward good maintenance control and is expected to continue improving as long as resources are maintained.

SOURCE: Ferriter et al. (2008).

an alternate state which is stabilized by a new set of feedback relationships (van de Koppel et al., 2004). Furthermore, there can be a pronounced hysteresis, in which reestablishment of the initial environmental conditions fails to move the system back to its original state due to the stabilizing feedbacks present in the alternate state (Beisner et al., 2003). The response of each ecosystem regime to environmental change is highly nonlinear (Mayer and Rietkerk, 2004) and can show time lags and unexpected, even surprising, patterns of change (Groffman et al., 2006).

There are important implications of a system of multiple dynamic regimes for the Everglades restoration effort. Theory and observation suggest that even a small environmental change could create an alternate ecosystem configuration that could prove very difficult to reverse even if managers "get the water right," and such changes could occur very rapidly. With the current system showing declines and losses of resilience in many components, together with increasing threats from invasive species—as described in this section—there is concern that further environmental changes could result in a degraded system that could be very difficult to restore.

CONCLUSIONS AND RECOMMENDATIONS

The Everglades ecosystem is one of the world's greatest ecological treasures, but for more than a century it has been subject to widespread changes resulting from the installation of an extensive water control infrastructure. Culminating in the C&SF Project completed in the 1970s, canals, dikes, and gates to control flows have changed the geography of South Florida and have facilitated extensive agricultural and urban development. These changes have had profound ancillary effects on regional hydrology, vegetation, and wildlife populations, resulting in an extensive decline in the vitality of all components of the ecosystem, including not only the central "River of Grass" but also Lake Okeechobee and the coastal estuaries. The CERP, a joint effort led by the state and federal governments and authorized in 2000, seeks to reverse the general decline of the ecosystem in the midst of a changing human and environmental context.

Population growth and associated development will make restoration more difficult. Increasing water demands from an expanding, more densely settled population in Florida could create competition with ecosystem restoration when supplies are limited. Agriculture and other undeveloped lands face an uncertain future in South Florida. The EAA in particular intervenes directly in the flow of water between Lake Okeechobee and Everglades National Park and influences the movement of water, sediment, and nutrients for the rest of the system. The use of "smart growth" principles that integrate the needs of environmental restoration with human demographic changes can lessen the negative impacts of population growth, if cities, counties, the state, and CERP planners are all involved.

Human-induced climate change is likely to impact the effectiveness of CERP projects, and CERP planners should assess and factor into planning and implementation the most recent projections of the impacts of climate change in South Florida. Precipitation, evapotranspiration, and the intensity of rainfall events in South Florida are expected to change during the current century due to climate change. Impending climate change should not be an excuse for delay or inaction in the restoration, but instead provides further motivation to restore the resilience of the ecosystem. The CERP Guidance Memorandum on climate change recommends consideration of sea-level rise and changes in precipitation quantity, distribution, and evapotranspiration in all CERP planning, but new analysis of impacts based on higher sea-level rise assumptions are needed. Among those possible changes that should be assessed and factored into planning and implementation are: changes in the water budget, including increasing human demands for water; changes in the return frequency and intensity of hurricanes; the effects of climate change on the distribution of biota in the Everglades ecosystem; and impacts of projected sea-level rise on the hydro-geomorphology of the estuaries and the mangrove zone.

Ongoing delay in South Florida ecosystem restoration has not only postponed improvements to the hydrologic condition but also has allowed ecological decline to continue. Recent water management strategies have not produced conditions that are conducive to restoring the Cape Sable seaside sparrow and appear to be negatively impacting the snail kite. Tree islands have undergone a multidecadal decline in both number and surface area—a trend that appears likely to continue until significant CERP and non-CERP restoration progress has been made. In the past decade, Lake Okeechobee has experienced continued water quality and habitat degradation. Meanwhile, the number and area of influence of invasive species are increasing and represent very real challenges to restoration efforts.

In the face of these numerous challenges, Everglades restoration efforts are even more essential to improve the condition of the South Florida ecosystem and strengthen its resiliency as it faces additional stresses in the future. If ecological resilience is not restored, the possibility exists that environmental changes could precipitate rapid and deleterious state changes that might be very difficult or impossible to reverse. Unless near-term progress is achieved on major restoration initiatives, including CERP and non-CERP efforts, opportunities for restoration may close with further loss of species numbers and habitat deterioration, and the Everglades ecosystem may experience irreversible losses to its character and function.

3

Project Planning and Implementation

This committee is charged with the task of discussing significant accomplishments of the restoration and to assess "the progress toward achieving the natural system restoration goals of the Comprehensive Everglades Restoration Plan (CERP)" (see Chapter 1). The first National Research Council (NRC) review of restoration progress noted that in the first 6 years after the Water Resources Development Act of 2000 (WRDA 2000) was authorized, actual construction progress was limited. Instead, most of the CERP accomplishments were programmatic (e.g., land acquisition, project implementation reports [PIRs; see Box 3-1]) to lay the foundation for later project construction (NRC, 2007). In 2008, the beginning of construction for some CERP projects is encouraging, but many CERP and non-CERP restoration projects have been delayed and are far behind their planned completions, and for a variety of reasons, most CERP accomplishments remain programmatic.

In this chapter, an update to the NRC's previous assessment of CERP and related non-CERP project planning and implementation progress is provided (NRC, 2007). The chapter includes discussions of important issues related to CERP progress, such as funding and sequencing. Important challenges are addressed for restoration planning, including incremental adaptive restoration, endangered species, and current project planning impediments. In Chapter 6, details are provided on additional programmatic progress, including the monitoring and assessment plan, development of modeling tools, and other ways in which the foundations of adaptive management are being built in support of the restoration.

PROJECT IMPLEMENTATION

Actual progress restoring the natural system in the South Florida ecosystem will come about only through implementation of restoration projects. The analysis of implementation progress that follows is focused on the CERP, but

BOX 3-1
CERP Project Planning: Project Implementation Reports

Project implementation reports (PIRs) are decision documents that bridge the gap between the conceptual design contained in the Yellow Book (USACE and SFWMD, 1999) and the detailed design necessary to proceed to construction. PIRs for most CERP projects are sent to the U.S. Congress for approval as part of the project planning and authorization process (Figure 3-1). No federal funding in support of mid- to large-sized CERP project construction can be appropriated before PIR approval and project authorization. However, the Secretary of the Army can approve PIRs and proceed with construction for small CERP projects (projects under $25M, with a total not to exceed $206M) under program authority.

The final draft Guidance Memoranda (USACE and SFWMD, 2007a) describe the expected contents and supporting analyses required in the PIRs. The PIR includes an evaluation of alternative designs and operations for their environmental benefits in relation to costs, as well as engineering feasibility. Each PIR also includes detailed analyses that support the justification for a project being next in the queue for CERP implementation as opposed to being delayed to a later time. Each PIR must show conformance with the Savings Clause in WRDA 2000, including a statement of the water reservation for the natural system and for other uses. The Restoration, Coordination, and Verification (RECOVER) program reviews the draft PIR, evaluates the benefits of project alternatives, and assesses the contribution of the project to meeting the overall goals of the CERP. RECOVER also evaluates the project's contributions toward meeting the interim goals and interim targets.

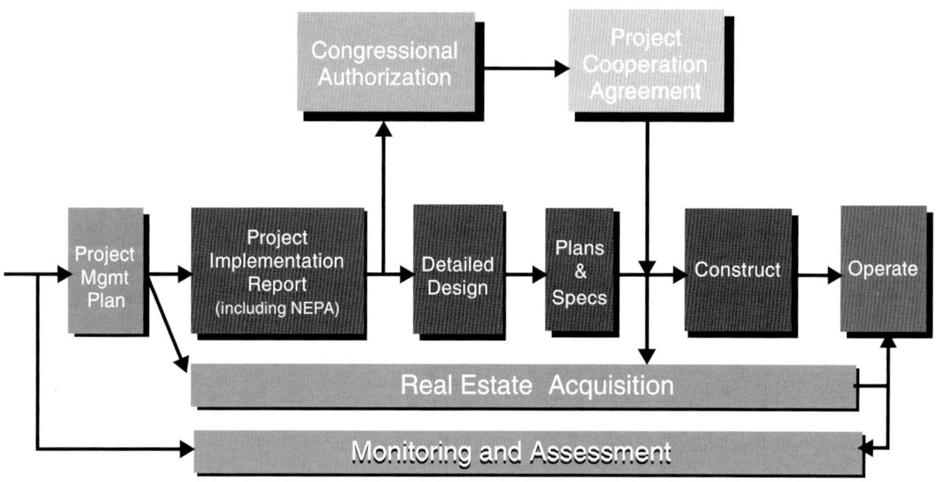

FIGURE 3-1 CERP project development process.

SOURCE: Adapted from Appelbaum (2004).

implementation progress for related non-CERP projects is summarized, with more details provided in Appendix C. A detailed analysis of progress in one critical non-CERP foundation project, the Modified Water Deliveries to Everglades National Park Project (Mod Waters), is provided in Chapter 4. Additional detail on implementation progress can be found in the CERP Annual Report (Williams, 2008), Tracking Success (SFERTF, 2007a), and the Government Accountability Office (GAO) report on the South Florida ecosystem restoration (GAO, 2007).

CERP Project Implementation

The original Yellow Book plan (USACE and SFWMD, 1999) defined 68 projects and identified a schedule for implementation. An updated implementation schedule called the Master Implementation Sequencing Plan (MISP) (USACE and SFWMD, 2005a) was released that organized CERP projects into seven 5-year bands, according to their estimated completion dates. The status of the earliest planned CERP projects—those that were expected, as of 2005, to be completed between 2005 and 2010—are summarized in Table 3-1 (also called Band 1 projects; see Figure 3-2). MISP Band 1 includes those CERP projects expedited by the state of Florida through its Acceler8 program, launched in 2004.[1] Band 1 includes projects with the primary purposes of habitat restoration, water storage, seepage management, and improved future project design (see Appendix D).

As of July 2008, at least four CERP restoration projects are under construction, and four CERP pilot projects are in an installation and testing phase. Many more projects are in planning and design phases (see Tables 3-1 and 3-2). However, not a single CERP project has been completed as of the production of this report. Many of the early projects are already well behind schedule, and the 2005 MISP is already outdated (see Figure 3-3). The GAO notes that some CERP projects are behind schedule by up to 6 years (GAO, 2007). Reasons for these delays are discussed later in this chapter.

A few project components expedited by the state of Florida through the Acceler8 program or through other initiatives are proceeding on or ahead of the original Yellow Book schedule: C-51 and L-8 Basin Reservoir, Phase 1 (projected completion 2008 versus 2011 in the Yellow Book) and Biscayne Bay Coastal Wetlands, Phase 1 (projected completion 2011 versus 2018 in the Yellow Book) (Table 3-1). The remaining Acceler8 projects show delays similar to the other MISP Band 1 CERP projects (Figure 3-3).

The Committee on Independent Scientific Review of Everglades Restoration (CISRERP) has found it difficult, however, to obtain reliable projected completion

---

[1]For more information, see *http://www.evergladesnow.org*.

TABLE 3-1 Updated CERP Band 1 Project Status

Project or Component Name	Yellow Book (1999) Estimated Completion Date	MISP 1.0 (2005) Estimated Completion Date	Current (2008) Estimated Completion Date
Caloosahatchee River (C-43) ASR Pilot (Fig. 3-2, No. 1)	2002	2006	2012
Hillsboro ASR *Pilot* (Fig. 3-2, No. 2)	2002	2006	2009
Lake Okeechobee ASR *Pilot* (Fig. 3-2, No. 5)	2001	2007	2012
L-31 (L-30) Seepage Management *Pilot* (Fig. 3-2, No. 4)	2002	2008	2010
Melaleuca Eradication and Other Exotic Plants	2011	2007	2026
Winsberg Farms Wetlands Restoration (Fig. 3-2, No. 3)	2005	2008	2010
Biscayne Bay Coastal Wetlands (Phase 1) (Fig. 3-2, No. 6)	2018	2008	2011
Picayune Strand Restoration (Formerly called Southern Golden Gate Estates) (Fig. 3-2, No. 7)	2005	2009	2015
Indian River Lagoon - South (Fig. 3-2, No. 8)			2023
- C-44 Reservoir*	2007	2009	2014
- Natural Areas Real Estate Acquisition (Phase 1)	Not specified	2009	Not specified
Broward County WPAs			2017
- C-9 Impoundment* (Fig. 3-2, No. 9)	2007	2009	2014
- Western C-11 Diversion Impoundment* (Fig. 3-2, No. 10)	2008	2009	2014
- WCA 3A-3B Levee Seepage Management* (Fig. 3-2, No. 9,10)	2008	2008	2017
Acme Basin B Discharge (Fig. 3-2, No. 11)	2006	2007	2009
Site 1 Impoundment* (Phase 1) (Fig. 3-2, No. 2)	2007	2009	2013
C-111 Spreader Canal* (Fig. 3-2, No. 12)	2008	2008	PIR#1: 2011 PIR #2: TBD

Project Implementation Report (PIR) and Authorization Status	Planning/Design	Construction Status (or Installation and Testing Status for Pilots)	Original Cost (in millions of 1999 dollars)	Estimated Cost (in millions[a]) from 2007 IFP
NA	Ongoing	Ongoing	6	8.3
NA	✓	Ongoing	9	9.3
NA	✓	Ongoing	19	27.4
NA	✓	Ongoing	10	12.0
In development	Ongoing	NA	5.8	8.3
In development	Ongoing	Phase 1: ✓ Phase 2: Not begun	14.1	18.1
In development	Phase 1: ✓	Not Begun	300	438 (Phases 1 & 2)
Submitted to Congress in Sept. 2005; Authorized in WRDA 2007	✓	Ongoing	46	393
Submitted to Congress in Aug. 2004; Authorized in WRDA 2007	✓	Ongoing	823	1,497
	Unknown	NA		
Final April 2007			314	691
	Ongoing	Not Begun		
	Ongoing	Not Begun		
	Ongoing	Not Begun		
Discontinued[b]	✓	Ongoing	20	28
Submitted to Congress in Dec. 2006; Authorized in WRDA 2007	Ongoing	Not Begun	39	84
PIR #1: In development	Ongoing	Not Begun	94	370

continued

TABLE 3-1 Continued

Project or Component Name	Yellow Book (1999) Estimated Completion Date	MISP 1.0 (2005) Estimated Completion Date	Current (2008) Estimated Completion Date
North Palm Beach County – Part 1			
- C-51 and L-8 Basin Reservoir, Phase 1 (PBA) (Fig. 3-2, No. 13)	2011	2008	2008[c]
Everglades Agricultural Area Storage Reservoir (Fig. 3-2, No. 14)			
- Part 1, Phase 1*	2009	2009	2010[d]
Lake Okeechobee Watershed			2015
- Lake Istokpoga Regulation Schedule* (Fig. 3-2, No. 15)	2001	2008	Not specified
Modify Rotenberger Wildlife Management Area Operation Plan (Fig. 3-2, No. 16)	Not specified	2009	2009
Lakes Park Restoration (Fig. 3-2, No. 17)	2004	2009	TBD
C-43 Basin Storage Reservoir (Fig. 3-2, No. 1)	2012	2010	2013

NOTES: Gray shading reflects projects being expedited and/or carried out entirely with state funding as of 2007. In most cases, construction of these projects was moving forward prior to the finalization of the PIR. Some of these projects are still considered CERP components, while others are now considered outside of the CERP. Recently, several state-expedited projects have reverted back to the USACE for remaining design and construction (e.g., Site 1 Impoundment, Broward County WPAs, Picayune Strand).

* Projects that were conditionally authorized in WRDA 2000, subject to approval of the PIR.

[a]Project costs in the Integrated Financial Plan (IFP) (SFERTF, 2007a) were reported as constant 2006 dollars, with the exception of the Acceler8 projects, which were reported as the present day value at the time the estimate was performed (~2007).

[b]The SFWMD has decided to work with local interests to complete the design and construction of the Acme Basin B Discharge project and the Lakes Park Restoration project outside of the CERP. Cost sharing under the CERP is not anticipated, thus work on these two PIRs has been discontinued, and CERP Planning/Design efforts have been ended.

[c]Although the IFP indicates that the C-51 and L-8 basin reservoir, Phase 1 (PBA), is scheduled to be completed by the end of 2008 with the use of temporary pumps, full capacity will not be available until construction of the final pump station, likely in fiscal year 2010. Construction of the permanent pump station was recently moved into Phase 2 of the project.

Project Implementation Report (PIR) and Authorization Status	Planning/Design	Construction Status (or Installation and Testing Status for Pilots)	Original Cost (in millions of 1999 dollars)	Estimated Cost (in millions[a]) from 2007 IFP
In development			437	608
	✓	Ongoing		
			437	594
Revised Draft in development	✓	Ongoing (but temporarily halted)[c]		
In development			456	643
	Ongoing	NA		
NA	Ongoing	NA	0	0
Discontinued[b]	Ongoing	Not Begun	5.2	6.6
Final Sept. 2007	✓	Not Begun	440 (Reservoir *and* ASR)	531 (Reservoir *and* ASR)

[d]The EAA Storage Reservoir project is on hold, pending the resolution of two lawsuits underway (USA, et al. v. SFWMD, et al. 1:88-cv-01886-FAM; NRDC, et al. v. USACE, 9:07-cv-80444-DIV-Middlebrooks). If the state of Florida acquires large land holdings in the EAA from the U.S. Sugar Corporation, opportunities made available by this acquisition could affect future plans for completing the EAA Storage Reservoir.

SOURCES: DOI and USACE (2005); L. Gerry, SFWMD, personal communication (2008); G. Landers, USACE, personal communication (2008); Project Status Reports from www.evergladesplan.org; SFERTF (2007a); USACE and SFWMD (1999).

✓=Complete
NA= not applicable
TBD = to be determined

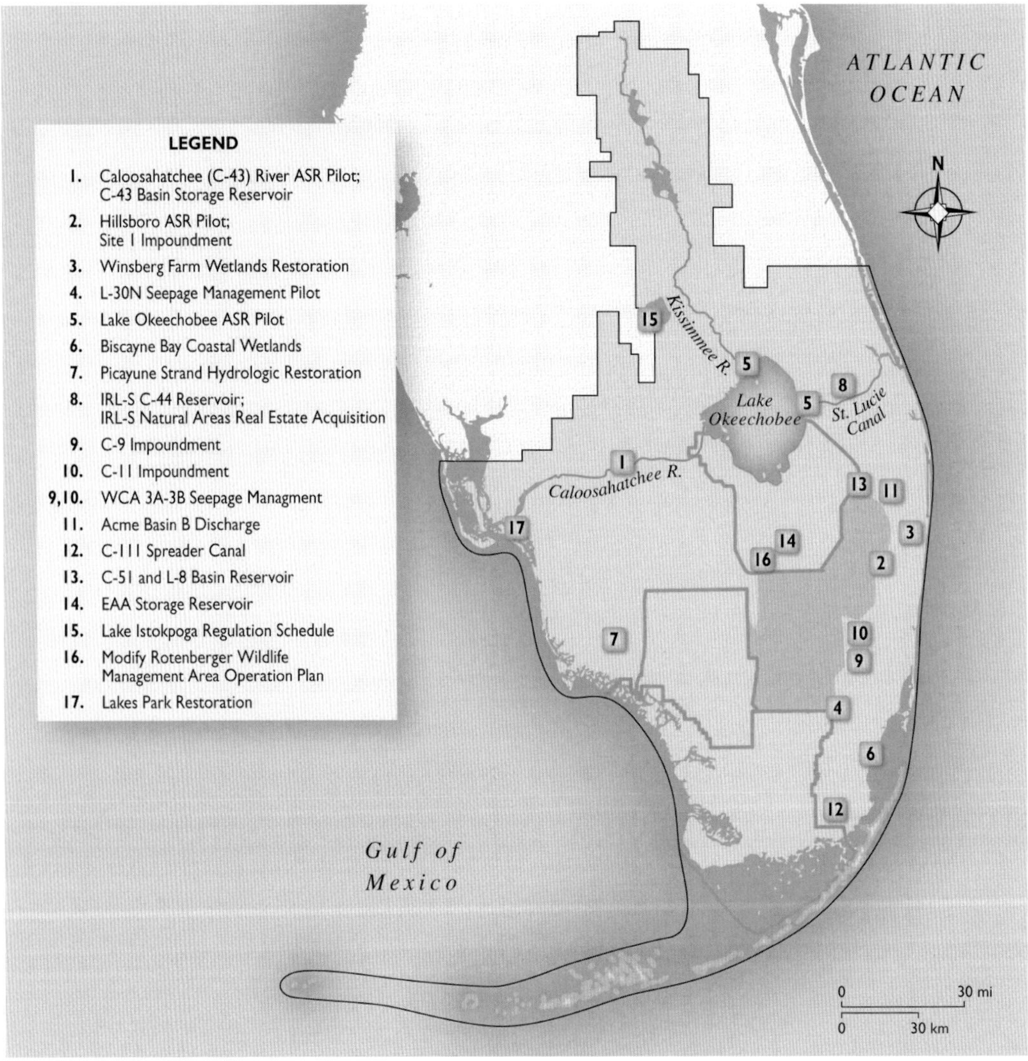

FIGURE 3-2 Locations of Band 1 CERP project components. © International Mapping Associates

TABLE 3-2 Status of 88 CERP and CERP-Related Restoration Projects

		Not Yet Implemented			
	Completed	In Implementation	Planning/ Design	Not Yet Started	Total
CERP	0	7	21	32	60
Related non-CERP projects	15	10	3	0	28

NOTE: Related non-CERP projects "when completed, will serve as the foundation for many of the CERP projects and are intended to restore a more natural water flow to Everglades National Park and improve water quality in the ecosystem" (GAO, 2007). See Appendix E for the names and status of the 88 projects. GAO (2007) also includes an assessment of 134 South Florida restoration projects that are not related to CERP and that do not serve as a foundation for the CERP.

SOURCE: GAO (2007).

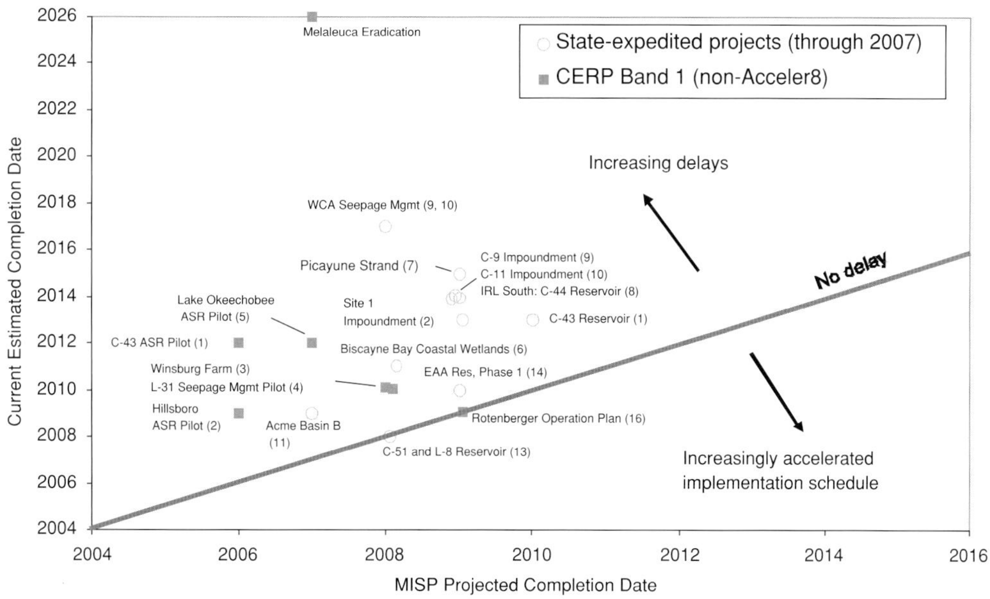

FIGURE 3-3 Project delays for CERP Band 1 projects, including state-expedited projects (e.g., Acceler8), based on projections from the Master Implementation Sequencing Plan (MISP) (USACE and SFWMD, 2005a) compared with estimated completion dates shown in Table 3-1.

NOTE: Some Acceler8 projects in 2008 are being returned to the USACE for completion as part of the federal government's share of the CERP, but the circles in this figure reflect those projects expedited by the state of Florida as of 2007. Numbers in parentheses reference project location on Figure 3-2.

dates from the CERP Web site and other information, probably in large part because the agencies themselves cannot overcome uncertainties in availability of funds. Available information on estimated completion dates in various documents is often contradictory. A more effective public communication mechanism is needed. The challenge for the CERP agencies is how to develop a realistic schedule in the face of these daunting financial uncertainties.

Natural System Benefits Derived from CERP Implementation

According to the MISP (USACE and SFWMD, 2005a), eight projects and four pilot projects were scheduled for completion by 2008, within the time-reporting range of this current NRC review. Installation for some pilot projects have been completed by 2008, but none have completed testing and analysis. No restoration projects are anticipated to be fully constructed by the end of 2008, although a few project subcomponents are nearing completion that will deliver some restoration benefits. These early benefits are described in the paragraphs that follow.

One project nearing completion is the L-8 Basin Reservoir (also called the C-51 and L-8 Basin Reservoir; see Figure 3-2 [No. 13] and Figure 3-4), a state-expedited project that includes an in-ground reservoir at the location of existing rock-mining pits with a storage capacity of about 48,000 acre-feet. The purpose of this project is to provide additional water storage that will increase water supply and reduce damaging high discharges to the Lake Worth Lagoon. Lake Worth Lagoon was historically a freshwater lake that was made estuarine by the creation of permanent inlets. Discharges from drainage canals result in occasional excessive releases of fresh water into the estuary. The L-8 Basin Reservoir project will also enhance hydroperiods in the Loxahatchee Slough and increase base flows to the Northwest Fork of the Loxahatchee River, which flows into the Loxahatchee Estuary (USACE and SFWMD, 2005b). The Loxahatchee Estuary has been heavily altered by inlet stabilization, channelization, and basin drainage. Drainage canals and disruption of natural flow patterns have resulted in salt water intrusion and loss of cypress swamp in the Northwest Fork and increased stormwater flows into the Southwest Fork. A future flow way project will direct excess water into the Grassy Waters Preserve (Figure 3-4). Construction of the permanent pumping infrastructure has been delayed until 2010, but the L-8 Basin Reservoir is scheduled to begin operating with a temporary pump by the end of 2008. Water reservations for this project have not yet been determined, as the PIR has not been completed for this state accelerated project. Sediment dredging from the C-51 canal is also under way in the vicinity of Lake Worth Lagoon (as part of the North Palm Beach County Phase 1 project) to reduce sediment discharged to the lagoon.

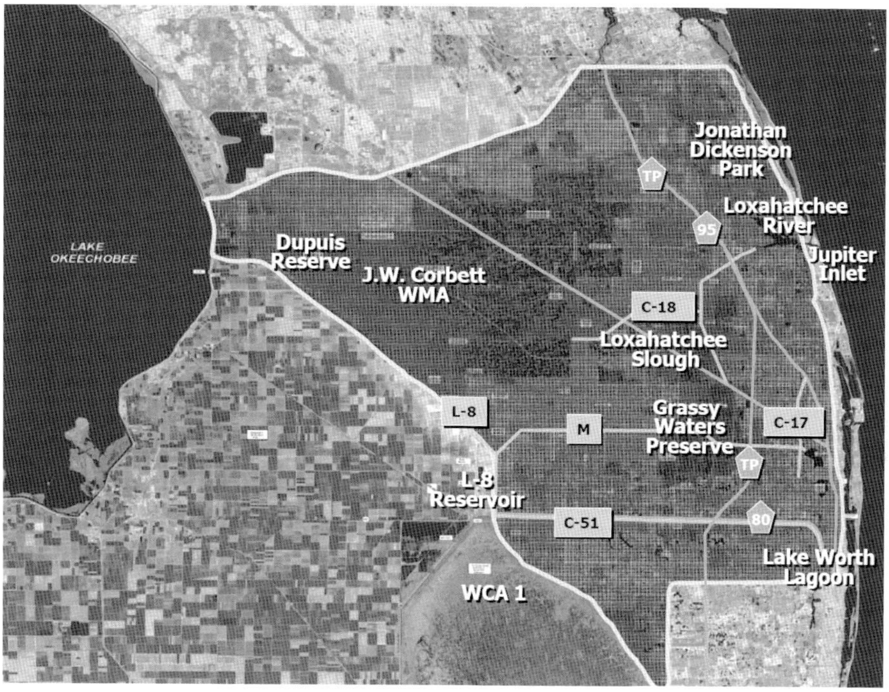

FIGURE 3-4 Location of the C-51 and L-8 Reservoir project.

SOURCE: USACE and SFWMD (2005b).

A number of other projects currently under construction are showing some phased benefits. For example, a new pump station constructed as part of the Acme Basin B project (Figure 3-2, No. 11) pumps stormwater into the C-51 canal and to STA-1E where it is treated before entering Water Conservation Area (WCA) 1. This state-expedited initiative eliminates the direct discharge of high-phosphorus stormwater from the urban area into WCA-1 (L. Gerry, SFWMD, personal communication, 2008). The Picayune Strand project (Figure 3-2, No. 7), currently under way, aims to restore and enhance over 55,000 acres of public lands by plugging and filling canals and returning sheet flow to the project site and adjacent natural areas, including the Fakahatchee Strand State Preserve, Florida Panther National Wildlife Refuge, Ten Thousand Island National Wildlife Refuge, and Collier Seminole State Park. Through this state-expedited project, 65 miles of roads were removed, and more than 160 structures and numerous trash sites have been demolished and removed in 2007. Cleanup of 25-plus acres of pesticide-contaminated soils (out of approximately 65 acres) has been completed. Seven miles of Prairie Canal adjacent to the road removal area have been plugged and filled, and native plants are becoming reestablished, thereby improving the habitat for native wildlife. Control of exotics within the construction footprints is also under way (SFWMD and FDEP, 2008a).

CERP Planning, Authorization, and Funding

Prior to appropriation of federal funding for CERP project construction, a PIR (see Box 3-1) must first be approved and for most mid- to large-sized projects, the U.S. Congress must authorize the project.[2] Ten CERP projects were conditionally authorized in WRDA 2000, subject to congressional approval of the PIR (those marked with asterisks in Table 3-1). PIR completion thus represents a major hurdle in the implementation process for all CERP projects. The Acceler8 program has expedited construction of some projects with state funding by moving forward with design and construction concurrent with development of the PIR. The state of Florida, through its Acceler8 program, however, takes the risk that the constructed projects may not actually meet final requirements, according to the final, approved PIR.

As noted in Table 3-1, of the 14 projects estimated (as of 2005) to be completed between 2005 and 2010, only 5 have PIRs that are considered final. Only 3 PIRs have been transmitted by the U.S. Army Corps of Engineers (USACE) to Congress for authorization: Indian River Lagoon-South (IRL-S), Picayune Strand, and Site 1 Impoundment (Figure 3-2, No. 8, 7, and 2; see Box 3-2). All three of these CERP projects were congressionally authorized under the Water Resources Development Act of 2007 (WRDA 2007). These projects provide significant restoration benefits, but they are expected to have only a minor effect on Everglades National Park. All three projects were originally among the earliest scheduled projects in both the Yellow Book and the MISP 1.0, although IRL project components were scheduled for MISP Bands 1–4, so restoration activities in that region were anticipated to continue for decades.

Original plans for the CERP rested on the assumption that key projects would be steadily and consistently authorized in Water Resources Development acts passed every 2 years and that congressional appropriations for approved projects would follow in due course. After the passage of WRDA 2000, however,

[2]Under program authority, the Secretary of the Army can approve PIRs and proceed with construction for CERP projects under \$25M total cost (\$12.5M federal cost), with a total not to exceed \$206M. Congressional action is not required for nonconstruction projects, such as operational changes.

BOX 3-2
Summary of Congressionally Authorized Projects with Approved PIRs

As of April 2008, three CERP projects with approved PIRs have been congressionally authorized—Indian River Lagoon-South (IRL-S), Picayune Strand Restoration, and Site 1 Impoundment. Not surprisingly, project plans for the three projects included substantial changes from the framework plans laid out in the Yellow Book.

Indian River Lagoon-South

The IRL-S project (Figure 3-2, No. 8), an approximately $1.2 billion component of the CERP (in 2004 dollars), is located northeast of Lake Okeechobee (Figure 3-2). The C-44 Basin Storage Reservoir is subsumed within the overall IRL-S project, to which are added the C-25 and C-23/C-24 North and South Storage Reservoirs. The original Yellow Book plan (USACE and SFWMD, 1999) was limited to these four storage reservoirs, but the project plans have since been significantly altered. The four storage basins are now proposed to provide 130,000 acre-feet of water storage, a substantial decrease in storage from the 389,000 acre-feet of storage proposed in the Yellow Book. An additional 65,000 acre-feet of storage are proposed through wetland restoration and utilization of three natural storage areas on 92,000 acres of land and in four new stormwater treatment areas (STAs). Finally, 7,900,000 cubic yards of muck will be dredged from the St. Lucie River and Estuary to provide 2,650 acres of clean substrate within the estuary for recolonization of marine organisms. The original Yellow Book plan aimed to reduce damaging flows to the St. Lucie Estuary and the IRL-S while also providing water supply for agriculture, thereby reducing demands on the Floridan aquifer. However, the PIR included added benefits for enhanced phosphorus and nitrogen reduction, improved estuarine water quality, restored upland habitats, increased spatial extent of wetlands and natural areas, and more natural flow patterns (USACE and SFWMD, 2004b). The 2004 cost estimates for this project have increased by $440 million (or 54 percent) above those in the 1999 Yellow Book, reflecting both inflationary increases and $240 million in project scope changes (DOI and USACE, 2005).

Picayune Strand Restoration

Located in western Collier County (Figure 3-2), the Picayune Strand Restoration project (Figure 3-2, No. 7) will restore and enhance more than 50,000 acres of wetlands in Southern Golden Gate Estates, an area once drained for development. The project will also improve the quality and timing of freshwater flows entering the Ten Thousand Islands National Wildlife Refuge, while maintaining flood protection for neighboring communities. The project includes a combination of spreader channels, canal plugs, road removal, pump stations, and flood protection levees. The project scope changes (e.g., additional road removal, larger pumps to provide additional flood protection), inflationary increases, and the failure to account for land acquisition costs in the original project cost estimates led to an increase in costs from $15.5 million in the original Yellow Book to $349 million (DOI and USACE, 2005; USACE and SFWMD, 2005c). This project is one of the most significant for increasing the spatial extent of natural wetlands.

continued

BOX 3-2 Continued

Site 1 Impoundment

Located in Palm Beach county south of the Arthur R. Marshall Loxahatchee National Wildlife Refuge (LNWR), the Site 1 Impoundment Project (Figure 3-2, No. 2; also called the Fran Reich Preserve) includes an aboveground reservoir adjacent to the Hillsboro Canal with a storage capacity of 13,280 acre-feet, an inflow pump station, spillways, and seepage management structures. The purpose of the project is to capture and store local runoff during wet periods and then use that water to supplement water deliveries to the Hillsboro Canal during dry periods, thus reducing demands for releases from Lake Okeechobee and the LNWR. According to the PIR, constructing and operating the impoundment will reduce the need for releases from LNWR during the dry season to meet local water demands and will facilitate the maintenance of more natural, desirable, and consistent water levels within the LNWR. The impoundment will also reduce groundwater seepage from LNWR. In addition there will be benefits to the downstream estuaries as a result of reducing peak freshwater flows from local stormwater runoff and pulsed releases from Lake Okeechobee. The cost of the recommended plan (October 2006 price levels) is $80.8 million, more than double the cost of $39 million estimated in the Yellow Book (USACE and SFWMD, 1999, 2006).

no such acts were passed until 2007, thus delaying project authorizations by up to 3 years after transmission to Congress (Table 3-1). Federal funding has not yet been appropriated to support these authorized projects, and on-the-ground restoration activity for these three projects has been deferred.

The original vision of the CERP was that it would be an equal partnership between the state and the federal governments, but this partnership has turned out to be distinctly unequal from a funding perspective. For the fiscal years (FY) 1999 through 2006, GAO (2007) reported that the federal government contributed $0.34 billion (in 2006 dollars) to the CERP compared to Florida's $2.0 billion. When the complete range of restoration-related initiatives in South Florida is considered, the federal government contributed $2.3 billion, while the state of Florida contributed $4.8 billion (Table 3-3). By 2006, total CERP funding had fallen short of original projected costs by about $1.2 billion. The overall shortfall between 1999 and 2006 was reduced to $1.2 billion only because Florida increased its contribution for CERP projects by $250 million during this period (GAO, 2007) as part of its expedited restoration initiatives, which advanced design and construction ahead of the CERP project planning and authorization process. Much of Florida's investment has been in the acquisition of land (see next section). Most of the shortfall in investment has been in the area of project construction.

TABLE 3-3 Project Purpose and Funding (in millions of 2006 dollars) Allocated Among CERP and All South Florida Restoration Projects and Activities, Fiscal Years 1999–2006

	CERP Projects		Total		
Type of Project	Federal	State	Federal	State	Total
Land acquisition	0	$1,788.6	$283.4	$2,274.1	**$2,557.5**
Project construction	0	25.7	835.7	1,123.1	**$1,958.7**
Support activities[a]	341.4	191.7	1,137.0	1,421.9	**$2,558.9**
Total	**$341.4**	**$2,006.0**	**$2,256.1**	**$4,819.0**	**$7,075.1**

[a]Support activities included RECOVER efforts, adaptive assessment and monitoring, the Interagency Modeling Center, program coordination, and science- and mission-related activities that indirectly benefit the restoration, such as invasive species control. In addition, for the USACE and SFWMD, support activities include some funding for project design, pilot project design, and feasibility studies.

SOURCE: GAO (2007).

The original CERP plan was developed based on estimated state and federal CERP funding of about $200 million each per year (in 1999 dollars) (USACE and SFWMD, 1999). In fiscal years 2005 and 2006, the USACE received only $64 million for the CERP, largely to support programmatic functions, such as monitoring and assessment, program planning and design, and interagency coordination (Trulock, 2007). Some of this funding was also directed toward previously authorized pilot projects (Table 3-1). In FY 2008, the USACE received $131 million for Kissimmee River restoration, the Mod Waters project, *and* the CERP combined. The comparable budget request for FY 2009 is $185 million (Department of the Army, 2008). Federal funding has thus fallen far short of that originally envisioned.

The South Florida Water Management District (SFWMD) has accepted a proposal by the USACE to complete the design and construction of two Acceler8 restoration projects—the Broward County Water Preserve areas (consisting of C-11 and C-9 impoundments, and WCA-3A/3B Seepage Management Area; Figure 3-2, No. 9 and 10) and the Site 1 Impoundment (also called the Fran Reich Preserve; Figure 3-2, No. 2)—as a part of the federal government's share of the CERP. The USACE has also accepted an SFWMD request for the federal government to construct the remaining components of the Picayune Strand Restoration project and anticipates a similar request for the Caloosahatchee River (C-43) Storage Reservoir project in order to better balance overall CERP cost sharing. Considering the lengthy CERP project planning and authorization process which Acceler8 handled parallel to construction, and the slow pace of federal funding to date, this transfer of Acceler8 projects to the USACE will likely further delay the construction of these projects.

While the planning process and subsequent delivery of funding for projects have lagged, the anticipated total costs have continued to increase (NRC, 2007). CERP costs have increased from $8.2 billion (in 1999 dollars) to $10.9 billion (in 2004 dollars) as a result of inflationary increases and scope changes (DOI and USACE, 2005). In general, restoration cost increases have been driven by inflation (not unexpected), changes in project scope, a tangled and complex federal design and planning process, increases in land costs, and unexpected rises in building costs. Land costs, the result of intense development pressure driven by population growth and redistribution, increased 50 to 88 percent in a single year between 2004 and 2005 (Reynolds, 2006). Building costs in Florida rose rapidly in the wake of several hurricanes in the state during 2003–2005, putting a strain on building supplies and skilled labor. Current CERP estimates do not include the costs for remaining land acquisitions and final project design cost estimates, which are not yet known. Therefore, GAO (2007) concludes that restoration costs are likely to continue to increase in the years ahead.

Land Acquisition

Land management for an effective CERP depends on acquiring particular sites within the project area and protecting more general areas within the South Florida ecosystem that could help meet the broad restoration goals. Nearly 56 percent of the land for the CERP has been acquired as of September 2007 (SFWMD and FDEP, 2008a), up from the approximately 51 percent reported by the last committee review (NRC, 2007). Approximately 99 percent of the land needed for the Acceler8 projects has been acquired (SFWMD and FDEP, 2008a). As of September 2007, 217,584 acres had been acquired for the CERP, up from 207,000 acres reported in 2005 (DOI and USACE, 2005). Meanwhile, current estimates for land needed to meet specific project plans has been reduced to 387,154 acres (down from 406,000 acres reported in DOI and USACE, 2005). Thus, the increase in the percentage of CERP project land acquired represents both continued rigorous land acquisition efforts by CERP partners—primarily the state of Florida—and a reduction in the estimated acreage needed for CERP projects. The land acquisition to date is ahead of schedule when compared to the MISP land acquisition strategy (SFERTF, 2007b; USACE and SFWMD, 2005a).

The committee is impressed with the very effective efforts by the state of Florida to acquire the land needed to complete CERP. The state has shown leadership in dealing with the expensive and often difficult process of land acquisition. The result is an emerging foundation for success for many CERP components.

Restoration partners have invested at least $1.3 billion[3] in land acquisition for CERP projects: the federal government provided $276 million; the state of Florida provided $1 billion; and local governments provided $72 million. An estimated $1.65 billion is needed to complete acquisition for the CERP (SFERTF, 2007b). The state anticipated spending more than $100 million per year over the next 5 years on land acquisition needed to implement the CERP (SFWMD and FDEP, 2008b), but the current state budget provided only $50 million toward Everglades restoration in FY 2009 (Kam, 2008). The recent real estate downturn, however, may provide some opportunities for cost savings in CERP land acquisition in the near term. The committee considers land acquisition a wise up-front investment which ensures that required project lands are purchased before development forecloses certain restoration alternatives and before land costs rise even further.

Despite the state's budget pressures, in June 2008 the governor of Florida announced a plan for the state to purchase 187,000 acres of land in the Everglades Agricultural Area (EAA), the complete land holdings of the U.S. Sugar Corporation, at a cost of $1.75 billion. If the proposed acquisition occurs, the land would continue to be farmed for the next 6 years, after which time the land would be turned over to the state for restoration purposes. This land acquisition was not part of the original CERP plan, but it could be extremely important to help meet the broad restoration goals by providing large land tracts for future but as-yet-unidentified CERP and non-CERP restoration needs (e.g., additional surface storage, STAs) and by preventing land use conversions that might negatively impact the restoration. As the state finalizes the acquisition and identifies the funding mechanism to support the purchase, restoration planners are expected to develop plans for how best to utilize this land for the benefit of South Florida ecosystem restoration.

Non-CERP Foundation Projects

Some of the largest accomplishments and some of the greatest challenges in South Florida restoration are associated with non-CERP projects that are directly related to the success of the CERP in achieving its restoration goals (Table 3-2). Progress in some of the major non-CERP foundation projects, including documented natural system restoration benefits (where feasible), is summarized in

[3]Unlike Table 3-3, these cost data have not been corrected for inflation. The figures cited here also include land acquisitions prior to 1999. As a result, these land acquisition totals differ from those reported in Table 3-3.

Appendix C (see Box 2-2 for descriptions of the major projects and Appendix E for a complete listing of all related non-CERP projects). Completed non-CERP projects are of the form of habitat acquisition and improvement, construction of STAs, drafting of water supply plans, and some exotic species control projects. The Kissimmee River Restoration continues to be the restoration activity with the most documented natural system benefits (NRC, 2007; SFWMD and FDEP, 2008a). STAs in the EAA continue to remove large quantities of phosphorus from surface waters, although they are suffering from damages from 2006 Hurricane Wilma and from the 2007 drought conditions (SFWMD and FDEP, 2008a) and removal efficiency may be declining (see also Chapter 5). Additional state funds for Lake Okeechobee have sparked plans for water quality improvements (e.g., STAs, improved land management practices) and enhanced water storage north of the lake (see Chapter 5). Nevertheless, non-CERP projects, like CERP projects, are experiencing delays and escalating costs (see Appendix C).

While the non-CERP foundation projects themselves are intended to provide restoration benefits, they are also important as necessary precursors for many CERP projects. Effective STAs are critical for improving the quality of water delivered to natural areas. The Mod Waters project, including Tamiami Trail modification, is essential for decompartmentalization and the restoration of flows in Everglades National Park. WRDA 2000 recognized this precedence relationship and required completion of Mod Waters before appropriations are made for other restoration projects in the east Everglades, including the WCA-3A Decompartmentalization and Sheet Flow Project (Decomp) (§601[b][2][D]). Delays and funding concerns for the Mod Waters project are discussed in detail in Chapter 4.

Assessment of Restoration Delays

The delays afflicting CERP and foundational non-CERP projects are creating increased concern that the Everglades ecosystem may suffer irreparable losses before major restoration actions are taken to reverse the ecosystem decline. A variety of reasons can be cited for CERP and non-CERP project delays.

- A complex project planning and approval process is required for CERP projects and other federal restoration projects. WRDA 2000 added unforeseen requirements and complexity, such as the Programmatic Regulations, the Savings Clause, water reservations, and assurances.
- Resolutions to major concerns or agency disagreements must be negotiated in the planning process.
- Unresolved engineering and ecological uncertainties can stall the project

planning and authorization process, especially for complex or contentious projects.

- Legal challenges (e.g., land acquisition, endangered species, water quality) and resulting litigation have delayed projects such as Mod Waters.
- New environmental issues and restoration risks (e.g., instability of Hoover Dike at high lake levels) have been recognized that need to be included in the planning.
- Personnel turnover and general staffing limitations can delay planning and development of modeling tools (see Chapter 6).
- Project authorization has caused a lag in submission to Congress by up to 3 years, because the Water Resources Development Acts have not been passed every 2 years as anticipated.
- Escalating costs due to changes in project scope, inflation, land cost, and construction costs have increased the overall cost of CERP, leading to funding challenges and schedule extensions.

Among these various reasons for delay, the inadequacy of the project planning, authorization, and funding process is noteworthy for the CERP, as previously discussed. It has been 8 years since the signing of WRDA 2000, yet only three projects have been submitted to Congress for authorization. This fact highlights the challenges faced in the project planning process. Meanwhile, the state has moved forward with the construction of some projects with its own funding, while the draft PIRs are contested among agencies and stakeholders.

The committee judges that the lack of federal funding in the first 8 years of the CERP is more of a symptom of the problems caused by the complex and lengthy CERP planning and authorization process, rather than the sole, or even the most serious, cause of the pervasive CERP delays. However, federal funding limitations have caused notable delays in non-CERP projects (see Chapter 4) and in some CERP pilot studies (NRC, 2007). And now that three CERP projects have both congressional authorization and approved PIRs, funding limitations will certainly create additional constraints to CERP progress in the years ahead, as non-CERP and CERP projects compete for limited state and federal resources.

Anecdotal information provided to this committee suggests that morale of dedicated agency personnel is also suffering from delays in CERP implementation. Hundreds of scientists, engineers, and other professionals have devoted a significant portion of their professional careers to the CERP. When implementation continues to be delayed, it becomes more difficult for these public servants to maintain their enthusiasm, and creativity in support of the overall effort suffers. The ultimate success of the CERP depends upon the considerable skills, knowledge, dedication, and ingenuity of the agencies' staff. If experienced staff increasingly become fatigued with the CERP process and choose to work elsewhere, the restoration will suffer a tremendous loss of knowledge and continuity. Loss of morale will also impact the recruitment of talented young scientists, engineers, and planners to replace them. When CERP projects that lead to documented ecosystem restoration are well under way, however, it should provide a needed boost to staff morale.

The delays and lack of predictability of funding that have hampered CERP progress are also stressing the CERP partnership in addition to driving up costs. The committee suggests that senior management work to develop approaches to reduce administrative roadblocks, which will benefit not only the CERP but also other major federal ecosystem restoration efforts as well.

An alternative to normal project-by-project review, authorization, and yearly funding is needed to allow a more scientifically, managerially, and economically efficacious approach to fund comprehensive restoration programs that assures funding over a multiple-year period. This problem is not unique to CERP; indeed, it affects other multicomponent ecosystem restoration projects (NRC, 2004). For example, a Working Group for Post-Hurricane Planning for the Louisiana Coast (2006) recommended an authorization and financing process separated from the WRDA that provided project flexibility and a reliable appropriations stream within a programmatic authorization.

PROJECT PLANNING ISSUES

Even with adequate funding, the CERP still lacks good mechanisms to prioritize and expeditiously plan and design projects. In this section, major project planning issues that affect the rate of CERP progress are discussed, including restoration scheduling, incremental adaptive restoration, creative approaches to minimize endangered species conflicts, and problems created by the USACE project benefits analysis process termed the "next added increment" (NAI).

Identifying Restoration Priorities

Developing a realistic schedule and project prioritization sequence is a critical need for the restoration effort. CERP planners have recognized that their target project implementation schedules were inconsistent with the realities of the planning process and the federal funding stream. Meanwhile, stakeholder frustration, budget constraints, and cost escalation highlight the need to focus on projects that deliver meaningful restoration benefits to the natural system in the near term.

Proper sequencing of projects is critical to meet hydrologic constraints. For example, restoring sheet flow through the Decomp project requires greater flow of water into the northeast area of Everglades National Park and the completion of some form of the long-delayed Mod Waters project. At the same time, additional seepage controls are needed to minimize flooding in the eastern urbanized areas, and Tamiami Trail modifications are needed to protect the road base and to prevent flooding of this hurricane evacuation route. However, GAO (2007) noted that there were no overarching sequencing criteria used for decision making, that implementation decisions were mostly driven by available funds, and that the MISP is not consistent with project-sequencing criteria established by the USACE.

The cumbersome planning process also advances projects with large stakeholder support or minimal contention over large and complex projects with vast stakeholder disagreements, regardless of their potential benefits. The three Everglades restoration projects authorized in WRDA 2007—Indian River Lagoon, Picayune Strand, and the Site 1 Impoundment—represent projects with strong local stakeholder support (or minimal stakeholder opposition), even though they may not represent the highest-priority projects with the greatest potential for achieving system-wide restoration benefits, assuming that limited federal funds are available. Without clear priorities for project planning and funding, projects with large potential restoration benefits may see lengthy restoration delays while other, less-contentious projects that address only isolated portions of the ecosystem may tie up available funding. Unless CERP planners exercise leadership while confronting the difficult budgetary decisions, these tough decisions will effectively be made for us by future budget constraints.

In response to advice from GAO (2007) and NRC (2007), CERP planners are now developing a revised project implementation schedule, termed the "integrated delivery schedule" (IDS), for the South Florida ecosystem restoration that would reprioritize the timing and funding of future restoration activities. The goals of an integrated schedule are to:

- Update existing project schedules to provide current status and practical time lines for implementation;
- Focus on delivering meaningful restoration benefits as early as possible;
- Phase large projects as necessary to provide early benefits and learning;
- Include related non-CERP projects as well as CERP projects in program sequencing; and
- Include new programs such as Northern Everglades restoration (Appelbaum, 2008a).

Including non-CERP foundation projects in an integrated schedule is advantageous to highlight the precedence relationships among these projects.

As of May 2008, this project scheduling and prioritization effort was still under development. CERP planners evaluated how various projects addressed the following five objectives of Everglades restoration:

- Reestablishing sheet flow to the central Everglades,
- Optimizing water storage,
- Improving water quality,
- Restoring the northern estuaries and Lake Okeechobee, and
- Restoring/enhancing wetlands and natural areas.

Planners are currently working to develop various draft project implementation schedules, with expected progress dependent upon different federal funding scenarios. Preliminary drafts show a large number of CERP projects being pushed back beyond the 2020 time frame. However, Appelbaum (2008a) noted that "no CERP projects are being taken off the table" (see Box 3-3). The planning efforts to date recognize the sizable existing restoration commitments, and there appears to be a strong emphasis on "finishing what's on our plate," including authorized but unfunded projects (Appelbaum, 2008a). This approach would continue to advance those projects with the strongest stakeholder support and least opposition, possibly above projects with greater potential system-wide restoration benefits. CERP planners should certainly prioritize critical non-CERP foundation projects that are necessary for the CERP to achieve its restoration goals, but the sequencing of all CERP projects (including already authorized projects) should be carefully reevaluated to advance those projects that can obtain the greatest restoration benefits given the anticipated budget constraints.

In response to these challenges, the Quality Review Board[4] asked an ad hoc group of senior Everglades scientists to provide advice on restoration priorities. The Quality Review Board specifically asked, where are "…the places in the natural system where we can get the strongest ecosystem restoration response (i.e., where should we direct our water and resources to achieve the maximum level of Everglades restoration)?" The committee commends the Quality Review Board for asking these important questions and beginning a dialogue on this issue. Based on a September 2007 workshop, the ad hoc group of scientists recommended that a fully successful Decomp project would "provide the stron-

[4]The Quality Review Board (QRB) is a group of senior CERP agency managers that was formed by USACE and SFWMD leadership as a means to resolve issues across agencies, improve collaboration, and provide common direction to CERP staff. The QRB is not a decision-making body, although QRB participants include most senior CERP decision makers.

BOX 3-3
Guiding Principles for Development of the Integrated Delivery Schedule

The following list summarizes the guiding principles for a new integrated delivery schedule (IDS) that were developed at a South Florida Ecosystem Restoration Task Force workshop held in February 2008:

1. No CERP projects are being taken off the table; this reevaluation is merely to update the project sequencing and develop a more realistic implementation schedule for the initial set of CERP projects to be constructed.
2. The IDS acknowledges the federal and state agencies' commitment to complete the implementation of key ongoing projects. The term *commitment* refers to projects currently authorized, under construction, or both. It includes the "Foundation Projects" (i.e., Modified Water Deliveries, Kissimmee River Restoration, C-111 South Dade, C-51/STA - 1East, etc.) and other projects for which the federal and state agencies have committed to accelerate implementation.
3. The IDS should include all projects related to the restoration of the Everglades. These projects include both state and federal initiatives such as Herbert Hoover Dike Rehabilitation, the Northern Everglades Plan, and the Long-Term Plan for Achieving Water Quality Goals in the Everglades Protection Area.
4. The IDS federal funding scenarios will include only those projects under the South Florida Everglades Ecosystem Restoration Program (SFEER). The SFEER projects include: MWD, Kissimmee, Central & South Florida (includes C-111 South Dade, C-51, and CERP), and Everglades & South Florida.
5. Projects should be implemented in a sequence that achieves restoration objectives at the earliest practicable time, consistent with funding constraints.
6. As appropriate, projects should be broken into multiple project implementation reports to facilitate the incremental adaptive restoration (IAR) approach recommended by the National Research Council. Each separable element will conform to National Environmental Policy Act (NEPA) guidance, as well as other federal and state laws.
7. The IDS will provide the basis for the MISP update for the CERP, as currently required by the Programmatic Regulations. The updated MISP, in turn, will be a major component of the wider-ranging IDS.
8. Project and component interdependencies will drive the sequencing order for constructing projects (e.g., pilot projects must be completed prior to a full-scale project).
9. As appropriate, the interim goals and targets should be used to measure restoration progress
10. Key points in implementation will be defined by new system operating manuals.

SOURCE: Appelbaum (2008a).

gest and most desirable ecosystem restoration responses in the system of South Florida wetlands, and should be the highest priority for achieving maximum ecosystem restoration benefits." The group came to this conclusion because of the prospective project benefits for the southern estuaries and mangrove transition zones and for restoring historic landscape patterns of production and biodiversity in the Everglades ecosystem. The ad hoc group concluded that moving large quantities of water from the northern basin would benefit the system as a whole, but it may result in sustained degradation of some plant communities due to existing levels of nutrients. The senior scientists acknowledged this "trade-off" but noted the greater good that comes from fully restoring flows through the Everglades and southern estuaries (Ad Hoc Senior Scientists, 2007).

As the ambitious program for restoring the Everglades has advanced, CERP planners have been hindered by deficiencies in the framework for making difficult decisions. Difficult restoration decisions include advancing projects in the face of conflicts, adjusting the sequences of projects under constraints of authorization and funding, and deciding among priorities when confronted with limited resources. Although the incremental adaptive restoration (IAR) approach described in NRC (2007) promises a means for making decisions in the face of both scientific uncertainty and resource limitations, even within an IAR approach, challenging decisions must be made in project scoping and sequencing and in implementing projects where there are stakeholder conflicts. How effective the new IDS process will be in making difficult decisions remains to be seen, particularly because the expectations for the program seem to be increasing. By prioritizing a smaller number of projects given current fiscal constraints, CERP planners risk diminishing support from some stakeholders. If important steps are to be made in the next decade toward the CERP's system-wide restoration goals, given the probable fiscal constraints, CERP planners will need to consider alternate mechanisms for decision making and evaluating trade-offs rather than relying upon mechanisms that favor projects with uncontested stakeholder support.

Adoption of Incremental Adaptive Restoration

A key recommendation from the committee's first biennial review (NRC, 2007) was the call for an alternative framework for advancing natural resource restoration in the Everglades, which it termed incremental adaptive restoration (IAR). The aim of IAR is to resolve decision-critical scientific uncertainties and to address project-sequencing constraints to improve the pace of restoration. As conceived, an IAR approach makes investments in restoration project increments that are large enough to secure significant environmental benefits, while simultaneously testing hypotheses selected to resolve important scientific uncertainties about the response of the system to management interventions. Such steps would likely be smaller than the CERP projects because the purpose of the IAR is to take actions that promote learning that can guide the remainder of project design, although an IAR approach might combine increments of several projects. NRC (2007) noted that IAR is likely to be of particular value in devising management strategies for dealing with complex and contentious ecosystem restoration projects in which probable ecosystem benefits are difficult to predict. This approach would also help address current constraints on restoration progress, including Savings Clause requirements for water supply and flood-control obligations. As an application of adaptive management, IAR would require rigorous monitoring and assessment to test hypotheses, yielding valuable information that could expedite future decision making and improve future project design. Existing authorization and budgeting processes may need to be modified to accommodate the IAR process.

The recommendation of the first biennial review (NRC, 2007) to apply IAR has been widely embraced by CERP staff at all levels of organization (e.g., Grosskruger, 2007) as well as by various stakeholders, but the concept has not yet been fully applied. The concept of IAR appeals to many who want to realize more restoration benefits; who are attracted by a more achievable, incremental approach; or who seek to resolve scientific, engineering, and policy uncertainties that are obstacles to progress.

CERP agencies have set about applying this approach to a number of CERP projects. Planning efforts are under way that consider application of IAR for the C-43 Water Storage Basin, projects in the EAA, the C-111 Spreader Canal, and the Biscayne Bay Coastal Wetlands Project. And an IAR approach was also considered for an interdependent array of southern Everglades projects (see Box 3-4). The committee is gratified to witness such earnest efforts to apply IAR in CERP projects.

The committee discussed these IAR efforts with senior agency managers, technical managers, scientists, and stakeholders. All of these individuals felt that IAR offered opportunities for significantly advancing CERP restoration and had already succeeded in bringing CERP participants together in an attempt to break through logjams in multiproject planning. Some, however, cautioned that it is still unclear whether or not IAR will be successful. Nevertheless, it was clear that those from different agencies or with different roles had disparate views about the primary benefits and requirements for IAR. Some emphasized the need to achieve tangible ecosystem restoration benefits as soon as possible, either because of fear of loss of public and political support or of risks of further degradation of the ecosystem. Others stressed the practical requirements to

BOX 3-4
Preliminary Draft IAR Plan for the Southern Everglades Restoration Projects

The initial IAR concept for the southern Everglades restoration projects (USACE and SFWMD, 2008a) pursued an integrated, comprehensive, and watershed-based approach to plan and implement restoration, incorporating four projects: Modified Water Deliveries (Mod Waters), WCA-3A Decompartmentalization and Sheet Flow (Decomp), C-111 Spreader Canal (C-111 SC), and Everglades National Park Seepage Management (ENPSM). The southern Everglades area was selected as an area that historically supported the diverse and abundant biota characteristics of the Everglades and adjacent estuaries, where restoration efforts could potentially return the most benefits with limited expenditures. The proposal considered the four projects as interdependent and their integration as yielding a CERP regional plan formulation. It sought "to advance early ecosystem restoration while addressing high-risk uncertainties" and thus to alleviate the time-delaying effects these uncertainties have on project design and implementation. A risk-assessment ranking was applied to identify the highest-ranking ecological and hydrological uncertainties:

- The rates and patterns of marsh sheet flow needed to achieve the desired topographical and ecological responses,
- The interaction of flow and depth on ridge and slough communities, including tree islands,
- The relationship between soil and water nutrient loads and ecological responses as volumes and rates of flow increase,
- The volumes and rates of flow into the southern estuaries needed to achieve desired levels of production in the estuaries,
- The interaction of depth and duration of marsh flooding on lower east coast groundwater and seepage patterns, and
- The operational, storage, and structural plans that will best meet the water quality, flow, and volume targets for Everglades restoration.

Reducing these uncertainties would affect important decisions about the design, location, and capacity of upstream water storage and design and operation of projects to meet flow rate and volume requirements of the system. The southern Everglades IAR team posited that these uncertainties can be reduced by learning based on careful observation of the outcomes of the integrated implementation of components of the four projects (Mod Waters, Decomp, C-111 SC, and ENPSM), while achieving the early restoration benefits for the Shark River Slough and Taylor Slough basins. It was further argued that collective implementation of the four projects would be at a spatial scale large enough to provide measureable restoration benefits for both the freshwater Everglades Basin and the northern Florida Bay and adjacent estuaries, as well as significant learning benefits.

The proposal also identified the various policy constraints and implications that need to be addressed, including impacts to existing projects, implications for USACE planning policy, and cost implications. However, after consideration of this proposal, the Chief of the Everglades Division of the USACE Jacksonville District directed that the Draft Incremental Adaptive Restoration (IAR) Proposal for Southern Everglades Restoration Projects would not be used in the potential IAR application efforts.

implement affordable restoration by breaking large projects into effective, but less-expensive components, even reasoning that significant half-measures might be all that can be accomplished. Scientists tended to emphasize the importance of the experimental approach to adaptive management (active adaptive management) to advance learning about natural system response that would benefit the larger project, other projects, or the CERP as a whole. Still others stressed the opportunity to overcome planning and procedural conflicts, such as flooding risks, that have been obstacles to restoration projects. Most fundamentally, there seemed to be inherent tension between applying IAR as a means to finally achieve restoration benefits versus as a means to narrow uncertainties and advance learning. Some managers observed that, despite the official adoption of adaptive management as the guiding process for the CERP, convincing agency and political leadership and the public to invest in learning is a hard sell, while the prospect of significant restoration is appealing.

In the committee's earlier report (NRC, 2007), IAR was developed more as a concept than a prescriptive methodology; therefore, there are no established criteria to judge whether the various potential applications comport with IAR. While the CERP participants will define what IAR becomes in practice, it is appropriate to evaluate whether the IAR applications being considered are likely to be successful in advancing measureable restoration while promoting decision-critical learning.

The proposed integration of components from four projects to achieve early and significant ecosystem restoration in what is the defining portion of the Everglades ecosystem was a laudable goal, and the committee does not know what led to the decision not to use the southern Everglades IAR proposal (see Box 3-4). Nevertheless, there are lessons that can be learned from this bold concept. Each of the four projects brought its own set of obstacles and costs that have already stymied planning, implementation, and funding. Collectively, the proposed effort was more substantial than incremental. This meant that it might not have been practically, politically, or fiscally achievable in the short term. As recognized in the proposal, the integrated approach posed major challenges with regard to USACE planning policies and project sequencing, even though it offered opportunities to overcome some of the serious limitations of the next added increment analysis through collective consideration of the benefits of interdependent components (see the following section). Further, the collective approach reduced the opportunity to take risks in pursuit of active adaptive management experiments compared to more incremental and less costly approaches to IAR.

The size and complexity of the IAR Proposal for Southern Everglades Restoration Projects also posed dilemmas for resolving uncertainties through adaptive learning. On one hand, it was argued that such a substantial and integrated implementation is required to elicit measureable results, particularly for scale-dependent hydrologic and ecological responses. On the other hand, the more factors that are affected through large-scale modifications, the more difficult it is to sort out cause-and-effect relationships. Also, many topographical and ecosystem responses develop over decades, making it difficult to incorporate knowledge into adaptive learning on the shorter time frames required to improve operations and subsequent project design. Finally, the project components considered are themselves very large projects, and there may be a reluctance to fully admit the uncertainties embedded in them or to commit the necessary resources to resolve the uncertainties as they progress.

While the committee recognizes that the dual attraction of early restoration benefits and learning has stimulated creative approaches for more-effective integration of projects in the southern Everglades as well as incremental implementation of other CERP projects, it suggests that the most-effective applications of the IAR concept will probably be in the incremental execution of project components that produce significant outcomes but are of a scope and scale that can be feasibly implemented and assessed. It is in such more-limited applications of IAR where the opportunities for testing hypotheses to resolve uncertainties are greatest. In addition to the continued efforts to apply adaptive management in the Decomp project there might be an IAR focus on design and outcome assessment for the C-111 Spreader Canal, or as it is now being conceived, the Taylor Slough Enhancement, within the broader integrated planning of southern Everglades restoration projects. Furthermore, because many desired ecological changes are likely to take many years or decades to respond to IAR actions, emphasis should be placed on assessing variables, such as sediment flow and water quality, that are leading indicators of likely long-term ecological responses.

Endangered Species Act Impacts on Restoration

The original vision of the CERP was that, through a combination of land acquisition and ecosystem restoration, the project would increase the extent, quality, and connectedness of habitats for multiple threatened and endangered species, thereby reducing the need for intensive single-species recovery efforts and avoiding the U.S. Endangered Species Act (ESA) conflicts and litigation. In practice, conflicts over endangered species have delayed CERP and related foundation projects such as Mod Waters (Rizzardi, 2001). In this section, alternative management approaches are discussed to minimize legal conflicts over endangered species management that threaten to further delay Everglades restoration.

More than 65 species of plants and animals in Central and South Florida are listed under the ESA as either endangered or threatened. Lawsuits have been filed over water management or construction impacts on the Cape Sable seaside sparrow, snail kites, and wood storks (Rizzardi, 2007). The potential for conflicts only grows with continued delays in ecological restoration because of the sensitivity of many of the listed species to water management regimes.

Much of the recent ESA-related litigation has been over water management in the WCAs and associated hydrologic regimes in Everglades National Park. Emergency water management for Cape Sable seaside sparrows under the interim operational plan (IOP) illustrates the failure of species-by-species management. The resulting water regimes have led to unwanted flooding of tribal lands and probably have contributed to declines of snail kites and tree islands in WCA-3A. Benefits to the sparrows under IOP have been mixed at best (SEI, 2007). Until flow is partially restored through Mod Waters and Decomp, water managers will continue to debate trade-offs "over which species will be allowed to suffer the most from ongoing ecosystem degradation" (SEI, 2007).

The single-species focus of the ESA, coupled with CERP planning deficiencies, is creating substantial problems in a multispecies environment. The CERP currently lacks scientifically credible, conceptual, and operational bases for managing multiple species at risk (SEI, 2007). The *South Florida Multi-species Recovery Plan* considers 68 listed species and 23 plant communities, but it addresses each individually and provides practically no guidance on how to manage trade-offs among species, to set priorities, or to deal with regulatory requirements and conflicts. Effective multispecies management will require three main initiatives: a higher level of interaction among species specialists; continued development and application of ecological models, like the Across Trophic Level System Simulation (ATLSS), that can be used to examine potential species trade-offs in space and time (e.g., Curnutt et al., 2000); and formulation of a robust decision framework for managing multiple species under uncertainty (e.g., Nicholson and Possingham, 2007). At the same time, the ESA should be applied less reactively and without such extensive procedural delays (Bean, 2006).

The ESA provides various mechanisms to avoid litigation and facilitate the management of multiple-listed species during the transition period of CERP implementation. These include incidental-take permits for multiple species, multispecies habitat conservation plans (HCPs) for incidental take of species on nonfederal lands, and safe-harbor agreements for those restoration activities that might provide ephemeral local benefits to species where subsequent activities might result in indirect take of those same species (Rizzardi, 2007). As a last resort, the Endangered Species Committee can approve an exemption to the ESA.

Given the advanced stage of CERP planning and the large amount of federal land, pursuing a multispecies HCP for the Everglades is probably ill advised. Successful multispecies HCPs at the scale of CERP have been established in California, Arizona, and elsewhere, but not without considerable time, cost, and controversy (Scott et al., 2006). However, there may be opportunities for programmatic permitting that could reduce single-species ESA-driven conflict. Under the ESA, the U.S. Fish and Wildlife Service (FWS) can authorize permits to "enhance the propagation or survival" of listed species. This provision has been used in restoration activities elsewhere to obtain incidental-take permits covering multiple species. Bean (2006) described the example of the Peninsula Open Space Land Trust near San Francisco, which obtained a programmatic permit allowing incidental take of listed species during prescribed burns as part of a large-scale, long-term grassland restoration program.

The Archbold Biological Station in the upper Kissimmee River basin has a similar agreement with the FWS permitting incidental take from a controlled burning program designed to mimic the historical fire regime, to enhance biological diversity, to promote threatened and endangered species, and to reduce fire hazards (H. Swain, Archbold Biological Station, personal communication, 2007). Obtaining such permits is not simple: The Peninsular Open Space Trust required 2 years to obtain their permit and had to endure a number of setbacks by the local FWS office (Bean, 2006). The Department of the Interior (DOI) should take a leadership role by convening a high-level group of science and policy experts to explore the available approaches and produce a multispecies adaptive management strategy to accompany the existing South Florida Multispecies Recovery Plan.

As pointed out by Goble (2006), the ESA has evolved since 1973 from a prohibitive statute into a somewhat more flexible permitting system. ESA conflicts of a scale comparable to CERP have been resolved through scientifically based conservation planning and private and public cooperation, particularly at local and state levels. The CERP benefits from a depth of biological data and scientific understanding that far exceed these other efforts. In principle, a scientifically strong and administratively streamlined multispecies management plan for endangered (and nonendangered) species is possible.

Programmatic permits could facilitate CERP adaptive management and restoration but would require flexibility by staff in the FWS regional office and a high level of trust and cooperation among the relevant CERP agencies (Bean, 2006; Burnham et al., 2006; Swain, 2006). Meaningful stakeholder engagement will also be essential, as difficult choices and compromises will inevitably have to be negotiated. The recent revised critical habitat designation for the Cape Sable seaside sparrow (USFWS, 2007) shows that the FWS and other parties can and are willing to make such difficult decisions, in this case trading off possible increased risk for sparrow subpopulation A (see Figure 2-15) to improve restoration prospects for several other species including wood storks and snail kites.

Problems Created by the Next Added Increment Requirement

Elements of the complex CERP project planning process are contributing to delays in restoration progress. Presentations to this committee identified key procedural problems, some characterized as "self-inflicted wounds," contributing to this protracted process. One particular concern is the absence of a systematic approach to analyze the costs and benefits across multiple projects in support of project planning. Fundamentally, the CERP is designed as a system of related projects (i.e., components) that work together in the aggregate to produce overall restoration benefits. Without a system-wide planning process, it is not clear how system benefits can be optimized for any one project without any systematic consideration of other projects. The next added increment (NAI) is a notable example of the lack of system-wide considerations in the CERP planning process.

In the Programmatic Regulations (see 33 CFR 385.1; 385.26(b)(3)), the NAI is defined as

> the evaluation of an alternative as the next project to be added to a system of projects already implemented. For the purposes of this part, this means analyzing an alternative as the next project to be added to a system of projects that includes only those projects that have been approved according to general provision of law or specific authorization of Congress and are likely to have been implemented by the time the project being evaluated is completed.

The primary objective of the NAI evaluation criteria is to ascertain whether sufficient ecosystem restoration benefits could be attributable to a specific project to justify the cost, if no additional CERP projects (other than those already existing or authorized) were built. The NAI analysis also illuminates the dependence of other CERP projects on the project under evaluation (USACE and SFWMD, 2007a). The NAI analysis does meet these objectives, but as currently implemented, it undermines system-wide CERP planning, including the capacity to implement a project-sequencing plan.

Difficulties arise with the NAI because the CERP consists of roughly 50 interrelated projects, and some sequencing constraints require the implementation of particular projects prior to other projects (e.g., increased water storage needs to be in place to support removal of barriers to flow). Analyzing these projects individually, based on the NAI, appears to minimize the benefits predicted to arise from these early CERP projects, because the benefits that derive from linkages between the proposed project and future, as yet unauthorized projects are not considered in the benefits analysis. Additionally, this benefits quantification approach requires the effects of a single project to be calculated on a system-wide basis, although the Everglades system covers in excess of 16,000 square miles, an area nearly twice the size of New Jersey.

In Guidance Memorandum #2 (USACE and SFWMD, 2007a), project teams are advised to avoid this problem by "combining the tentatively selected plan with other CERP components to identify an alternative that can be justified on a NAI basis or to consider delaying the implementation of the tentatively selected plan…" However, project teams have avoided combining interrelated CERP components into a single PIR because there is a perception that Congress and/or the Office of Management and Budget (OMB) is unwilling to authorize PIRs that include more than a billion dollars in projects. As a result, the reduced benefits predicted for individual projects by NAI methodology hinder specific project approval by the USACE and submission to Congress for authorization and subsequent funding. For multicomponent large-scale restoration programs authorized on a project-by-project basis, the NAI methodology appears designed to fail unless Congress and OMB express willingness to approve larger, interrelated suites of projects.

The EAA storage reservoir project, Phase 1 (A-1), provides an example of the difficulties created by the NAI. There is wide recognition that additional water storage can make contributions to Everglades restoration and to the reduction of adverse impacts of lake water releases into the Caloosahatchee and St. Lucie estuaries. Nevertheless, the project is entangled in evaluation methodology and project approval procedures that are contributing to delays in the project's approval and authorization. Not surprisingly, consideration of Phase 1 (A-1) of the EAA reservoir project (190,000 acre-feet of storage) in isolation, consistent with the NAI approach, is making it difficult to demonstrate sufficient benefits for it to go forward. Downstream benefits of the EAA reservoir to the WCAs and Everglades National Park contingent upon the Decomp project cannot be considered because that project is not yet authorized. Also, the EAA reservoir itself only adds 3 percent to Everglades-wide storage capacity; thus, it is difficult to demonstrate that—on its own—it makes an appreciable beneficial contribution to restoration on an Everglades system scale. Meanwhile, benefits analyses are hindered by limitations in the models themselves. For instance, the South Florida Water Management Model has embedded within it operational rules that distribute water between the environment and agricultural purposes in ways that cannot easily be adjusted, and the model output appears to emphasize water supply and flood control rather than restoration, even though decisions have not been made regarding formal water allocations for the natural system.

This mix of procedural requirements, operational considerations, and statutory directives, when applied to the EAA storage reservoir project, yields evaluations predicting less-than-expected and relatively small benefits to the Everglades ecosystem. These results are at odds with the conclusions of an array of experts and observers that are persuaded that additional water storage is a necessary and critical component of the restoration effort. The CERP project planning and approval process, especially NAI, fails to recognize that CERP's purpose is to restore an ecosystem rather than build a particular project. The CERP is designed as a suite of interacting projects or components to provide an aggregated set of ecological benefits, and there appears to be a fundamental tension between this objective and the isolated nature of the project approval and authorization process, of which the NAI methodology is just one component. As CERP planners embrace the concept of IAR, more CERP projects may be constructed incrementally, which may lead to additional project approval problems if the issue of NAI is not addressed. Furthermore, NAI is symptomatic of difficulties arising from trying to retrofit traditional USACE project- and engineering-oriented processes on a large-scale ecological restoration program.

An opportunity does exist to correct this problem. The Programmatic Regulations, promulgated in 2003, are slated to undergo a 5-year review in 2008. Following review, USACE may propose revisions to the Programmatic Regulations that facilitate improved project planning and evaluation as part of an overall integrated ecological restoration program.

In Chapter 2, the committee concluded that ongoing restoration delays have contributed to ecological decline in the Everglades and that major restoration efforts are needed to strengthen the resiliency of its ecosystems to help avoid irreversible losses. Yet, the current planning mechanisms, including NAI, do not help resolve the central questions that affect system-wide prioritization and the delivery of restoration benefits. Given the limited funding environment and the lack of certainty that the CERP will be fully constructed, what are the best opportunities for restoration that will produce the most near-term ecological benefits? Effective mechanisms for alternatives evaluation and project prioritization and sequencing are essential to CERP planning, and CERP planners should develop mechanisms to improve system-wide planning and decision making for the CERP.

CONCLUSIONS AND RECOMMENDATIONS

The CERP is one of the most ambitious, detailed, and comprehensive blueprints for managing an integrated built and natural environment ever planned, and the attempt to restore an ecosystem as large and complex as the Everglades is an unprecedented challenge. Despite programmatic accomplishments and the beginning of construction for some projects identified in the CERP, natural system restoration has been delayed. The South Florida ecosystem continues to suffer as a result of a complex and sometimes contentious planning process, funding uncertainties, lack of clear restoration priorities that are central to restoration, and statutory and regulatory impediments such as the Endangered Species Act and next added increment rules.

It is too early to evaluate the response of the ecosystem to CERP projects because none have been completed. Construction completion for the first CERP components has not been achieved through mid-2008, and key foundational pre-CERP projects, such as Mod Waters, remain far behind schedule. Furthermore, the natural system in South Florida continues to decline. If limited natural system restoration progress continues, frustration will further increase among stakeholders and agency staff, and public support for restoration is likely to diminish. Actual construction and implementation of key non-CERP and CERP projects are the only means to arrest the degradation and assure that natural system restoration begins. State efforts to construct projects in spite of funding limitations and other serious obstacles to progress are commendable. Some partial benefits have been produced from phased construction in the Picayune Strand Restoration (wetland restoration) and Acme Basin B (stormwater treatment) projects. Additionally, several non-CERP activities are positive harbingers of future CERP programs and indicate that when project implementation does occur, bona fide ecological restoration benefits will be demonstrated. For example, the success of the Kissimmee River Restoration effort continues to be the most important piece of evidence that restoration of a natural system is possible in the Everglades region.

The state of Florida should continue its active land acquisition efforts, accompanied by monitoring of and regular reporting on land conversion patterns in the South Florida ecosystem. Land management for a successful CERP depends on purchasing particular sites within the project area and protecting more general areas within the South Florida ecosystem that could help meet the broad restoration goals. The committee commends the state of Florida for its aggressive and effective financial support for acquiring important parcels, including the announcement to enter into negotiations for the potential purchase of 187,000 acres of land in the EAA for $1.75 billion. The acquisition of this large amount of land has the potential to alter basic CERP plans, but because of the structure of the purchase and the possibility of numerous land exchanges made after the purchase, direct effects of the deal may not be seen for a decade or more.

The complex project planning and approval process has been a major cause of delays for CERP projects to date. The greatest challenge in the project planning process has been developing technically sound project plans that are acceptable to the many agencies and stakeholders involved. The process of resolving disagreements among agencies and stakeholders has led to lengthy delays in the development of some PIRs that can be submitted to Congress for authorization. The infrequent and unpredictable federal authorization mechanism for CERP projects has caused some additional problems and attendant delays. The committee judges that the lack of federal funding in the first 8 years of the CERP is not the most serious cause of the CERP delays. Instead the slow pace of federal funding has largely been a symptom of the problems caused by the complex and lengthy CERP planning and authorization process for each project. However, now that three CERP projects have been approved for their PIRs and congressional authorization, funding limitations will certainly create additional constraints to CERP progress in the years ahead. Non-CERP and CERP projects will increasingly compete for limited state and federal funding, while project costs increase due to inflationary pressures and scope changes. Both state and federal partners are facing budget constraints, and dramatic state budget cuts in FY 2009 will affect the speed of restoration progress.

Deficiencies in CERP system-wide planning are affecting future natural system restoration benefits. The CERP lacks a systematic approach to analyze the costs and benefits across multiple projects in support of project planning. Fundamentally, the CERP is designed as a system of related projects (i.e., components) that work together in the aggregate to produce overall restoration benefits. Without a system-wide planning process, it is not clear how system benefits can be optimized for any one project without any systematic consideration of other projects. The next added increment is a benefits evaluation method that considers benefits only from the proposed and previously authorized projects, and as currently implemented in the Everglades, it undermines system-wide restoration planning and sequencing. The current planning process also appears to reward the least-contentious projects, regardless of their potential contribution to ecosystem restoration. Without clear priorities for project planning and funding, projects with large potential restoration benefits may see lengthy restoration delays while other, less-contentious projects that address only isolated portions of the ecosystem may tie up available funding. During the 5-year review of the Programmatic Regulations, the USACE should address deficiencies and impediments in the CERP planning process that are affecting restoration progress. CERP planners should also develop mechanisms to improve system-wide planning and decision making for the CERP.

Developing a realistic schedule and sound project sequence is a critical need for the restoration effort. In this time of increasing fiscal pressures, it is critical that CERP planners find a means to prioritize and properly sequence restoration projects so that public funds are allocated by the degree to which the projects are essential to restoration of the South Florida ecosystem, rather than by local stakeholder support or the order of authorization. Public Web-based reporting on project progress, delays, and anticipated completion dates should be more transparent than is currently the case.

The executive and legislative branches of the federal government should consider departing from traditional project-by-project review, authorization, and yearly funding to benefit both the CERP and other multicomponent ecosystem restoration projects across the nation. It may be far more efficacious—scientifically, managerially, and economically—to design a different approach for comprehensive restoration programs that provides assured funding over a multiple-year period.

The incremental adaptive restoration (IAR) concept proposed in NRC (2007) has stimulated creative restoration approaches to Everglades restoration but has not yet been fully applied. The prior committee's recommendation to apply IAR has been widely embraced by implementing agencies at all levels of organization as well as by various stakeholders, but an effort to apply IAR to an integrated group of Southern Everglades restoration projects was discontinued. CERP planners, however, are using the IAR concept in planning the Biscayne Bay Coastal Wetlands and C-111 Spreader Canal projects. The most-effective applications of the IAR concept will probably be in the incremental execution of project components that produce significant outcomes but are of a scope and scale that can be feasibly implemented and assessed. Because most desired ecological changes are likely to take years or decades to respond to IAR actions, in developing IAR strategies, the emphasis should be placed on assessing variables, such as sediment transport and water quality, that are leading indicators of likely long-term ecological responses.

To reduce restoration delays, CERP planners should develop a stronger conceptual basis for multispecies recovery planning and management. Although implementation of the ESA has become focused increasingly on single species management, the statute does provide various mechanisms that can reduce the threat of legitimate litigation and facilitate the recovery and management of multiple-listed species. However, effective multispecies management under the ESA requires a high level of integration of scientific knowledge about individual species and species interactions to understand risks and trade-offs during construction and under alternative water management regimes. It also requires strong federal leadership and a high level of trust and cooperation among the regulatory and management agencies and other stakeholders to allow for learning, compromise, and decision making under uncertainty.

In addition, jeopardy determinations for endangered species and associated litigation are a significant, unresolved challenge for adaptive management and IAR. There currently is no scientifically credible operational plan for managing multiple species at risk in South Florida. To expedite multispecies restoration under the ESA, DOI should immediately initiate and lead the development of a South Florida multispecies adaptive management strategy, including both science and policy dimensions, to accompany the existing South Florida Multi-species Recovery Plan.

4

Mod Waters

The Modified Water Deliveries to Everglades National Park (Mod Waters) project was authorized nearly 20 years ago to substantially restore more natural water flows into Everglades National Park. Nearly 50 years of water blockages and diversions have altered and impaired Shark River Slough, in the heart of Everglades National Park, making the northeastern side too dry and the western side too wet. Mod Waters had a simple goal: offset the continuing adverse effects of these flow diversions by restoring more natural water flows into Everglades National Park and, thereby, set the stage for the Comprehensive Everglades Restoration Plan (CERP) (Figure 4-1).

Nothing else about Mod Waters has been simple. A variety of political, legal, and financial objections have stymied the project, and the attendant delays and resulting changes in scope have dramatically increased its cost. The failure to achieve modified water flows into Everglades National Park has also undermined the learning opportunities associated with the first steps to "get the water right."

Everglades restoration is at a crossroads: completion of the Mod Waters project would put in place a cornerstone for CERP, while continued failure to implement Mod Waters will delay critical components of the CERP and allow the Everglades ecosystem to continue to degrade. The significance of Mod Waters motivates its review in this progress report.

Accordingly, the project is described, obstacles overcome are identified, and remaining impediments examined. Lessons learned from the Mod Waters project are also included in this chapter. The chapter concludes with a recommendation that the remainder of Mod Waters be implemented promptly so that the first steps toward achieving restoration in Everglades National Park can begin. While the Mod Waters project as currently planned will not provide as many environmental restoration benefits as other project alternatives, it is critical for the overall restoration effort to achieve some environmental benefit from Mod Waters after several decades of effort.

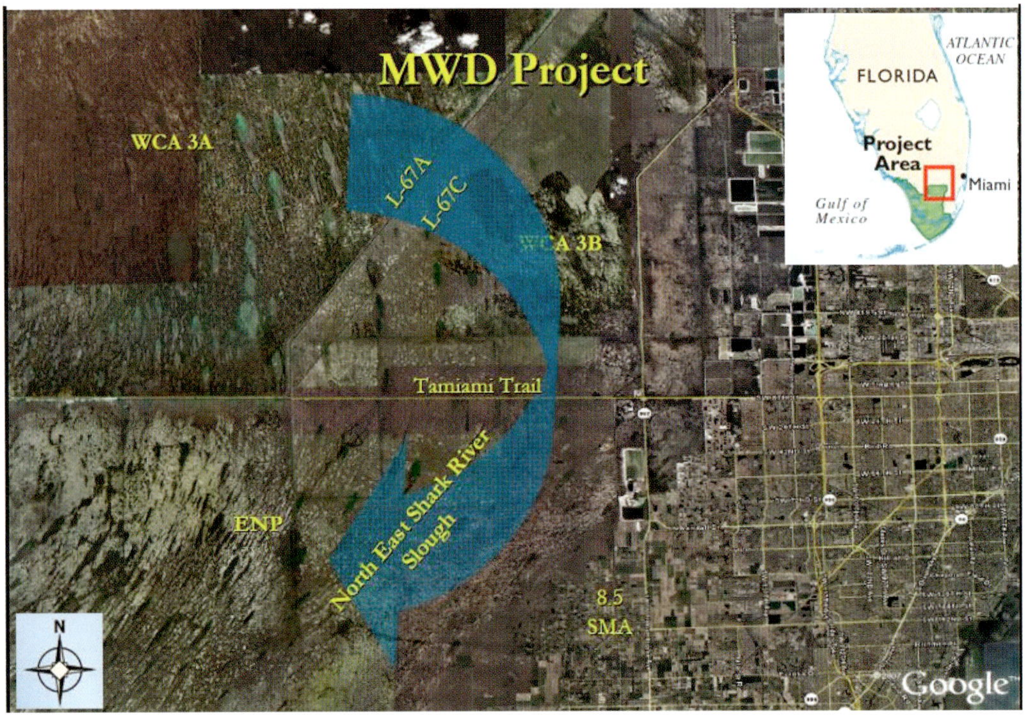

FIGURE 4-1 Overall goal of the Modified Waters project. Water would be diverted from north of the Tamiami Trail (SR 90), particularly from Water Conservation Area (WCA) 3A. This water would enter Everglades National Park (ENP) from the northeast and flow through the Shark River Slough system. Several levees (L-67A, L-67C, L-29) and the Tamiami Trail itself currently hinder this flow. Seepage controls in the 8.5-square-mile area (8.5 SMA) would prevent flooding into the developed areas of the eastern Everglades.

SOURCE: Adapted from USACE and DOI (2008); inset map © International Mapping Associates.

OBJECTIVES OF THE MODIFIED WATER DELIVERIES PROJECT

Roads and canals have long affected water flow into the Shark River Slough in Everglades National Park, but the construction of the water conservation areas (WCAs) in 1963 dramatically reduced the flow to what is now the northeastern portion of the park (see Figure 4-2). The Central and South Florida (C&SF) Project primarily channeled flow through West Shark River Slough, reducing mean flow volumes into Northeast Shark River Slough from 65 to 10 percent of the total flow southward across Tamiami Trail, while frequently flooding the park's western

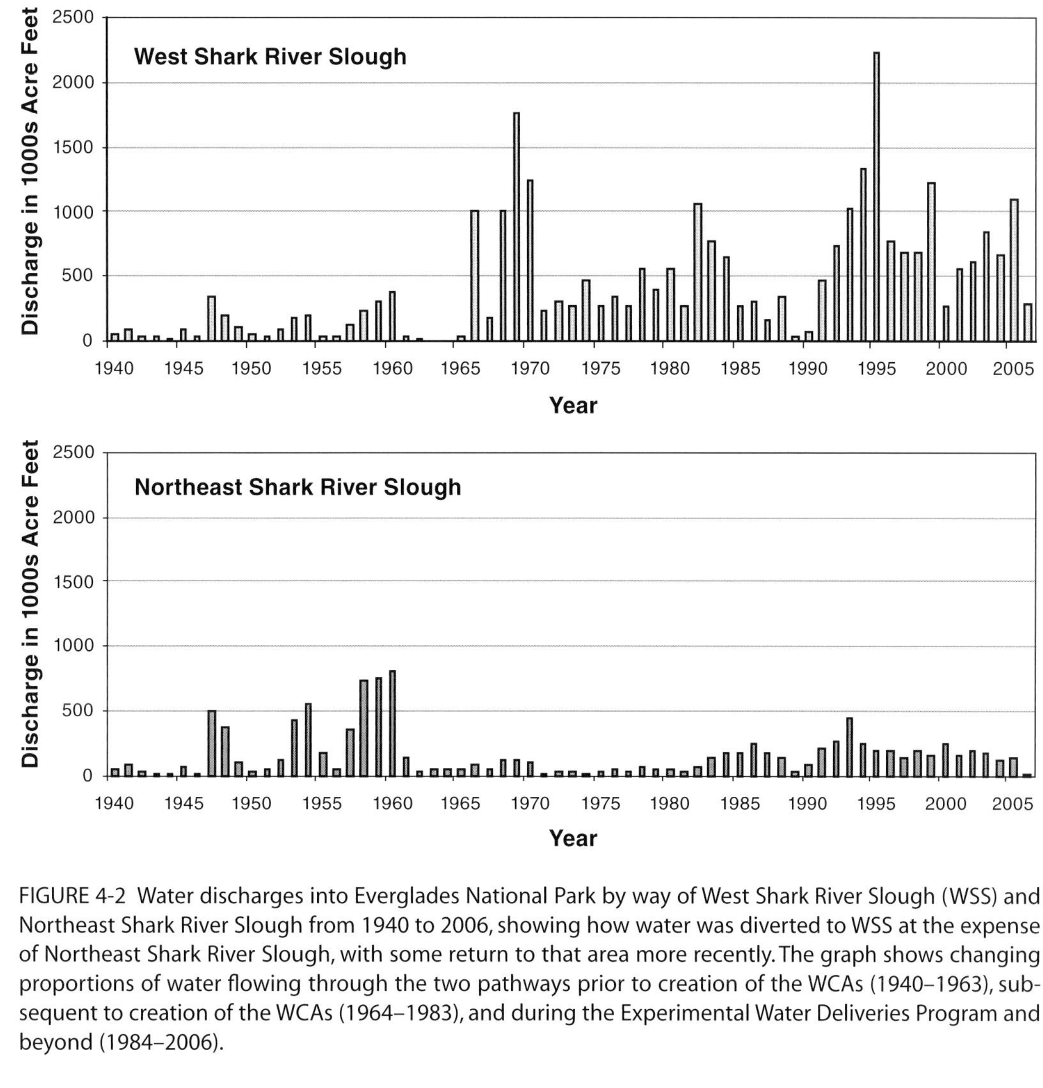

FIGURE 4-2 Water discharges into Everglades National Park by way of West Shark River Slough (WSS) and Northeast Shark River Slough from 1940 to 2006, showing how water was diverted to WSS at the expense of Northeast Shark River Slough, with some return to that area more recently. The graph shows changing proportions of water flowing through the two pathways prior to creation of the WCAs (1940–1963), subsequent to creation of the WCAs (1964–1983), and during the Experimental Water Deliveries Program and beyond (1984–2006).

SOURCE: Data from R. Johnson, NPS, personal communication, 2008.

habitats. The project also dramatically reduced flow velocities and water depths in WCA-3B. Decades of altered flow volumes and velocities have led to destruction of habitat through loss of soil and deterioration of tree islands, decreased native species diversity, and increased invasion by exotic species.

The Mod Waters project was authorized in 1989 as part of the Everglades

Expansion Act after a series of measures implemented in the 1970s and 1980s to alleviate/reverse the effects of an extended, historic drought on the ecosystems in Everglades National Park proved to be inadequate (see Figure 4-3 and Table 4-1). The Everglades Expansion Act authorized modifications to the C&SF Project to improve water deliveries into Everglades National Park and, to the extent practicable, to take steps to restore natural hydrologic conditions in the park while maintaining flood protection of and water supply to the built environment (PL 101-229, Section 104).

If successfully completed, Mod Waters will provide significant restoration in Everglades National Park by increasing the flow of water into Northeast Shark River Slough. Deterioration of ecological conditions in the natural system will continue until the Mod Waters project is operational as described in the authorizing legislation. Examples of ecological decline include:

- Loss of soil elevation and the characteristic ridge and slough pattern in Everglades National Park due to soil oxidation and more intense and more frequent fires (McPherson and Halley, 1996),
- Loss of tree islands in WCA-3A as a result of unnaturally high water levels during wet years (drowning) and unnaturally low water levels during dry years (soil oxidation) (Schortemeyer, 1980; Zaffke, 1983),
- Movement of wading birds and other large animals away from Everglades National Park because of a lack of food as the heterogeneity of the landscape is lost (Frederick and Spalding, 1994; Hoffman et al., 1994), and
- Continued population decline for the snail kite in WCA-3A and lack of recovery for the Cape Sable seaside sparrow in Everglades National Park (SEI, 2007).

Halting the decline soon is an essential and urgent step on the path to restoration. It took biological, geological, hydrologic, and climatological processes thousands of years to build the soils and tree islands (Stone et al., 2002; Willard et al., 2002), whereas an inch of soil can be lost in a single fire or by microbial oxidation of peat in a single year (McPherson and Halley, 1996). Once the soil is lost and tree islands drown, it may not be possible to restore these characteristics. It is not clear how long restoration of such losses might take, but the replacement of continued decline with gradual improvement is one step on the path to restoration.

The Mod Waters project is also a critical precursor to the CERP. That is, critical elements of the CERP, especially the WCA-3 Decompartmentalization and Sheet Flow Project (or Decomp), a centerpiece of the CERP, are designed to be built upon this cornerstone. Although Mod Waters does not have the visibility

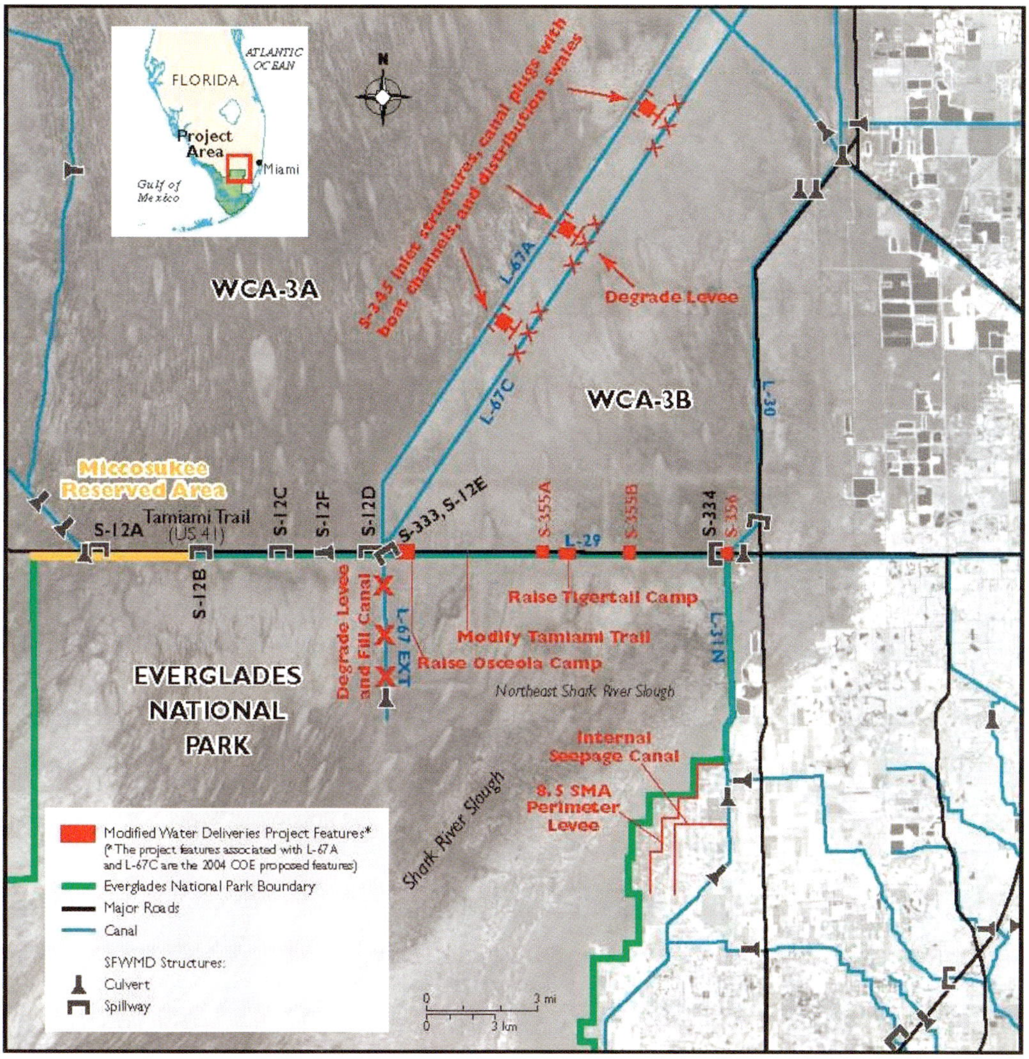

FIGURE 4-3 Mod Waters features. © International Mapping Associates

SOURCE: Modified from B. Gamble, National Park Service, personal communication, 2006, and K. Taplin, U.S. Army Corps of Engineers, personal communication, 2008.

TABLE 4-1 Time Line Leading to Initial Authorization and Preliminary Plans for the Mod Waters Project

Date	Event	Purpose
1960s	Extended historic drought affects Everglades National Park (ENP)	
1968	ENP South Dade Conveyance System (Flood Control Act)	Enlargement of the L-31N and C-111 canals to supplement water deliveries to south Dade and ENP
1970	Minimum Water Delivery Schedule (PL 91-282)	Required a minimum of 315,000 acre-feet of water deliveries to ENP each year, with a fixed monthly allotment[a]
1983	ENP Seven-Point Plan issued	Reduce the impacts of high S-12 regulatory flows on West Shark River Slough[b]
1983-1999	Experimental Water Deliveries Program (PL 98-181)	Test different water delivery schedules to restore more normal flow, especially in the eastern ENP
1989	ENP Expansion Act (PL 101-229)	Acquire 109,000 acres in Northeast Shark River Slough; authorized Modified Water Deliveries and C-111 (South Dade) Projects[c]
1992	General Design Memorandum (GDM) for Mod Waters finalized (USACE, 1992)	Restore historic flow way between WCA-3A, WCA-3B, and Northeast Shark River Slough and relieve high flows from WCA-3A to Western Shark River Slough. The plan aimed to deliver 55 percent of the total flow volume east of L-67 to reflect historic flow paths

[a]260,000 acre-feet delivered to West Shark River Slough, and 55,000 acre-feet delivered to Taylor Slough and Eastern Panhandle basins.

[b]Seven points included: (1) fill in the L-28 and L-67 extension canals and remove the levee (promote sheet flow); (2) gap the L-67A and L-67C levees (promote sheet flow and restore flows through WCA-3B); (3) redistribute West Shark Slough inflows along the full length on the Tamiami canal (L-28 to L-30); (4) establish a bimonthly water quality monitoring program for ENP; (5) defer implementing a proposed drainage district in the East Everglades; (6) field test a rainfall-based water delivery schedule for the WCAs and ENP; and (7) suspend the minimum water delivery schedule (Light and Dineen, 1994).

[c]Objectives of C-111 (South Dade) project were to: (1) restore historic hydrologic conditions in the Taylor Slough, Rocky Glades, and Eastern Panhandle basins of ENP; (2) protect the natural values associated with ENP; (3) eliminate the damaging freshwater flows to Manatee Bay/Barnes Sound and increase flows to northeast Florida Bay from the lower C-111; (4) maintain the level of flood damage reduction associated with the 1994 C-111 (South Dade) recommended plan; (5) ensure that C-111 (South Dade) project waters diverted to ENP meet all applicable water quality criteria.

or political support that the CERP enjoys, it is of special interest because of the prominence and importance of the Shark River Slough for the ecosystem and because appropriations for the Decomp project are predicated on completion of the Mod Waters Project. As noted in the Water Resources Development Act of 2007 (WRDA 2007) Conference Report, "Without a change in water delivery to the Park, restoration of the Everglades, and many of the projects authorized as components of the Comprehensive Everglades Restoration Plan (CERP) in 2000, will not succeed."

Since congressional authorization nearly 20 years ago, the Mod Waters project has been plagued by complex and difficult obstacles. Among these obstacles are:

- planning and implementation being driven by litigation[1] and threats of litigation rather than optimal restoration of the natural system;
- Congressional authorization and appropriation processes with limited ability to cope with a project of the scope of Mod Waters[2];
- limited experience of agencies working together resulting in lack of coordination and weaknesses in strategic planning;
- deep differences in stakeholder goals resulting in protracted conflicts over project design (see Box 4-1);
- large, unanticipated scope changes; and
- dramatic cost escalations over time ($81 million in 1990 to $398 million estimated in 2006 and $523 million in 2008; Sheikh, 2005; USACE and DOI, 2008).

Working through these obstacles has resulted in significant delays in project design and implementation. The project was originally scheduled for completion in 1997, but the projected completion is now estimated for 2013. The cost increases and delays have threatened the viability of the project.

[1]Litigation has involved issues such as property rights (Garcia v. United States, No. 01-801-CIV-Moore, 2002; see later section "Flood Mitigation in the Eastern Everglades Developed Areas"), endangered species (Natural Resources Defense Council v. U.S. Army Corps of Engineers, No. 99-2899, 2001 U.S. District), and procedural requirements (e.g., Miccosukee Tribe of Indians v. U.S. Army Corps of Engineers, No. 08-21747-CIV-King, 2008), among others.

[2]Mod Waters was initially treated as a National Park Service (NPS) project. However, since the total NPS annual construction is approximately $200 million ($218 million in FY 2008), Mod Waters—even as initially approved in 1989—would use the entire construction budget or more. The U.S. Congress has been unwilling to commit this kind of large percentage of the NPS construction account to a single project.

BOX 4-1
Stakeholder Conflicts in the Combined Structural and Operational Plan

The Combined Structural and Operational Plan (CSOP) describes how water for the Mod Waters and C-111 projects will be managed to provide benefits to the natural system and for flood control, recreation, and water supply while meeting applicable water quality standards. It exemplifies the conflicts among stakeholders that have plagued the Mod Waters project and those that may face the CERP as it moves forward. Stakeholders with the potential to be impacted by the CSOP include Everglades National Park (ENP), environmental groups, urban, agricultural, and recreational interests, as well as tribal, state, regional, and local government resources.

Stakeholder conflicts were so significant in the development of the CSOP that the U.S. Institute for Environmental Conflict Resolution stepped in to assess opportunities to facilitate a multistakeholder Environmental Impact Statement process at the request of the U.S. Army Corps of Engineers (USACE), ENP, U.S. Fish and Wildlife Service), and the South Florida Water Management District (SFWMD). Specific substantive issues of greatest concern to stakeholders included concerns about flooding east of the L-31N and C-111 canals, water supply, and prestorm drawdown of canals by urban and agricultural interests, hydrologic conditions in natural areas by tribal and environmental interests, the focus on single species management by tribal interests, access for recreational purposes especially by sport fishing interests, and comprehensive restoration of the Everglades by nearly all parties (Alverez et al., 2002).

Four major process issues were identified as significant stakeholder concerns related to preparing the CSOP: lack of trust among the stakeholders, the perceived need to implement CSOP quickly, skepticism about agency commitment to a collaborative process, and process/meeting fatigue (Alverez et al., 2002). The most difficult of the issues to resolve was that of mistrust among stakeholders. Alverez et al. (2002) describe five scenarios through which mistrust developed among stakeholders:

- The perception that an agency has gone back on a commitment or a promise made in the past.
- The perception that persons or agencies were working behind the scenes contrary to public pronouncements.
- The perception that agency action was not driven by legal requirements and technical data but rather that persons within the agencies manipulated legal requirements and technical data to advance a preferred outcome.
- The perception that certain agencies or groups never considered their concerns or act against the interests of the stakeholder.
- Perception about delays, disputes among agencies, or errors that have occurred during the planning and implementation of the Mod Waters and C-111 (South Dade) projects. Consequently, there is an assumption that the responsible agencies—ENP, the SFWMD, USACE, and USFWS—cannot be trusted to competently complete these projects.

Alvarez et al. (2002) identified the following specific obstacles to collaboration among the stakeholders: "...the long history of polarized relationships, ongoing litigation involving CSOP-related issues, the highly technical nature of the CSOP, extreme process fatigue, Federal Advisory Committee Act (FACA) restrictions regarding nongovernmental participation on advisory bodies, the constraints posed by congressionally-authorized purposes for MWD and C-111 (South Dade) projects, severe time factors, and continuing uncertainty about the agencies' commitment to collaboration." Each of these issues, if not addressed through collaborative approaches, also has the potential to negatively impact the CERP.

OVERVIEW AND STATUS OF MOD WATERS PROJECT COMPONENTS

The Mod Waters project has four interrelated components (Figure 4-3).

1. Tamiami Trail modifications to a 10.7-mile section of U.S. Highway 41 from Structure 333 (S-333) on the west to Structure 334 (S-334) on the east near Krome Avenue;
2. Flood mitigation in the eastern Everglades residential areas (8.5-square-mile area);
3. Conveyance and seepage features in WCA-3B; and
4. Program implementation, including management, operations (e.g., the Combined Structural and Operational Plan [CSOP]), and monitoring.

Each of the components contributes to the overall project goals in different, yet critical ways. And each of the Mod Waters components has been controversial enough that each has encountered significant delays in planning and/or implementation. The progress and major issues associated with each component are discussed below.

Tamiami Trail Modifications

Tamiami Trail (U.S. Highway 41) cuts across the central portion of the Everglades ecosystem along the northern boundary of Everglades National Park (Figures 4-3 and 4-4). The highway links the eastern and western coasts of Florida, providing a critical transportation route for commerce and emergency evacuation. Tamiami Trail and the adjacent L-29 levee impede the slow flow of water southward in a region that was historically sawgrass prairies, marshes, and tree islands. Modifications to the eastern 10.7 miles of Tamiami Trail are necessary to increase sheet flow in Northeast Shark River Slough because higher water levels in the L-29 canal and/or larger openings in the trail (e.g., bridges rather than culverts) are required to allow greater water inputs to Northeast Shark River Slough, and higher water levels can adversely affect the stability of the road base given modern traffic loads and speeds.

Much of the controversy surrounding the Tamiami Trail component centers on the road design necessary to accomplish hydrologic restoration "to the extent practicable" while protecting the integrity of the roadbed and also minimizing the likelihood of flooding of tree islands within WCA-3B. The original 1992 design increased flows from WCA-3B into Northeast Shark River Slough by passing water through two weirs in the L-29 levee and then through existing culverts beneath the road, and only short sections of the roadway were targeted for eleva-

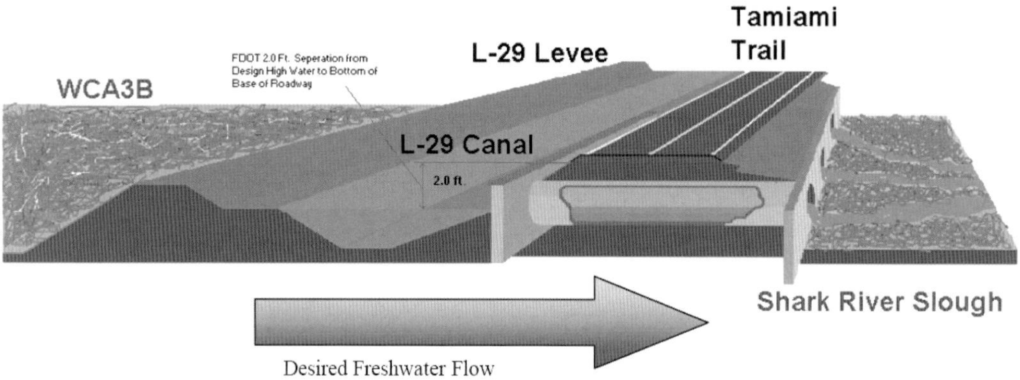

FIGURE 4-4 Built in the 1920s, the two-lane Tamiami Trail (and the adjacent levee L-29) interrupts the natural north-south flow of water through Big Cypress National Preserve and Everglades National Park. Desired freshwater flow under Tamiami Trail is currently limited by the elevation (stage) of the water in the L-29 canal, which is restricted to prevent flooding of the Tamiami Trail roadway and damage to the subbase under modern traffic loads and speeds.

SOURCE: USACE and DOI (2008).

tion. However, subsequent analyses revealed that water levels in the L-29 canal to accommodate target flows could adversely affect the road base, given current traffic loads, and could also possibly overtop the road. Furthermore, the Tamiami Trail is a designated hurricane evacuation route, which makes the consequences of potential damage more severe. A hydraulic rating curve—relating stage (i.e., water elevation) in the L-29 canal on the upstream (north) side of Tamiami Trail to flow through the culverts beneath the highway—is shown in Figure 4-5. Given sufficient head (or water elevation above the downstream elevation of Shark River Slough) in the L-29 canal, a variety of desired flows beneath the highway could be achieved. The 1992 General Design Memorandum (USACE, 1992) described maximum target flows of 4,000 cfs based on water flow that could be accommodated through four inflow structures (i.e., S-333, S-355A, S-355B, and S-356) based on their maximum capacities. The U.S. Army Corps of Engineers (USACE) and the Department of the Interior (DOI) (USACE and DOI, 2008) note that 4,000 cfs peak flow is important "because it allows for a discharge sufficient to create the physical changes to the landscape." However, a flow of 4,000 cfs would require a stage of over 9 feet according to Figure 4-5, and the Florida Department of Transportation (FDOT) has currently set a maximum

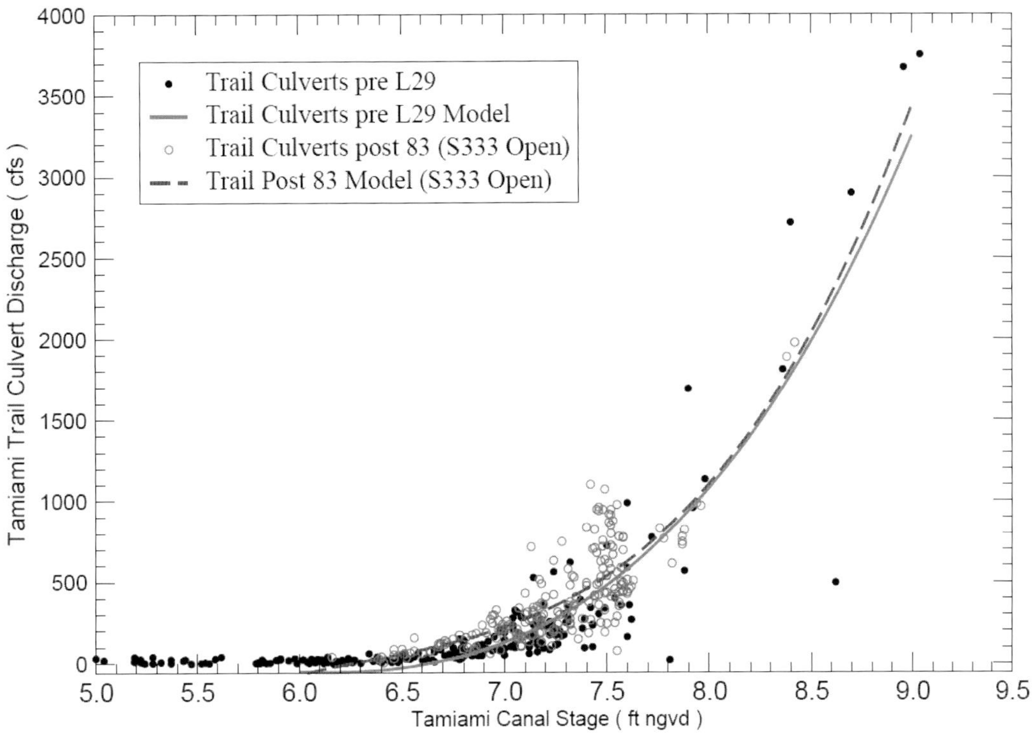

FIGURE 4-5 Hydraulic rating curve for flow through existing culverts beneath the Tamiami Trail, both before the construction of the L-29 (Tamiami) canal (circa 1963) and with current operational options (post-1983).

SOURCE: K. Kotun, ENP, personal communication (2008).

allowable water level of 7.5 feet in the L-29 canal to avoid impacts to the road. Thus, the main impediment to higher flows into Northeast Shark River Slough is the required L-29 canal stage to pass the desired flow through the existing culverts.

Wider openings under the road (e.g., new bridges) would allow the required Mod Waters flows to pass under the road at a lower stage in the L-29 canal than with the culverts alone, thereby reducing the required height of and improvements to the road and minimizing adverse high-water-level effects on tree islands once Northeast Shark River Slough is reconnected with WCA-3B. USACE and DOI (2008) demonstrated that there is a significant difference in the water stage (or head loss) across Tamiami Trail when the opening size is less than 5,000 feet (circa one mile). This differential represents the additional height of water necessary to move water from the L-29 canal into Everglades National Park, due to the interaction of the bridge opening size and the resistance of the downstream marsh to flow. With openings smaller than 1 mile, much of the increase in stage of the various alternatives would be consumed by the head loss and little would be left to increase flows. (The trade-off of canal stage versus width of opening is shown in Figure 4-7 of the USACE and DOI, 2008.) Bridges would also benefit wildlife by improving the ecological connectivity between the lands north and south of Tamiami Trail. However, adding bridges to the design plan raises the costs of the project significantly. Sheet flow can still be achieved downstream of an unbridged road, although there would be some as-yet-unspecified distance of unnatural flow downstream of the culverts that may alter native habitats.

As CERP projects, particularly Decomp, were being developed in the late 1990s, it became clear that reevaluation of Tamiami Trail modification plans was needed to ensure consistency between Mod Waters and CERP objectives (USACE, 2003). In 2003, a plan for the road was developed and reported in the General Reevaluation Report (GRR; USACE, 2003) (see Table 4-2) that recommended a single 3,000-foot bridge and allowed some potentially damaging high water levels along the rest of the road in return for funding placed in escrow that FDOT could use for any necessary road repairs. However, increased FDOT safety requirements resulted in reconsideration of the 2003 plan. A 2005 plan was developed that consisted of two bridges (a 2-mile western bridge and a 1-mile eastern bridge), with the remaining 7 miles of roadbed raised and widened to accommodate L-29 canal stages of 9.7 feet (USACE, 2005). These design changes and an escalation in construction costs led to dramatic increases in project costs. In 2007, the projected cost of the two-bridge 2005 Tamiami Trail modification plan was estimated to be approximately $430 million (USACE and DOI, 2008).

2008 Limited Reevaluation Report

In November 2007 the U.S. Congress rejected the 2005 preferred two-bridge plan as too expensive and directed the USACE to complete a Limited Reevaluation Report (LRR) of Tamiami Trail alternatives by July 2008. The USACE was tasked to reexamine prior reports and "to pursue immediate steps to increase flows to the Park of at least 1,400 cubic feet per second, without significantly increasing the risk of roadbed failure." The WRDA conference report also stated that flows to the park should have "a minimum target of 4000 cubic feet per second so as to address the restoration envisioned in the 1989 Act," while

TABLE 4-2 Post Authorization Progress of the Mod Waters Project

	Tamiami Trail Modifications		8.5-Square-Mile Area (SMA)		Conveyance, Seepage, and Operating Plan	
Date	Event	Purpose / Outcome	Event	Purpose / Outcome	Event	Purpose / Outcome
1992	Design proposed in General Design Memorandum (GDM)	Includes raising the road in the vicinity of S-334, but no additional culverts or bridges	Design proposed in GDM	Includes a pump station, canal, and a levee around entire residential area	Design proposed in GDM	Includes gated culverts in the L-67A, spillways in L-29, raising of 2 Miccosukee villages, and seepage control. Operations plan to be developed by the USACE in consultation with DOI and the state. If no water control plan is developed, the Modified Rain Driven Operational strategy will be implemented once construction is complete.
1994			Congress amended ENP Expansion Act (1989)	Authorized federal assistance for land acquisition in 8.5 SMA		
1995			Governor's East Everglades 8.5 SMA Study, because of technical concerns about 1992 plan	Detailed evaluation of 8.5 SMA alternatives		

continued

TABLE 4-2 Continued

	Tamiami Trail Modifications		8.5-Square-Mile Area (SMA)		Conveyance, Seepage, and Operating Plan	
Date	**Event**	**Purpose / Outcome**	**Event**	**Purpose / Outcome**	**Event**	**Purpose / Outcome**
1998			SFWMD Review Team Study	Recommends total buyout of 8.5 SMA		
			Legal challenge to SFWMD review team study	Suspends further action		
1999			FL Governor replaced SFWMD governing board		FWS jeopardy opinion on Cape Sable seaside sparrow	Ends Experimental Water Deliveries program at iteration # 7
			SFWMD and DOI requested USACE to examine alternatives for 8.5 SMA			
2000			USACE finalizes and approves revised project design, Alt. 6D	Plan includes protective levee with acquisition of the western third of the 8.5 SMA	Interim Structural and Operational Plan (ISOP 2000)	Emergency Deviation in water deliveries to ENP to protect CSSS

2001		Legal challenge to Alt. 6D filed by residents of 8.5 SMA over USACE use of eminent domain authority to acquire the property interests		USACE begins consultation with U.S. Institute for Environmental Conflict Resolution	Suggested by Council on Environmental Quality to resolve long-standing interagency conflicts related to the Interim Operational Plan (IOP)
2002		Federal Judge rules USACE lacks authority to implement Alt 6D	Halts progress on 8.5 SMA component	Interim Operational Plan EIS Record of Decision signed	IOP implemented
				Injunction on IOP filed by Miccosukee Tribe	Litigation continues through 2007
				Combined Structural and Operational Plan (CSOP) EIS initiated	Examines environmental consequences of integrating operation of Mod Waters and C-111 (South Dade)

continued

TABLE 4-2 Continued

Date	Tamiami Trail Modifications		8.5-Square-Mile Area (SMA)		Conveyance, Seepage, and Operating Plan	
	Event	Purpose / Outcome	Event	Purpose / Outcome	Event	Purpose / Outcome
					U.S. Inst. For Env. Conflict Resolution explores opportunities for multi-stakeholder collaboration	Assessment published in Alverez et al. (2002)
2003	General Revaluation Report (GRR) for Tamiami Trail completed	Recommended 3,000-foot bridge with flowage easement on Tamiami Trail and compensation to FDOT for damages	Consolidated Appropria-tions Resolution of 2003 (PL 108-7)	Congress directs USACE to construct Alt 6D; resolves legal issues associated with litigation.	South Florida Ecosystem Task Force establishes the CSOP Advisory Team	To develop consensus multi-stakeholder recommenda-tions on CSOP
	Tamiami Trail GRR withdrawn because no agreement could be reached with FDOT	Halts Mod Waters progress				
2005	Tamiami Trail Revised GRR (RGRR) completed	Recommended modifications to roadway allowing 9.7-ft water levels; bridge extent expanded to 2-mile and 1-mile sections	8.5 SMA Record of Decision approved	USACE begins construction of protective levee in 8.5 SMA	CSOP Advisory Team submits consensus recommenda-tions on Alternative 5	Leads to some modifications

2006	Record of Decision signed for RGRR				Supplemental Environmental Impact Statement for IOP completed	Alternative 5R proposed by project delivery team. Record of decision never signed
	Congress finds costs of RGRR unacceptable					
2007	Congress directs USACE to prepare a Limited Revaluation Report (LRR) for Tamiami Trail	Addresses cost increases of Tamiami Trail construction				
2008	USACE recommended 1-mi eastern bridge with road modifications to allow water levels of 8.5 ft		Expected construction completion		USACE recommends 1-mi eastern bridge in Tamiami Trail modifications	Necessitates that the CSOP be revisited

avoiding modifications that are duplicative or incompatible with the CERP (U.S. Congress, 2007).

The limited reevaluation assessed 27 alternatives, generally grouped into five categories based on the maximum height (or stage) of the L-29 conveyance canal north of and parallel to the Tamiami Trail (Figure 4-4). The five categories are listed below. (Note that the category number also serves as a prefix to the identification numbers for the 27 alternatives as listed, for example, in Table 4-3.)

1. Stage 7.5 feet National Geodetic Vertical Datum: No roadway elevation or stage increase.
2. Stage 8.0 feet: Minimum roadway improvement. Minimum roadway crown 11.05 feet.
3. Stage 8.5 feet: Moderate roadway improvement. Minimum roadway crown 11.55 feet.
4. Stage 9.7 feet: Major roadway improvement. Minimum roadway crown 12.75 feet.
5. Alternative roadway realignments.

With higher stages in the L-29 conveyance canal, there would be greater water flow into Everglades National Park. The USACE conducted a simple spreadsheet analysis using 24 years (1983–2006) of water flow, dividing the flow from WCA-3A, assuming that 55 percent of this flow would be moved into Northeast Shark River Slough. The specific mechanism of conveyance was not considered; thus, the results were not limited by the capacity of current conveyance structures (T. Ferguson, USACE, personal communication, 2008).

Some benefit measures and preliminary cost estimates for 5 of the 27 alternatives considered in the reevaluation are summarized in Table 4-3. A 10.7-mile bridge (or skyway) would allow the most flow, but the estimated cost by the USACE would be $1.7 billion (USACE and DOI, 2008). A 10.7-mile bridge was considered the "environmentally preferred alternative" by the USACE (2005) without consideration of cost constraints.

The USACE gave the study team explicit guidance to identify an alternative with a lower cost than the 2005 recommended plan (USACE and DOI, 2008), even though the WRDA conference report did not include such explicit restrictions. The reevaluation study team then rejected any alternative with a cost in excess of $400 million, although such alternatives had significant increases in water flow and habitat unit increases. Examining incremental differences between two recommended alternatives, the 2005 recommended alternative (alternative 4.2.3 in Table 4-3) had more than double the restoration benefits (calculated in habitat units) for a 75 percent increase in cost relative to the 2008 recommended alternative. This

TABLE 4-3 Characteristics of Five Alternatives Considered in the Limited Reevaluation of Tamiami Trail Modifications

Alternative No.	Action	Design Stage (ft)	Peak Flow (cfs)	Average Annual Volume (acre-ft/yr)	Habitat Unit Increase (HU)	Preliminary Total Cost Estimate ($M)	Total Cost per Habitat Unit ($/HU)	Expected Completion Date (Year)
1.1	Do Nothing	7.5	1250	177	0	0	n/a	—
1.3	Add Swales and Culverts	7.5	1371	188	238	73	306,723	2011
3.2.2.a[a]	Raise road, add 1-mile E Bridge	8.5	1848	340	13109	319*	24,334	2012
4.2.3[b]	Raise road, add 1-mile E Bridge and 2-mile W Bridge	9.7	2331	436	28361	557	19,640	2014
4.2.4[c]	10.7-mile Bridge	9.7	4036	472	53010	1,648	31,088	2020

[a]USACE (2008) recommended alternative. After value engineering reported later in the report, final estimated cost is $227M. However, the value engineering process was not applied to all scenarios, so the updated cost value is not useful in comparing alternatives.
[b]USACE (2005) recommended alternative.
[c]Alternative with largest environmental benefit.

SOURCE: USACE and DOI (2008) Table 4-3.

incremental analysis suggests that the 2005 recommended alternative would be a more cost-effective alternative without the imposed budget constraint.

The USACE and DOI (2008) recommended plan (alternative 3.2.2a) has a single 1-mile eastern bridge (Figure 4-6) that would achieve the immediate goal of permitting peak flows in excess of 1,400 cfs at approximately half the cost of the 2005 two-bridge plan (alternative 4.2.3). However, the eventual goal of 4,000 cfs required to achieve the desired ecological effects would not be achieved. Maximum flow under the recommended plan is 1,848 cfs (Table 4-3).

This committee recognizes the importance of completing this initial step of increased flows in Mod Waters. If completed, this plan will provide important steps toward restoration in Everglades National Park by increasing the volume of water entering Northeast Shark River Slough by at least 163,000 acre-feet per year over the current level (or an increase of at least 92 percent over cur-

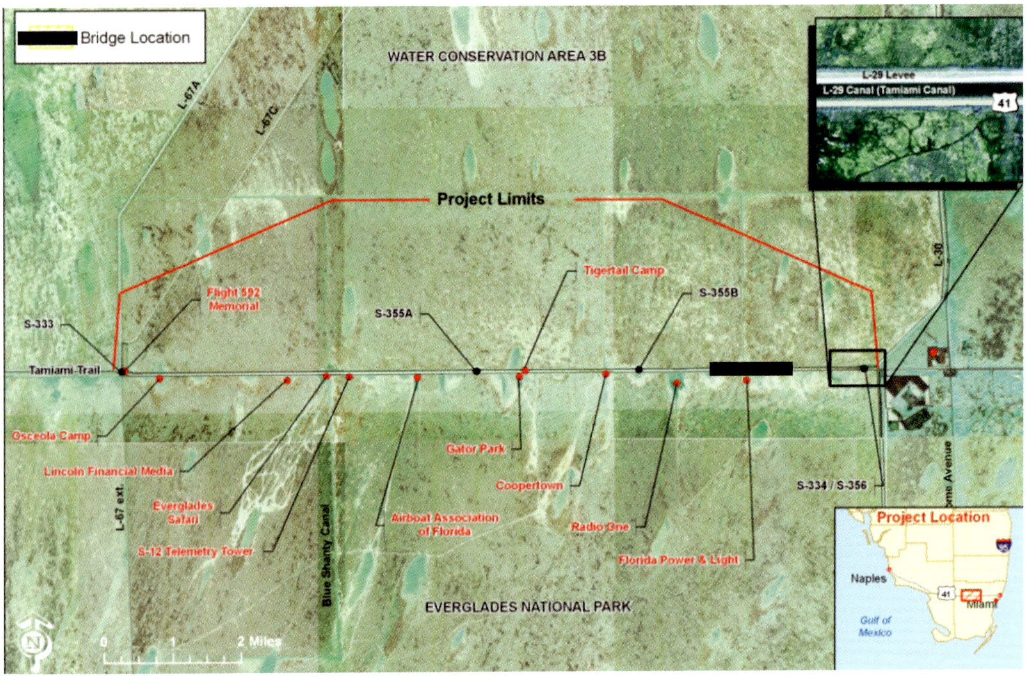

FIGURE 4-6 Eastern 1-mile bridge option 3.2.2a proposed in the 2008 LRR.

SOURCE: USACE and DOI (2008).

rent mean flow volumes; USACE and DOI, 2008). Achieving this increment of flow would provide more ecological restoration benefits than the alternatives of do-nothing or culvert and swale additions. Other Mod Waters alternatives would have even greater restoration benefits, but they are not consistent with the cost constraints imposed by Congress and the USACE. Imposing these cost constraints may well increase the eventual costs of the entire CERP effort and will delay achieving restoration benefits. However, after two decades of effort, achieving some benefits from Mod Waters is critical for the overall CERP program, especially in light of continuing deterioration of the Northeast Shark River Slough ecosystems.

The recommended plan does not provide *more* water into Everglades National Park than before; it only redistributes the 1983–2006 water coming out of WCA-3A according to a 45–55 percent distribution in West Shark River

Slough and Northeast Shark River Slough. Analysis of historic Shark River Slough flows makes it clear that restoration of significantly higher annual flows cannot occur until southerly flows from Lake Okeechobee are restored (ENP, 2008). Hydrologic modeling is needed to determine whether or not the 2008 recommended plan (alternative 3.2.2.a) will provide adequate hydraulic capacity to pass higher overall flows into Everglades National Park while maintaining the 45–55 percent distribution between West Shark River Slough and Northeast Shark River Slough.

Real estate issues and higher construction costs at the western location were two main factors that influenced selection of the eastern bridge location (alternative 3.2.2.a) instead of the western location (alternative 3.2.2.b). Detailed planning has already been performed for the eastern bridge; therefore, construction can begin sooner, with attendant cost savings. Some concerns have been raised, however, that an eastern bridge may result in much greater loss of flow via seepage to the east when compared to a western bridge. Inflows to Northeast Shark River Slough through either a western or an eastern bridge are driven by the modestly increased head in the L-29 canal. A western bridge would likely have a lesser effect on seepage losses, because water introduced at the western end of the 10.7-mile highway segment would have more of an opportunity to flow due south into Shark River Slough via south-flowing sloughs near the L-67 extension. But some of this water, too, would eventually be subject to seepage losses. The implication is that an additional western bridge will enhance water deliveries to Northeast Shark River Slough if one is constructed as a future restoration increment beyond the currently proposed Mod Waters alterations.

Other issues also point toward the importance of additional Mod Waters and CERP construction. While delivering a greater annual volume to Northeast Shark River Slough than occurs now, the LRR-recommended one-bridge plan (alternative 3.2.2.a) does this mainly by increasing dry-period flows rather than enhancing the passage of higher, wet-period flows (USACE and DOI, 2008, Appendix D, Figure 12). Additional wet-period flows require higher heads in L-29 than the 8.5-foot maximum of alternative 3.2.2a. High wet-period flows—for instance, up to an approximate 10-year return period flow of about 4,000 cfs—better mimic historic maximum flows according to the Natural Systems Model. The historic range of flows over seasons and years drives the creation and restoration of landscape and is thus important. But high flows cannot pass under Tamiami Trail until restrictions on the head in the L-29 canal are raised to at least 9.5 feet. Furthermore, under alternative 3.2.2a, the *average* wet-season (October) head in L-29 is planned to be 7.89 feet to protect the roadway integrity (USACE and DOI, 2008, Section 6.1.3). Operational procedures will be instituted to cease inflows to L-29 if storms are forecast that might drive the head above the 8.5-foot maximum.

Ultimately, to achieve CERP goals for the restoration of Everglades National Park, stages in the L-29 canal will need to be higher than current constraints. Thus, it should be recognized that the Tamiami Trail modifications proposed by the USACE in the current plan (alternative 3.2.2a) should be viewed as a necessary but partial first step toward restoration. Additional Tamiami Trail modifications will be needed in the future to move a sufficient quantity of water south. The recommended Tamiami Trail design alternative in the LRR would therefore leave some benefits to be completed in the CERP, such as the ability to pass 4,000 cfs. Unless additional restoration steps are taken, this design will further delay the delivery of crucial ecological benefits from the Mod Waters project until the completion of Decomp, a project already plagued by delays and with an uncertain target completion date.

Flood Mitigation in the Eastern Everglades Developed Areas

Control of seepage and flooding in the eastern Everglades developed area, especially the 8.5-square-mile area, has been a highly controversial component of the Mod Waters project. This component provides flood mitigation to the urban and agricultural areas within the 8.5-square-mile area associated with the higher water levels created by Mod Waters in the restored Northeast Shark River Slough of Everglades National Park (Figures 4-3 and 4-7). A series of legal challenges and court decisions associated with this project component (e.g., challenging the USACE's use of eminent domain authority to acquire the property interests) slowed progress toward completion and escalated the cost of the Mod Waters project (Table 4-2).

The issues centered on land acquisition and flood protection of the land in this relatively small tract with 1,500 residents. Modeling results suggested that implementation of Mod Waters would flood most of the 8.5-square-mile area because the area was once part of the original Everglades with low elevation and lack of drainage. The 1992 General Design Memorandum contained plans for mitigation of flooding, but these plans were deemed inadequate by a South Florida Water Management District (SFWMD) review team established by Florida's governor in response to stakeholder concerns about the technical feasibility of the plan. In 1998, the SFWMD recommended a complete buyout of the area and establishment of a flow way. This alternative, and others, were rejected in legal challenges brought by stakeholders in the 8.5-square-mile area. In 2000, USACE completed a general revaluation report and supplemental environmental impact statement recommending a plan referred to as Alternative 6D, which proposed: (i) acquisition of 2,100 acres of the 5,600 acres in the area and purchase of 77 residential tracts (Sheikh, 2005); (ii) construction of an interior

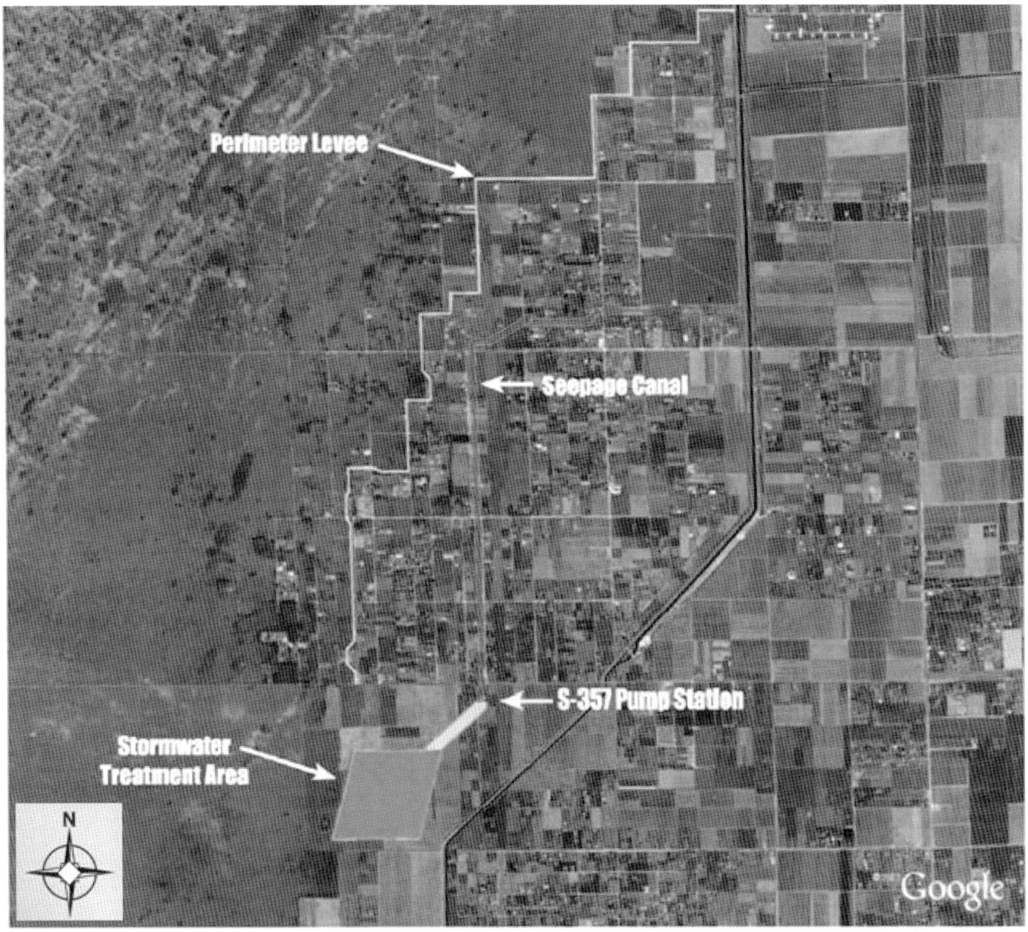

FIGURE 4-7 Flood mitigation efforts in the 8.5-square-mile area in the eastern Everglades residential area.

SOURCE: Adapted from *http://www.saj.usace.army.mil/dp/mwdenp-c111/docs/8-5SQMILE_2008Jan.pdf.*

canal and three internal levees to provide mitigation for the remaining area; and (iii) construction of a pump station and an associated stormwater treatment area (STA). After additional court challenges from 8.5-square-mile area residents, the USACE was directed by Congress to acquire the property necessary to carry out Alternative 6D and to provide flood protection in the remaining residential area with a protective levee. Construction of the protective levee, the seepage canal, and the STA has been completed. Land acquisitions are nearly complete with the exception of two parcels.

Another issue that arose in the 8.5-square-mile area involved the discovery of hazardous wastes in the form of buried lead-acid motor vehicle batteries. The resulting hazardous waste removal and disposal further increased costs for the project.

Conveyance and Seepage Control

The conveyance and seepage control features of Mod Waters were planned to move water from WCA-3A to WCA-3B and, subsequently, into Northeast Shark River Slough, to control seepage, and to eliminate the barriers to natural flow patterns between Northeast Shark River Slough and West Shark River Slough (see Figure 4-1). As of December 2007, the following project components have been completed (see Figure 4-3):

- Gated spillway structures 355-A and 355-B in the L-29 levee to connect the L-29 canal to Northeast Shark River Slough have been constructed;
- A portion of the L-67 extension (4 of 9 miles) has been backfilled;
- The S-356 pump station, which returns seepage from Northeast Shark River Slough to the L-29 canal has been constructed;
- S-333, a gravity-driven spillway, has been modified to allow greater flow from WCA-3A into the L-29 canal; and
- Tigertail Camp has been raised to prevent flooding (see Figure 4-3; USACE and DOI, 2008).

Negotiations with the Miccosukee Tribe are ongoing concerning details of raising the Osceola Camp to prevent flooding. Contracts have been signed for construction of S-331 (Command and Control Building). Engineering Documentation Reports will be completed in 2008 for the S-345 flow structures and S-349 spillways that would allow flow through the L-67 levees between WCA-3A and WCA-3B, and weirs in L-29 (Appelbaum, 2008b).

The general specifications for the conveyance features in the L-67 levees (i.e., WCA-3B inlet structures [S-345s] and the canal plugs with boat channels [S-349s]; see Figure 4-3) were determined as part of the CSOP negotiations (see next section). However, the maximum inflow for the recommended LRR alternative (approximately 1,850 cfs into the L-29 canal) can be achieved through existing inflow structures into the L-29 canal (S-333, S-355); therefore, it is unclear whether additional conveyance features will be added (e.g., S-345s, weirs in the L-29 levee) because these features may be difficult to justify as part of Mod Waters.

Without the L-67 conveyance features, the Mod Waters project will not serve to reconnect WCA-3A to WCA-3B and Everglades National Park, thereby eliminating important sheet-flow restoration benefits of the project for the WCAs. Failure to reestablish sheet flow perpetuates the unnatural ponding of water in WCA-3A and WCA-3B, while other areas of the system are overdrained. These conditions are in stark contrast to those which formed and maintained the ridge and slough pattern for over millennia (Ross et al., 2006) in what is now WCA-3A, WCA-3B, and Everglades National Park.

Program Implementation

The final piece in implementation of the Mod Waters project is to develop an operating plan for the pumps and gated structures to move water from WCA-3A to Northeast Shark River Slough while ensuring the safety of the Tamiami Trail roadbed. The CSOP is an integrated structural and operational plan for two modifications to the C&SF Project: the Mod Waters project and the C-111 (South Dade) project (see Figure 2-3). The purpose of CSOP was to develop an operations plan for the Mod Waters and C-111 (South Dade) projects that would be consistent with their project purposes and provide the most environmental benefits. There is a need for a combined operational plan because the C-111 (South Dade) and Mod Waters projects are hydrologically linked to each other and the larger regional water management system, even though the two projects were authorized by separate congressional legislation.

An interagency and multistakeholder CSOP advisory team spent more than 2.5 years in facilitated negotiations to develop a consensus operational plan, based on the two-bridge alternative proposed in the GRR (USACE, 2005). The CSOP advisory team examined a wide range of alternative operational plans, and while none of the alternatives was acceptable to all parties, all agreed that the CSOP should move forward with a modified rain-driven option (Alt 5R), to be implemented as a test unless a preferred operating strategy could be developed before completion of Mod Waters.

The proposed Tamiami Trail LRR plan (USACE and DOI, 2008), however, represents a significant change in the operational assumptions underpinning the CSOP, particularly because stage constraints in the L-29 canal will affect operations.

Thus, the CSOP is no longer valid and cannot go forward without major changes. The operational strategy to achieve a water distribution of 55 percent east of S-333 and 45 percent to the west, as simulated in the spreadsheet model of the LRR, is only now being reconsidered. Until a plan for Tamiami Trail modifications receives congressional approval, it is premature to establish consensus operations for unknown project conditions and structures.

Nevertheless, the negotiated CSOP operating criteria could potentially be applied to existing and recently completed structures for operation of the redesigned Mod Waters project. Additional delays in implementing Mod Waters may occur if consensus on the details of the new operational plan is difficult to achieve.

Cost Escalation for the Mod Waters Project

The estimated costs of the Mod Waters project have increased dramatically with the project delays, difficulties in land acquisition, and changes in project scope. By 2008, the estimated cost for the total project is nearly six times the original estimate made in 1990 (Table 4-4). In 2008, estimated project costs totaled $523 million, including $227 million in Tamiami Trail modifications and additional costs for protection of the 8.5-square-mile area, conveyance and seepage controls, design costs, and project implementation support (Table 4-5). In contrast, the general increase in the USACE civil works cost index was only 50 percent during this same 18-year period (USACE, 2007a). Florida has seen greater cost increases over time, particularly for land acquisition (FDOT, 2007a). Since 2004, the rate of construction cost inflation has increased, in Florida and elsewhere.

For the Tamiami Trail portion of the Modified Waters project, the major cost escalation occurred when the necessity for increasing the elevation of the roadway to protect its integrity was recognized. The early plans for the project simply included additional culvert volume under the existing roadway, at a cost originally estimated to be $2.9 million (USACE, 1992). The USACE recommended an option with two bridges and other roadway improvements (USACE, 2005) that was originally estimated at $144 million in 2005. By 2007, the same plan was estimated to cost $430 million (USACE and DOI, 2008). After the limited

TABLE 4-4 Escalation in Estimated Costs for the Entire Mod Waters Project, with Changing Scope

Estimated Total Cost ($ M)	Year	Tamiami Trail Modification	Source
81	1990	Flow through culverts	Sheikh, 2005
398	2006	2-bridge option	Sheikh, 2005
523	2008	1-mile eastern bridge	USACE and DOI, 2008, Table 6-3

TABLE 4-5 Estimated Costs and Funding Shares in 2007 Dollars for the 2008 Recommended Mod Waters Project, with the 1-Mile Eastern Bridge

Item	Expenditure Through FY07 ($ M)	FY08 Enacted ($ M)	FY09 and After ($ M)	Total Project Costs ($ M)
8.5-Square-Mile Area	170.4			170.4
Conveyance & Seepage	30.0		21.2	51.2
Tamiami Trail Modifications	45.5	18.4	161.6	225.5
Tamiami Trail Design	11.0	5.7		16.7
Project Implementation Support	41.5	0.0	17.8	59.4
Mod Waters Total	298.4	24.1	200.6	523.1
Funding Shares				
DOI	230.7	14.3		245.0
USACE	67.7	9.8		77.5
FDOT			4.5	4.5
Funding to be Determined			196.1	196.1

SOURCE: Developed from USACE and DOI (2008), Table 6-3.

reevaluation of Tamiami Trail alternatives, the Tamiami Trail modifications cost estimate for the recommended plan (alternative 3.2.2a) for a 1-mile bridge and related improvements is $227 million (USACE and DOI, 2008, p. C-10). This is 29 percent lower than the figure cited in Table 4-3 due to incorporation of potential value-engineering cost-saving options that could result from changes in asphalt placement, construction right-of-way needs, bridge inspection methods, and fill material sources, as well as from the elimination of spreader swales as part of the design.

The 2008 USACE estimate for the recommended Tamiami Trail modifications after cost savings from value engineering is shown in Table 4-6. This estimate includes a 28 percent escalation allowance for price increases in the 3-year construction period, suggesting a historically high expectation of continuing cost increases. In 2007, FDOT recommended a present-day cost inflation factor through 2011 of 14 percent, anticipating a reduction in recent high cost escalations (FDOT, 2007b). The USACE estimate is also intended to reflect a 90 percent confidence that costs would be at or below the $227 million figure. However, the 90 percent confidence level is only 3 percent above the 50 percent confidence level (USACE and DOI, 2008), reflecting a remarkable degree of certainty about the future increase in construction costs. Even with this cost escalation and 90 percent confidence level budgeting, the cost savings identified through

TABLE 4-6 Estimated Costs of the Recommended Tamiami Trail Modification (1-Mile Eastern Bridge)

Item	Cost Estimate After Value Engineering
Construction	$ 154.8 M
PED (planning, engineering, and design)	$ 1.5 M
EDC (2%) (engineering during construction)	$ 3.1 M
S/A (8.5%) (supervision and administration)	$ 13.2 M
Real Estate	$ 5.9 M
Escalation (28.1% based on Oct. 2008 Award)	$ 48.1 M
Total	$ 226.6 M

SOURCE: USACE and DOI (2008), Table 4, Appendix C, 90 percent confidence estimates.

value engineering reduced the estimated Tamiami Trail modifications cost from the preliminary $319 million estimate in Table 4-3 to the final estimate of $227 million.

The Mod Waters project is not yet complete. Through fiscal year (FY) 2007, $298 million has been spent on all aspects of the project, and $225 million is estimated for future costs, primarily to complete the proposed Tamiami Trail modifications (Table 4-5). The funding totals include an expected $4.5 million contribution from the state of Florida reflecting pavement maintenance savings on the Tamiami Trail. Compared to the entire CERP project, the expenditure is modest, but the project takes important first steps toward restoration within Everglades National Park and lays the groundwork for future modifications in the CERP. The project delivers an important increment of restoration benefits, albeit substantially less than the 2005 plan (Table 4-3).

Additional Tamiami Trail modifications will be needed as the CERP progresses and the Decomp project moves forward to allow greater water flow volumes into Everglades National Park. Therefore, it is essential to understand that the Modified Waters project proposed in the 2008 LRR represents only a first step in achieving the restoration goals of CERP. Nevertheless, progress in increasing the water flows into Everglades National Park is a critical first step to make. To maximize the usefulness of the Mod Waters project, critical uncertainties should be identified and a monitoring plan developed so that the Mod Waters project can inform the planning of future sheet-flow restoration projects.

LESSONS LEARNED FROM MOD WATERS

The challenges to completing Mod Waters have proven formidable, and it is likely that the same challenges will affect other projects connected to the CERP.

One of the tenets of the adaptive management framework of the CERP is learning while doing, and the struggles to overcome obstacles faced during planning and implementation of the Mod Waters project offer some important lessons for the CERP. In this section, the following lessons will be discussed, although there are undoubtedly many more:

- agreeing on the scope and operating targets early in the decision-making process;
- providing strong, unifying leadership;
- recognizing the need to adapt plans over time; and
- establishing clear project management structure.

Benefits from Early Agreement on Project Scope and Operating Targets

Agreement on project scope and operating targets is a widely recognized characteristic of successful projects (Diekmann and Thrush, 1986). With consensus agreements, design and construction can proceed in a straightforward, timely, and cost-effective fashion. The Mod Waters project illustrates the difficulties arising from an ill-defined scope and operating targets. After authorization as part of the 1989 Everglades National Part Expansion and Restoration Act, planning activities resulted in the 1992 General Design Memorandum that assumed existing culverts were sufficient to pass required water flows without damage to the road, and the original authorization limited land acquisition to only the original perimeter levee in the 8.5-square-mile area. The general management philosophy of the 1992 plan was to allow more unconstrained water flows into Everglades National Park. As time went on, the project scope and operating targets changed, with adoption of more active control of water flows, larger-scale land purchases, and significant change in the Tamiami Trail configuration. With scope changing so drastically, project costs increased dramatically. Indeed, issues such as the appropriate opening sizes for culverts and underpasses along the Tamiami Trail (US-41) are still controversial, and stakeholders and agencies continue to debate what amount of hydrologic restoration is, in fact, "practicable."

The alternative to consensus building is prolonged agency squabbling and litigation from stakeholders, a process that is slow and divisive. With litigation there are always winners and losers, and litigation promulgates an increasingly antagonistic environment through which the parties involved in restoration must negotiate. Conflict resolution efforts have offered the greatest hope for finding solutions that are at least partly amenable to all parties (NRC, 2003c). In the development of the CSOP, agencies and stakeholders used a facilitated conflict resolution process to reach agreement on the objectives of the operational plan and, subsequently, the details of the operations. Restoration planners hoped this approach would help to build trust among stakeholders and allow them to avoid litigation. Although it was a lengthy and challenging process, the facilitated conflict resolution process enabled the agencies and stakeholders to work collaboratively toward a common goal and to find a way to move beyond the planning stalemate. However, this team-based process will succeed only if all participants (and their agency leaders) agree upon the project goals and are committed to working together to accomplish these goals with a spirit of compromise, not confrontation. The entire process can be undermined when parochial interests prevail and without support from the policy level.

One of the major strengths of the CERP in the past has been the ability of state agencies, federal agencies, and stakeholders to combine their interests and present a united position in seeking authorization and funding for projects. Overall, the strategy was to ensure that everyone received some benefits from the large-scale project. The reality, however, is that every interest group may not benefit equally, and trade-offs are likely to be necessary. In many cases, the real consensus building occurs after project authorization, when specifics of design, construction, operation, and management sharpen differences among stakeholders. CERP leaders will need to find ways to move forward with CERP components even when there is a lack of consensus.

Leadership

Strong leadership remains essential to achieving Everglades restoration goals by overcoming the many potential conflicting interests involved. Elected state and federal officials, together with agency directors, can build and maintain support among stakeholders and maintain the system-wide vision for the restoration. Continuing efforts are needed to assure that interests coalesce around high-priority project components with the greatest restoration benefits for the South Florida ecosystem, to offset the seemingly natural inclination of many stakeholders or communities to take opposing positions.

This leadership is also important in holding coalitions together over the long time periods required for completion of projects even after they are authorized and funded. General agreement among stakeholders on priorities and trade-offs will make completion of CERP components possible; if such agreement is absent, it is unlikely that restoration will succeed. Strong political leadership resulted in the initial authorization of the CERP and remains critical to keep the CERP on track toward its restoration goals.

Recognize Need to Adapt Plans over Time

One of the most challenging aspects of large ecosystem restoration is that decisions need to be made despite some uncertainty in the supporting science and engineering knowledge base, and as a result, plans may need to be adapted over time as more knowledge is gained. For those well versed in adaptive management (see Chapter 6), this idea may seem obvious. Nonetheless, it carries with it the difficult challenge of how to maintain consensus and stakeholder support when the basis of that agreement and support changes. Planners recognize that they cannot anticipate all possible outcomes, but it remains to be seen how willing restoration decision makers will be to significantly alter project designs and/or operations once the projects come on line. Sometimes project design changes come at a significant cost, as with the Tamiami Trail, but the changes are essential to attaining the restoration goals. Once again, strong restoration leadership with clear communication with CERP scientists is needed to adapt to changing knowledge and make difficult but well-informed decisions.

Establish Clear Project Management Structure

The Mod Waters project has been managed by the DOI and the USACE with participation by the state and local agencies. A report by the Office of the Inspector General (DOI, 2006) raises concerns about the consultation and management role played by the DOI. The report recognized that many factors contributed to project delays and cost escalation beyond the participating agencies' control, but the report concluded that project management was ineffective due to a lack of communication among participating agencies. This stems, in part, from DOI's inexperience with implementing large water projects and lack of an institutional history of working with other federal, state, and local agencies to accomplish a major restoration project.

Undoubtedly federal policy makers also played a role in project delays. For example, the Mod Waters project was originally authorized as part of the construction budget of the NPS. The project represented a very large element of this budget, and it was a radical departure from business as normal. As a result, congressional appropriators exhibited great concern, and Congress was slow to provide funding for Mod Waters. In 2006, to ease concern about NPS oversight of Mod Waters and to facilitate completion of the project, the federal government shifted funding of some of the Mod Waters project to the USACE budget (Sheikh and Carter, 2006), although this shift has created some resistance to new funding for the project in Congress. However, as long as a significant fraction of the Mod Waters budget resides in the NPS's construction budget, there could be difficulties in implementing the project due to historic limitations on the magnitude of this budget.

With multiple managers and many affected parties, decision making for the Mod Waters project has been slow, in turn creating concern in Congress about continuing appropriations for a project already delayed 8 years and more than five times over the original budget. Similar delays are apparent for many CERP projects (see Chapter 3). Clear project management structure can facilitate decision making, thereby reducing delays and cost escalation. The state of Florida's experience with its Acceler8 program shows that clear lines of authority are also helpful in making timely progress.

INCREMENTAL ADAPTIVE RESTORATION AND MOD WATERS

The LRR supports a plan that would bridge a small (1-mile) portion of Tamiami Trail and elevate the road to 8.5 feet to accommodate additional flow into Northeast Shark River Slough, although this is less flow than was originally envisioned. These alterations should also be compatible with further modifications needed for Decomp. Some might describe this approach as incremental adaptive restoration (IAR) (NRC, 2007; see also Chapter 3), but the committee does not, primarily because there is no commitment to take the new knowledge and apply it to future project increments to move closer to the original project goals. Once the Tamiami Trail component of Mod Waters is completed, all future improvements to the project would need to be made through a separate authorization (e.g., the CERP), which would be funded and managed by entirely different mechanisms.

Instead, the primary motivation for moving forward with a small increment is financial. The political constraints that bounded the decision for the Tamiami Trail essentially eliminated the option for IAR, as envisioned by the committee. After 18 years with only moderate progress and massive increases in the scope and cost of restoration, legislators lost patience in the Mod Waters project and confidence that public funds were being well spent. The goal no longer was hydrologic restoration but to finish the project that was started, whether or not the full degree of hydrologic restoration envisioned in the 1989 authorization was achieved.

The current plans will result in some restoration benefits and may bring the Mod Waters project to a close so that CERP projects can move forward. However, the unfulfilled goals of the Mod Waters project and the unresolved challenges involving Tamiami Trail modifications appear likely to be shifted into the CERP and Decomp plans, a project already plagued by stakeholder conflicts and planning delays.

Nevertheless, there are important opportunities for learning in the proposed Tamiami Trail modifications. The committee encourages the USACE and DOI to develop appropriate hypotheses and supporting monitoring programs so that critical uncertainties affecting future restoration planning can be addressed and future sheet-flow enhancement projects improved. The LRR states that a monitoring program will support learning from this project, but the report does not identify any critical uncertainties that the project could address or metrics by which those uncertainties could be resolved. With an appropriately designed monitoring program, the committee hopes CERP planners can take what is learned from the Mod Waters project and use that information to help build consensus for Decomp.

CONCLUSIONS AND RECOMMENDATIONS

The history of the Mod Waters project is one of the most discouraging stories in Everglades restoration. The project, which would provide crucial first steps toward ecological restoration within Everglades National Park, has been plagued by changes in direction and scope, parochial interests, debilitating litigation, enormous cost escalation due both to inflation and plan modifications, unanticipated engineering constraints (e.g., Tamiami Trail integrity), and lack of coordinated leadership from the responsible agencies. How the project will be funded (i.e., involving the NPS, USACE, FDOT) is a further complicating factor. While some events may have been unavoidable, the outcome has been loss of support from Congress, which must fund the project, and loss of enthusiasm—or even understanding—from the public. Worst of all, the history of delay further damages Everglades National Park. Completion of Mod Waters is crucial to the success of Everglades restoration and the CERP projects that follow. If this relatively modest restoration project cannot proceed and provide some restoration benefits, the outlook for the CERP is dismal.

Without completion of Mod Waters, central components of the CERP cannot proceed and ecological conditions in the Everglades ecosystem will continue to deteriorate. Nineteen years have passed since the Mod Waters project was authorized, and the restoration of water flows has not occurred, even though it is a critical foundation project for the CERP. Serious doubt is cast on the prospects for successful Everglades restoration if this important project still languishes. Political leadership and the timely provision of funding are essential if progress on Mod Waters and the associated delivery of restoration benefits to Everglades National Park are going to occur.

Strong leadership focused on building and maintaining support among stakeholders and overcoming conflicts is essential for Everglades restoration projects

to achieve their restoration goals. If there is insufficient political leadership to align research, planning, funding, and management with restoration goals agreed upon by the stakeholders, the CERP will be likely to result in an abbreviated series of disconnected projects that ultimately fail to meet the restoration goals. Other lessons for the CERP that can be learned from the struggles faced during the planning and implementation of the Mod Waters project include the benefits of early agreement on project scope and objectives, the need for a clear project management structure, and the need to anticipate adapting project plans over time.

The reduced scope of Mod Waters attainable with the 2008 recommended plan for modifying Tamiami Trail (alternative 3.2.2.a) provides some environmental benefits but shifts increased responsibility (and cost) to the CERP to achieve authorized Mod Waters goals. The 2008 recommended plan represents a substantially smaller step toward restoration than was originally envisioned for Mod Waters. The recommended alternative is also less cost-effective than other alternatives when benefits are considered as habitat units per dollar spent (see Table 4-3). Although it is critical to move ahead and implement it quickly, the recommended alternative should be viewed only as a first step toward restoration. Moreover, it should be recognized that moving forward with the 2008 recommended plan increases the urgency to proceed more quickly to implement the additional necessary Tamiami Trail modifications through the CERP, or some other mechanism, so that the restoration benefits for Everglades National Park outlined in the WRDA 2007 conference report can be achieved as soon as possible.

5

Lake Okeechobee and Its Place in the Restoration of the South Florida Ecosystem

Lake Okeechobee (Figure 5-1) is at the heart of the hydrologic system, with connections to the east and west coasts as well as downstream to the southern Everglades. The lake was a critical part of the pre-drainage hydrology because it was a primary storage mechanism that modulated downstream flows by storing water in wet years and gradually releasing water during dry periods. After artificial connections to the St. Lucie Canal to the east and the Caloosahatchee River to the west, with a dike around its perimeter, the lake became a diversion point. Large amounts of water that once flowed southward were instead diverted to the ocean, and the lake became much less of a controlling factor for downstream flows to the south, while large, fluctuating flows to the Caloosahatchee and St. Lucie estuaries and the Lake Worth Lagoon have had adverse impacts on them (see Figure 5-2).

In addition, the lake's water quality has been degraded by the external loading of nutrients, especially phosphorus (Engstrom et al., 2006). As a result, although the lake is the largest potential natural source of storage for water in the system, its water cannot be delivered to the often-parched remnant Everglades ecosystem because today's stormwater treatment areas (STAs) do not have the capacity to treat increased volumes of nutrient-enriched water. Changes in water quantity and degradation of water quality (Havens and Gawlick, 2005; Johnson et al., 2007) have also adversely affected the lake's ecological value as a habitat for diverse biotic communities as well as the lake's recreational value.

As in many other complex water management problems (Alexander et al., 2007; Feldman, 2008), extensive research on the lake makes clear that water quantity and quality are inseparably intertwined and need to be considered together in planning and implementing restoration plans (James and Havens, 2005; RECOVER, 2007c). For example, increases in water level (quantity) directly affect the amount of submerged aquatic vegetation (SAV), which in turn allows phosphorus-rich sediments to be mobilized into the water column (quality) (Johnson et al., 2007). Although for organizational reasons water quantity and

FIGURE 5-1 Lake Okeechobee.

SOURCE: Adapted from *http://www.evergladesvillage.net/sat/everglades/thumbs.html.* © International Mapping Associates.

quality are considered separately in places in this chapter, readers should keep their close connection in mind.

This chapter explores several facets of the management of Lake Okeechobee and the potential role it might play in the Everglades restoration. The chapter also includes a discussion of the downstream effects of the disturbed lake, including

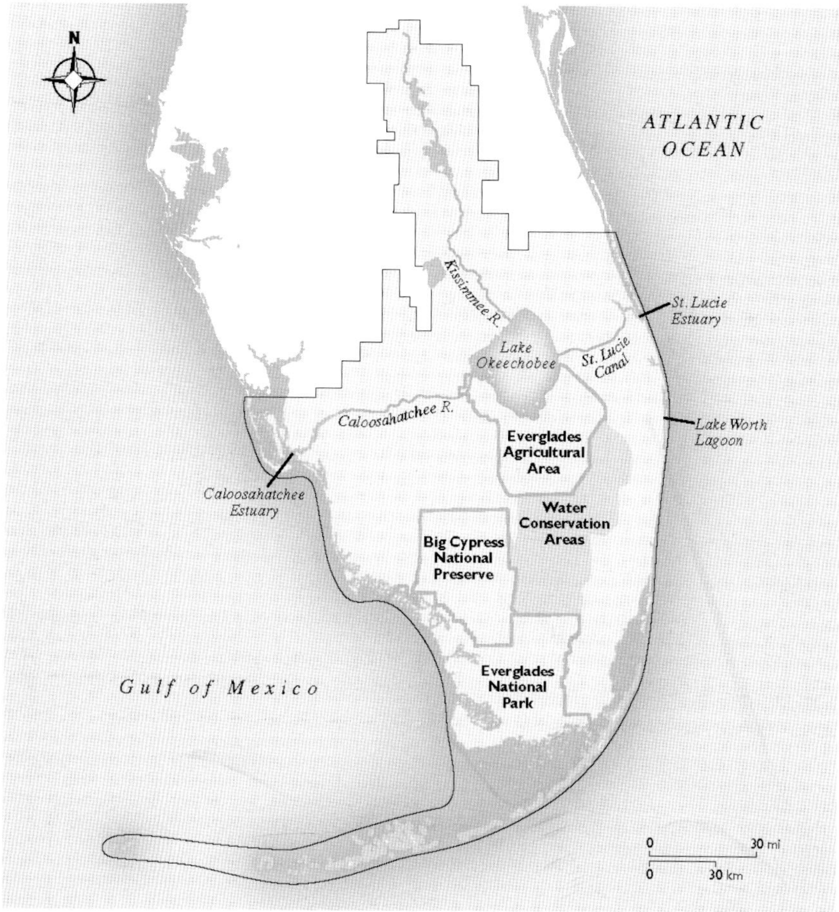

FIGURE 5-2 Lake Okeechobee within the South Florida ecosystem. © International Mapping Associates.

impacts on the estuaries that receive direct flows from it. Allowing the lake to function as the heart of the Everglades as it used to in the pre-drainage system requires large additional restoration efforts. Therefore, the final section of the chapter addresses current and proposed restoration efforts and discusses additional options for restoring the system.

THE CONDITION OF LAKE OKEECHOBEE

Lake Okeechobee presently is plagued by both high and, more recently, very low water levels as well as poor water quality. These conditions have adversely affected the lake's structure and functioning.

Water Quantity

Lake Okeechobee receives most of its inflow from Central Florida via the flows of the Kissimmee River. Until the late 1800s and early 1900s, no canals connected the lake to the Caloosahatchee or St. Lucie estuaries to the west and east, respectively (Blake, 1980; Brooks, 1974; Parker, 1974, Parker et al., 1955). In the rainy season, the lake levels sometimes increased to 21 or 22 feet above mean sea level, and lake waters also flowed southward through the central body of the Everglades via overland flow when the lake exceeded these levels (USACE, 1999). In rainy periods, some lake water also flowed west through wetlands into Lake Hicpochee, which served as the headwaters of the Caloosahatchee (Steinman et al., 2002a). Thus, the lake was a major source of water storage and supply for the entire Everglades ecosystem during periods of high water. Also, the lake, its watershed to the north, and the ecosystem to the south transmitted water fairly slowly. As a result, the seasonality of water level fluctuations in the lake and its watershed and the severity of most dry and wet periods in the ecosystem was considerably reduced compared to today's system (Beissinger, 1986; NRC, 2005).

The condition of Lake Okeechobee today differs distinctly from its historical condition. Lake Okeechobee has undergone major modifications; primary among these was diking that began with a small earthen dike in the 1910s, was expanded in the 1930s along the south shore of the lake, and gradually strengthened until the current Herbert Hoover Dike was completed in the 1960s (Blake, 1980; Brooks, 1974; Parker, 1974, Parker et al., 1955). The construction of the dike restricted the ability of the lake to expand in response to wet periods and reduced the total storage capacity of the lake. Today, the lake functions as a regional reservoir whose inflows and outflows are regulated based on water supply, flood control, and environmental needs (see Box 5-1). The area of the lake varies from about 300,000 acres at its historical low water level to about 470,000 acres (more than 730 square miles) when water levels reach 20 feet above mean sea level. When the lake is at an elevation of 9 feet, a very low level, it contains about 1.75 million acre-feet (MAF) of water; at the upper end of current operating policy of 17 feet, storage is about 4.8 MAF. Each additional

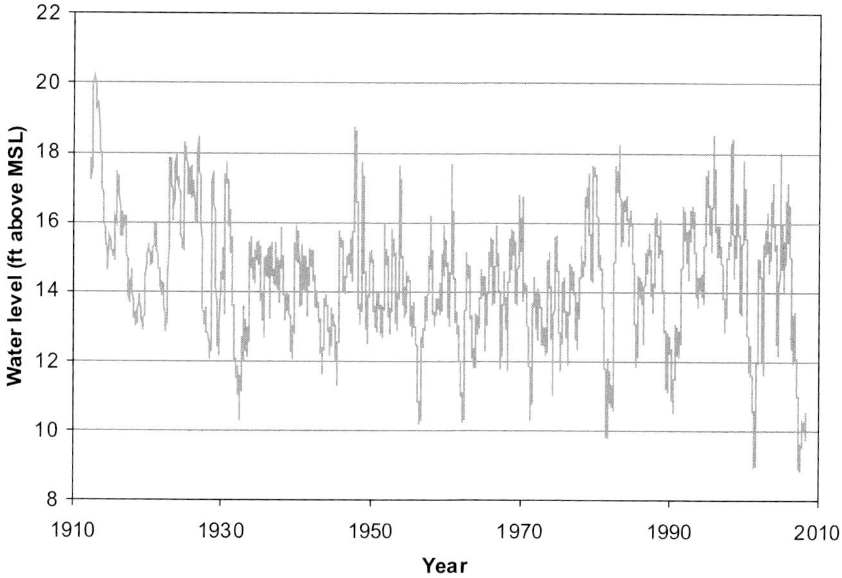

FIGURE 5-3 Elevations of Lake Okeechobee surface 1912–2006.

SOURCE: Daily water level data for April 12, 1912–May 22, 2008, accessed at *http://waterdata.usgs.gov/nwis/nwisman/?site_no=02276400&agency_cd=USGS.*

foot of elevation above 17 feet adds about 425,000 to 525,000 acre-feet of storage, up to 26 feet elevation (Abtew et al., 2007).

Natural lake functioning was also altered by the establishment of new connections to the east via the St. Lucie Canal and to the west via a canal to the Caloosahatchee River, and the construction of levees, water control structures, and locks (Rogers and Allen, 2008). Large releases of water that are frequently made through canals to the Caloosahatchee and St. Lucie estuaries during wet periods are adversely affecting the vitality of those ecosystems (as discussed later in this chapter).

Water levels are currently maintained at much lower levels than historical levels. The U.S. Army Corps of Engineers (USACE) estimated that before the first dike was constructed, the lake had a mean stage of 20.5 feet (USACE, 1999), but today the USACE aims to maintain the water level at about a 12-foot elevation to protect the integrity of the Herbert Hoover Dike (USACE, 2006). A 1999 report showed that at an elevation of 18 feet, 3 of the 13 components of the dike are

BOX 5-1
Average Annual Water Budget for Lake Okeechobee

The lake is fed from several watersheds north and west of the lake and the Everglades Agricultural Area (EAA) (see Figure 5-2). Total stream-flow input to the lake averages 1.62 million acre-feet (MAF) (based on data between 1965–2000), with an additional 1.67 MAF direct rainfall input resulting from an annual average rainfall of about 50 inches (4.2 feet) per year over the lake (see Figure 5-4). However, precipitation is more than offset by evapotranspiration, which amounts to 2.09 MAF per year. Not considering evapotranspiration, total outflow is 1.43 MAF, 29 percent of which is discharged to the Caloosahatchee River (as regulatory discharges and environmental releases), 12 percent to the St. Lucie River (as regulatory discharges and environmental releases), 7 percent to the lower east coast (as regulatory discharges), and 4 percent to the Water Conservation Areas (as regulatory discharges). Water supply applications (mostly agriculture) in these basins receive 38 percent of the outflows from Lake Okeechobee, and the remaining 10 percent is accounted for in other outflows. Inflows and outflows are highly variable within annual periods and from year to year.

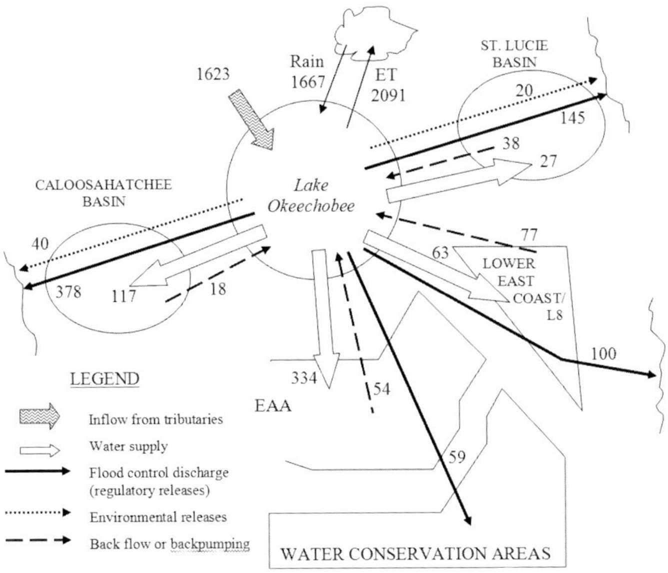

FIGURE 5-4 Average water balance of Lake Okeechobee based on the current Lake Okeechobee Regulation Schedule (LORS) and precipitation data from 1965–2000.

NOTE: All flows are in thousands of acre-feet. Diagram depicts all flows greater than 0.1 percent of the total water budget (including evapotranspiration), with the exception of 141,000 acre-feet in "other outflows." These other outflows include flow to small basins around the lake, the Seminole Tribe, and the Florida Power and Light Reservoir.

SOURCE: Data from J. Obeysekera, SFWMD, personal communication (2008).

classified as hazardous, with high probabilities of failure; at 21 feet, 8 of the 13 are hazardous, with 4 having probabilities of failure of 0.89 or higher; and at 26 feet, 11 of the 13 are hazardous, with 7 virtually certain to fail (USACE, 1999). In 2000, the U.S. Congress authorized the USACE to rehabilitate the dike, and initial construction of a 4.6-mile section of the 140-mile-long dike began in 2005. Completion of all improvements is scheduled over a 25-year period (USACE, 2006), contingent on congressional appropriations.

Water Quality

Historically, water flowing into Lake Okeechobee came primarily from the Kissimmee River, whose extensive wetland floodplain filtered nutrients from the water. This kept nutrients at extremely low concentrations throughout the system, particularly with respect to phosphorus, the contaminant of greatest concern. The extensive spread of agriculture in the upstream drainage basins, plus the channelization of the river and the creation of canals conveying storm-water from agricultural areas directly to the lake, resulted in high phosphorus loads to the lake (Engstrom et al., 2006). A large proportion of phosphorus loaded to the lake accumulated in sediments.

The role of phosphorus as a controlling factor in eutrophication of freshwater ecosystems has been recognized for several decades. High phosphorus concentrations in the lake adversely impact biota by altering the structure and functioning of both the lake and downstream ecosystems. The overall increase of phosphorus loading in the past decades has resulted in conversion of a phosphorus-limited system to a nitrogen-limited system. This has resulted in many changes in the lake, including increased frequency of algal blooms and an increasing abundance of nitrogen-fixing cyanobacteria (Havens et al., 2007). Unless phosphorus in Lake Okeechobee's water can be reduced, there are serious constraints to discharging large volumes of water to the south for use in Everglades restoration.

The phosphorus problem is exacerbated because much of the phosphorus accumulates in soils, ditches, wetlands, and lake bottoms where it can remain for a long time; such stored phosphorus is often referred to as *legacy phosphorus*. Legacy phosphorus is problematic both in the Lake Okeechobee watershed and in the lake itself, because the soil- and sediment-associated phosphorus can serve as steady sources of phosphorus to the water column. When it does so in the watershed, it contributes to the external phosphorus load to the lake; when it does so in the lake, it creates an internal phosphorus load to the lake water. The effects of legacy phosphorus on water quality can last several decades.

External Phosphorus Loads

Most of the current external phosphorus load to Lake Okeechobee comes from agricultural and urban activities in the watershed. Phosphorus is added to uplands in fertilizers, organic solids (e.g., sewage sludge, animal wastes, composts, crop residues), wastewater, and animal feeds. South Florida uses approximately 50 percent of the phosphorus fertilizer imported into the state of Florida (Reddy et al., 1999). Some of the phosphorus is exported from the drainage basin as agricultural products (i.e., harvested biomass), but a significant amount of the phosphorus applied to the land ends up in upland soils and sediments of ditches and streams, and a portion is then transported by river flow to the lake, where it accumulates in lake sediments (Engstrom et al., 2006) and contributes to eutrophication.

The Lake Okeechobee watershed consists of approximately 3.5 million acres, and primary land cover/land uses include: natural areas such as wetlands (37 percent), improved and unimproved pastures (20 and 4 percent, respectively), sugarcane (12 percent), citrus (7 percent), and urban use (11 percent) (SFWMD and FDEP, 2008a). The export of phosphorus to Lake Okeechobee was exacerbated by the channelization of the Kissimmee River in the 1950s and 1960s and the transport of large volumes of phosphorus-laden sediments. Approximately 10 percent of the phosphorus imported into the Okeechobee basin is eventually exported into the lake, although current estimates are based on limited data sets. The residual mass and annual load from legacy phosphorus in the watershed is not currently quantified (Reddy et al., 1996; SWET, Inc., 2008a, 2008b).

A total maximum daily load (TMDL) of 140 metric tons (mt)[1] of phosphorus per year was established for Lake Okeechobee using a goal of phosphorus concentration in Lake Okeechobee of 40 ppb (FDEP, 2000). Based on modeling studies, the Florida Department of Environmental Protection (FDEP) selected 40 ppb as a threshold concentration in nearshore waters for preventing imbalance in the composition of biotic communities (Havens and Walker, 2002). An estimated 35 mt per year of the load is atmospheric deposition (primarily as dust), leaving a target of 105 mt per year as the waterborne TMDL. Average total waterborne phosphorus loads to Lake Okeechobee in the past 5 years were 630 mt per year—six times greater than the target waterborne TMDL. The recent loads represented a decline from the previous 5-year average of 715 mt, largely due to the drought in 2007 when the total phosphorus load to the lake was 203 mt (Figure 5-5). Over the past 5 years, the average phosphorus concentration in the lake water column has been 179 ppb—4.5 times the target concentration (SFWMD and FDEP, 2008a). Intensive phosphorus-management strategies are

[1]One metric ton equals 2,200 pounds.

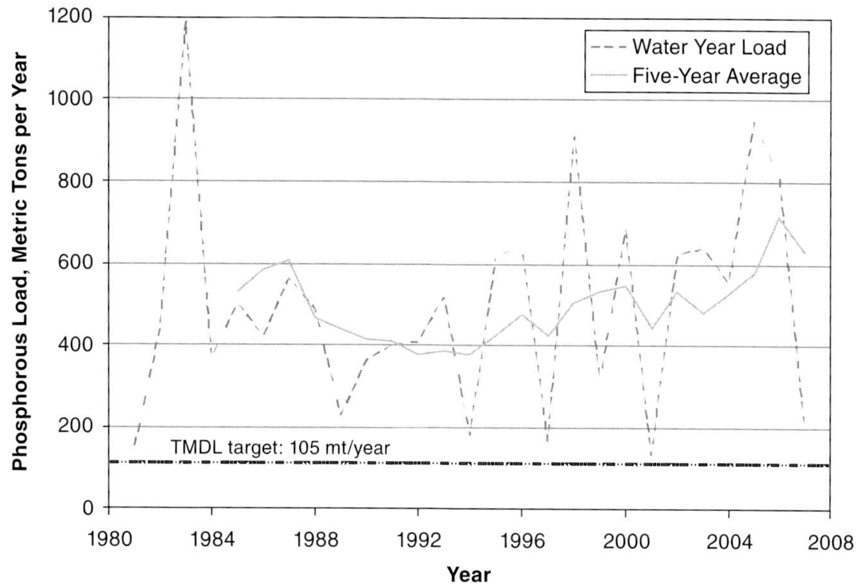

FIGURE 5-5 Annual phosphorus loads into Lake Okeechobee between 1981 and 2007.

SOURCE: Data for 1974–2005 from James et al. (2006); data for 2006–2007 from James and Zhang (2008).

needed to reduce loads from the basins and meet the current TMDL of 140 mt of total phosphorus in Lake Okeechobee by 2015.

Internal Phosphorus Loads

Excessive external phosphorus loads to the lake have accumulated in mud sediments in the center of the lake (Figure 5-6), and they create the current internal phosphorus loads to the lake water column. The phosphorus-rich mud sediment in Lake Okeechobee covers an area greater than 197,684 acres (40 percent of the lake bottom) and has a volume of approximately 162,142 acre-feet. Currently, there are nearly 30,000 mt of phosphorus that have accumulated in the upper 10-cm of these mud sediments (Fisher et al., 2001; see Figure 5-6), representing approximately 60 years' worth of external phosphorus loads. Phosphorus accumulated in sediments shows a dramatic increase in loading, beginning about 1950, coincident with elemental tracers of phosphate fertilizers (Engstrom et al., 2006).

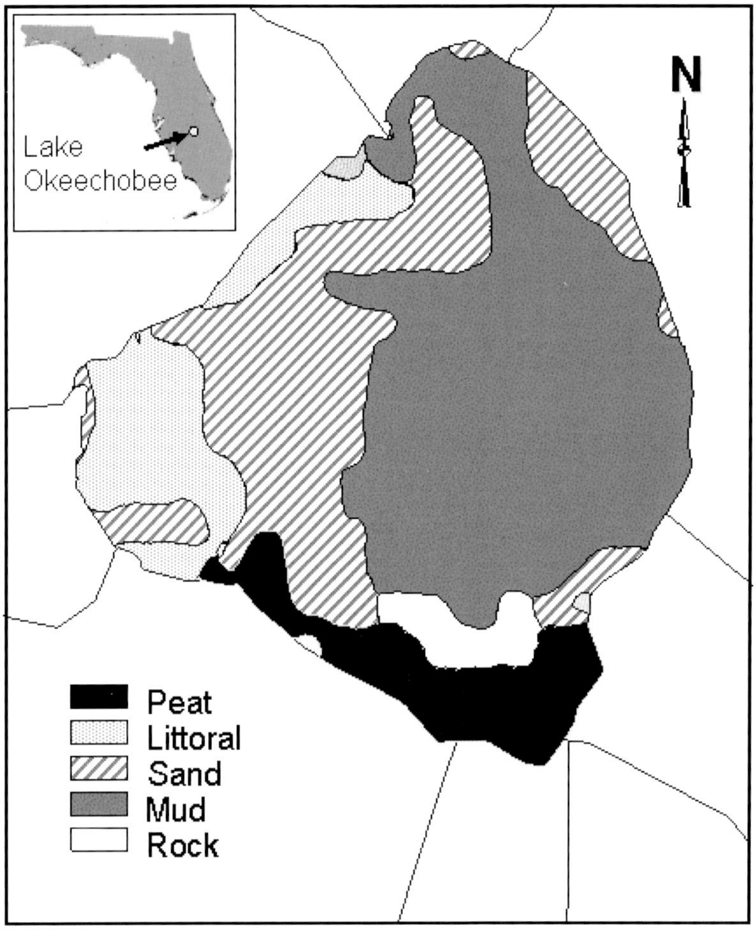

FIGURE 5-6 Lake Okeechobee showing the location of fine mud sediments on the lake bottom that can be resuspended by hurricanes and other wind events.

SOURCE: Adapted from Fisher et al. (2001).

Internal loads of phosphorus to Lake Okeechobee's water can be substantial (approximately 200 mt per year) and comparable to external loads (Fisher et al., 2005). These internal loads occur through diffusive flux of phosphorus from sediments to overlying water and during resuspension of surface sediments into the water column during wind events. After three hurricanes (Frances, Jeanne, and Wilma) moved directly over the lake in 2004 and 2005, sediments became resuspended, and nutrient budgets showed that these sediments became a source of phosphorus to the lake water rather than a sink. Resuspension of sediments also creates turbidity in the lake and prevents light penetration, resulting in poor establishment of SAV. In most hypereutrophic lakes, the bottom sediments are largely derived from deposition of planktonic materials, and the sediments are highly organic in nature. In contrast, one of the sources of Lake Okeechobee's bottom sediments is clay mineral matter derived from the Kissimmee River basin, and these low-density colloidal materials are easily redispersed following wind-driven mixing events and settle very slowly (Harris et al., 2007), which could have long-term impacts on water quality.

The origin and composition of the existing mud sediments and current sediment loads are pertinent to phosphorus management in the lake. At present, sediments delivered to the lake contain magnesium silicates. These colloidal sediments remain suspended in the water column for long periods. This may decrease the longevity of prospective dredging benefits by maintaining resuspended sediments in the water column. Also, the concentration of calcium in the water column can affect flocculation of suspended particles. At this time, very little is known about the reactivity of suspended particles with respect to phosphorus release and retention and the role of altered water chemistry on sediment resuspension.

Implications of Legacy Phosphorus

Once the external phosphorus loads from uplands are curtailed through the implementation of best management practices (BMPs) and other phosphorus management strategies in the drainage basin, the critical question concerns how the lake will respond to reduction of the external phosphorus load (Havens et al., 2007). Legacy phosphorus will continue to leach into the water even after other external loads have been reduced, extending the time required for the lake to meet environmental goals (Fisher et al., 2005). Given these conditions, how long will it take for Lake Okeechobee to reach background or alternate stable conditions? For example, in Lake Okeechobee, phosphorus accretion rates have increased about fourfold since the 1900s (from about 0.25 g P/m^2/year before 1910 to 0.85 g P/m^2/year in the 1980s; Brezonik and Engstrom, 1998), and most of that increase occurred since the 1950s (Engstrom et al., 2006). Although accretion of sediment-bound phosphorus suggests that particulate phosphorus flux is downward (i.e., from the water column to sediments), the dissolved reactive (bioavailable) phosphorus flux is upward (i.e., from sediments to the water column) in response to concentration gradients established at the sediment-water interface (Reddy et al., 2007). Average phosphorus flux from sediments is estimated to be at 0.3 g P/m^2/year, which remained approximately constant over a 10-year period (Fisher et al., 2005). At these rates, bottom sediments can be a source of phosphorus for several decades, unless the sediment phosphorus is stabilized through selective management strategies (e.g., chemical amendments).

Lake Biota

Lake Okeechobee represents an important ecological and recreational resource to the citizens of Florida, and it is an important ecological component of the South Florida ecosystem, both in its own right and as an integral part of the larger system. The lake is home to many species and biotic communities, including some that are (or were) not found elsewhere in South Florida. Some of those species—especially birds—use the lake for part of their life cycles but are important ecosystem components elsewhere at other times. The changes in Lake Okeechobee water quantity and quality, however, are threatening the condition of native lake biota, including vegetation fishes, and birds and exacerbating the spread of exotic species.

Vegetation

The plant communities of Lake Okeechobee are both the linchpin of the aquatic ecosystem and a sensitive indicator of the status of water quantity and quality in the lake. Historically, a series of plant communities occurred roughly in bands around the lake, with a distribution closely correlated with hydroperiod (Havens and Gawlick, 2005), although the littoral zone before the lake was diked is not as well documented. The size, community composition, and geographic arrangement of these communities have been strongly affected by the Herbert Hoover Dike and by water level fluctuations in the lake. Some bands have been entirely lost (e.g., pond apples) (James and Zhang, 2008), and exotic species that are highly tolerant of changes in water level have invaded and spread, altering the dynamic responsiveness of the vegetation to water changes.

SAV is a keystone indicator of many aspects of lake functioning. SAV stabilizes bottom sediments, provides essential habitat for fish and wildlife, and serves as a substrate for the periphyton community that removes nutrients from the water column (Havens and Gawlick, 2005; Havens et al., 2005; RECOVER, 2007a). SAV biomass and cover was markedly decreased after Hurricane Wilma, from approximately 54,000 acres in late summer 2004 (SFWMD and FDEP, 2005) to nearly 11,000 acres in August 2005 and less than 3,000 acres in 2006 (SFWMD and FDEP, 2008a). SAV biomass is highly sensitive to light penetration (Havens, 2003) and was affected by the high turbidity of the water, plus disturbance from wind action and high water levels, although it started to recover in 2007. Recovery requires high light penetration, which in turn results from low lake stages or lack of suspended solids or low concentrations of phytoplankton. However, the goal for shoreline water clarity (100 percent visibility to the lake bed from May through September) was met less than 10 percent of the time during the past 5 years (SFWMD and FDEP, 2008a). The effects of sediment and water quality (especially phosphorus) on both the total amount and the species composition of the SAV are less well understood and warrant inclusion in the research agenda.

A band of floating and emergent vegetation constitutes the littoral zone, which occupies approximately 98,842 acres along the western perimeter of the lake (Figure 5-1). This zone is characterized by high species diversity and a complex pattern of community occurrence, responding to small differences in water depth and hydroperiod. The littoral zone provides essential habitat for fish, wading birds, and other animals for nesting and feeding, and functions like the SAV zone as a keystone community to structure lake food webs and to affect water quality through uptake and stabilization or remobilization of P-rich bottom sediments (Havens and Gawlick, 2005; Johnson et al., 2007). The littoral zone has undergone dynamic change over the past three decades, largely in response to changes in hydroperiods due to variations in lake level (Havens and Gawlick, 2005). Very low lake stages in the past permitted the invasion and spread of two damaging exotic species (*Melaleuca* and *Panicum repens* [torpedograss]); the former species required an expensive and long removal program that has been successful. High lake stages reduced native bulrush stands and have enhanced the distribution of floating-leaved exotic species (*Eichhornia,* water hyacinth, and *Pistia,* water lettuce).

A recent extensive review of the response of both vegetation and fauna to lake levels (Johnson et al., 2007) clearly established two findings: first, most if not all native plant species are highly sensitive to small changes in stage and hydroperiod; second, an optimal range of water level fluctuations promotes a healthy vegetation mosaic that in turn supports a diverse and productive animal community. This review suggests that under the projected management of Lake Okeechobee in accord with Comprehensive Everglades Restoration Plan (CERP) planning, maintenance of lake levels between 12.1 and 15.1 feet above mean sea level should support extensive, dense, and diverse stands of SAV and fluctuating conditions in the interior littoral zone, with positive effects at least on largemouth bass and probably other fish species. However, hurricanes, droughts, and other considerations will influence the maintenance of the lake at those levels.

Even with optimal lake levels, exotic species management is crucial to rehabilitating the Lake Okeechobee ecosystem. More than 80 species of nonindigenous plants are found in the Lake Okeechobee region, of which 10 are considered important pests (James and Zhang, 2008). The South Florida Water Management District (SFWMD) implements an extensive monitoring and control effort for exotic plant species.

Fishes, Birds, and Exotic Animals

The changes in water quantity and quality—especially lake level and turbidity—and the related changes in vegetation have had substantial effects on the fishes and birds of the lake. In recent years, in part due to the four hurricanes that affected the lake in 2004 and 2005, piscivorous fishes declined, while omnivores and planktivores increased. There were marked declines in the fish species of greatest recreational and commercial interest, particularly largemouth bass (*Micropterus salmoides*) and various species of sunfish (*Lepomis*).

Lake Okeechobee was historically an important area for wading birds to nest and feed. White ibis (*Eudocimus albus*), great egret (*Ardea albus),* snowy egret (*Egretta thula*), glossy ibis (*Plegadis falcinellus*), and great blue heron (*Ardea herodias*) were the bulk of the species that historically used the lake (David, 1994a), but tricolored herons (*Egretta tricolor*) and little blue herons (*Egretta caerulea*) also occurred.

Before the 1940s, most of these species were considered so numerous that counts were not made. Numbers of wading birds appeared to peak in the early and mid-1970s due to a large increase in the number of white ibis, when 10,000-plus pairs nested. Increased lake levels that occurred after a change in the lake regulation schedule in 1978 seem to be related to the decline in wading bird numbers on the lake since the late 1970s (David, 1994b). Today Lake Okeechobee is still an important nesting area for wading birds, but the number of birds varies greatly from year to year. Nesting effort of wading birds has been greatest during moderate lake stages (between 13.6 and 15.5 feet for the mean January stage) and is typically very low during years with high or low lake stages (David, 1994b; Marx and Gawlik, 2007). Moderate lake stages probably increase productivity of fish and maximize the potential foraging habitat for wading birds (Smith et al., 1995). Although nesting effort in 2005 was below average, nesting effort in 2006 was near the historic high of 1974, with more than 10,000 nests on the lake (Marx and Gawlik, 2007). However, in the drought year of 2007 only 550 nests were found on the lake—the third lowest count on record (Marx and Gawlik, 2007).

Until recently, Lake Okeechobee had been one of the two most important areas for nesting of the endangered snail kite (*Rostrahamus sociabilis*) in Florida, along with Water Conservation Area (WCA) 3A. Kites nested frequently on the lake in the 1970s and 1980s (Snyder et al., 1989), peaking from 1991–1993 with 63–132 nests annually (Rodgers, 2007). Despite reaching a peak population size of more than 3,100 individuals, no kite nests were found on the lake from 1999 through 2002, and since then few nests have been initiated there (USFWS, 2007). When kites did nest on Lake Okeechobee, water levels directly affected the success of their nests. Kites on Lake Okeechobee from 1997–2007 produced an average of only about 3 young annually. This value is much lower than the mean of 87 young fledged per year in WCA-3A and 19 young fledged per year in the Kissimmee Chain of Lakes region (Martin et al., 2007b).

The factors responsible for reduced use of Lake Okeechobee by kites are not fully understood, as no quantitative analysis of causes has been conducted. However, reduced use appears to be associated with declines in kite foraging habitat from prolonged periods of both high and low water levels that have affected apple snail (*Pomacea paludusa*) populations (USFWS, 2007), which appear to be low and declining (Darby et al., 2006; Darby, 2007). The proportion of kite nests successfully producing young on Lake Okeechobee was positively related to stage levels, with low water levels having a more adverse effect on kites than other water levels (Beissinger and Snyder, 2002; Rodgers, 2007; Snyder et al., 1989).

In excess of 100 species of nonnative animals are found in and around the lake (James and Zhang, 2008), including channeled or island apple snails (*Pomacea insularum),* oscar (*Astronotus ocellatus*), sailfin catfish (*Pterygoplichthys multiradiatus*), and Cuban tree frogs (*Osteopilus septentrionalis*) (Ferriter et al., 2008). Although only a few of the many exotic fish species found in Lake Okeechobee have become established, and although they do not appear to have had large effects on the fisheries in the lake (D. Fox, Florida Fish and Wildlife Conservation Commission, personal communication, 2007), there is not enough information for this committee to evaluate the effects of exotic animals on Lake Okeechobee. Ferriter et al. (2008) conclude that although exotic animals could become, or might already be, problematic, "not enough is known about the population dynamics, reproduction, feeding habits and biology of any of these nonindigenous animal species to make evaluations of their current or future potential impacts to the Lake Okeechobee region." Given the experiences with exotic animals elsewhere, it would appear that the general matter of exotic animals deserves serious attention. Despite the existence of several efforts to monitor and manage specific exotic animal species, there is no coordinated effort to track the wide range of exotic animals in Lake Okeechobee or the South Florida ecosystem as a whole.

State Changes

As ecosystems are altered, they frequently undergo what is termed a regime shift, in which their physical characteristics, biogeochemistry, and biology change dramatically (Folke et al., 2004). Once a shift has occurred to a new alternative stable state, the new ecosystem regime may resist recovery, despite intensive restoration efforts (see also Chapter 2). Such state changes are particularly well known for lakes that shift from a clear-water regime—in which phosphorus inputs, phytoplankton biomass, and recycling of phosphorus from sediments are relatively low—to a turbid-water regime—in which these same variables are relatively high (Carpenter, 2003; Scheffer and van Ness, 2004). Under the enriched, turbid-water regime, submerged plants are reduced or eliminated and primary production is dominated by phytoplankton. Increases in bottom-feeding fishes destabilize the sediment substrate, further making it more susceptible to resuspension by winds and thereby increasing turbidity. Sediment resuspension not only decreases light penetration but also recycles phosphorus back into the water column. Anoxic conditions at the bottom can also release phosphorus from sediments (Fisher et al., 2005). Turbid phosphorus-rich conditions favor dominance by cyanobacteria, which tend to persist even when external nutrient loads are reduced. The conditions are even influenced by the trophic structure of the animal community, with increased fish populations consuming zooplankton that would keep the phytoplankton abundance in check and destabilizing vegetation beds. Current conditions in Lake Okeechobee correspond closely to the conditions generally described for regime shifts in lake ecosystems, making the general properties described in the literature of great relevance to the management of this lake.

The nexus of interrelationships involved in regime shifts in lakes complicates restoration efforts, including the reduction of phosphorus inputs (Carpenter, 2003; Søndergaard et al., 2007). The lakes may remain turbid and susceptible to algal blooms as a result of the remobilization of phosphorus in sediments due to anoxia or resuspension or because of low rates of grazing of phytoplankton. Controlling populations of planktivorous fishes that reduce zooplankton grazing or benthic fishes that cause sediment resuspension—either by removing them or by introducing larger fish to prey on them—has been used to assist recovery in some lakes. Drawing down water levels to remove sediments or to promote vascular plant growth also has been used, as is illustrated by a short-term experimental drawdown of the lake in 2000 (Steinman et al., 2002b). For most temperate lakes that have been studied, cessation of excessive phosphorus inputs has resulted in recovery over 10 to 15 years (Søndergaard et al., 2007), although the presence or absence of key plants and animals that affect water clarity may have an impact on recovery (Ibelings et al., 2007). Indeed, enough is known about physical and chemical changes that have occurred in Lake Okeechobee and their effects on its biota to indicate that any return to the pre-drainage state will not be quick or easy.

EFFECTS OF THE LAKE'S CONDITION ON THE SOUTH FLORIDA ECOSYSTEM

The construction of levees, canals, and other water control structures has significantly altered Lake Okeechobee and its interactions with the South Florida ecosystem. The Herbert Hoover Dike interrupted the southward flow of water, and the water quality difficulties described above pose considerable constraints to the movement of water from Lake Okeechobee to the southern part of the ecosystem in its current condition. As a result, excess water from Lake Okeechobee currently is released through constructed connections to the Caloosahatchee and St. Lucie estuaries, the lower east coast, and the WCAs.

Changing Flows to the Estuaries

The Caloosahatchee Estuary to the west and the St. Lucie Estuary and Lake Worth lagoon to the east (Figure 5-2) have been extensively altered by inlet opening, channelization, and wetland drainage. Most significantly, they have been greatly affected by canal drainage from Lake Okeechobee. Through the CERP, this latter effect is to be mitigated by reducing and modulating freshwater inflows to avoid ecologically harmful low and high flows.

The Caloosahatchee Estuary receives most of its freshwater inflow from the historically meandering Caloosahatchee River, which was supplied by rainfall in its watershed. Beginning in the 1890s, the river was channelized and connected to Lake Okeechobee to promote both navigation and drainage from the lake. An extensive network of canals was constructed to drain agricultural lands in the watershed, and the tidally influenced portion of the estuary was reduced by operation of the S-79 control structure. Approximately 55 percent of the regulatory (flood control) discharge from Lake Okeechobee is sent down the Caloosahatchee River (J. Obeysekera, SFMWD, personal communication, 2008), where it often dominates the wet-weather discharge to the estuary (Figure 5-7). Together, this discharge and the altered drainage patterns within the watershed have greatly changed the pattern, quantity, and timing of freshwater inputs, have caused abnormal salinity fluctuations, and have increased loading of nutrients and other materials (Doering and Chamberlain, 1999).

The St. Lucie Estuary was originally a freshwater system with intrusions of salt water during occasional opening of the ocean inlet. It became an estuary

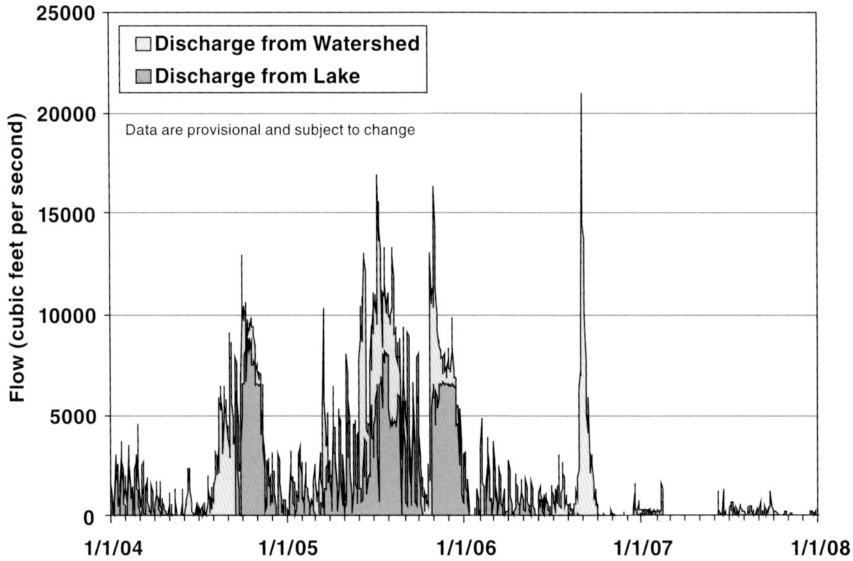

FIGURE 5-7 Total discharge rate into the Caloosahatchee Estuary (watershed releases) at S-79. The portion of the discharge rate accounted for by Lake Okeechobee releases is shown in blue and the portion from the C-43 basin is shown in yellow.

SOURCE: P. Doering, SFWMD, personal communication (2008).

when the permanent inlet was artificially established in 1898. Its watershed was expanded by agricultural and urban development and drainage; it was connected to Lake Okeechobee by the C-44 canal to provide for navigation and regulatory releases. It receives 21 percent of Lake Okeechobee's regulatory (flood control) discharge (J. Obeysekera, SFMWD, personal communication, 2008), and like the Caloosahatchee, the freshwater inflows are excessive at times and insufficient at others (Chamberlain and Hayward, 1996; Doering, 1996; Figure 5-8). Thick muddy deposits cover large areas of the bottom of the estuary, making it unsuitable for aquatic vegetation and other benthic life (Doering, 2007).

Lake Worth Lagoon was historically a freshwater lake, but the creation of permanent inlets has made it estuarine. Lake Okeechobee discharges from drainage canals toward the lower east coast result in occasional excessive releases of fresh water into the estuary.

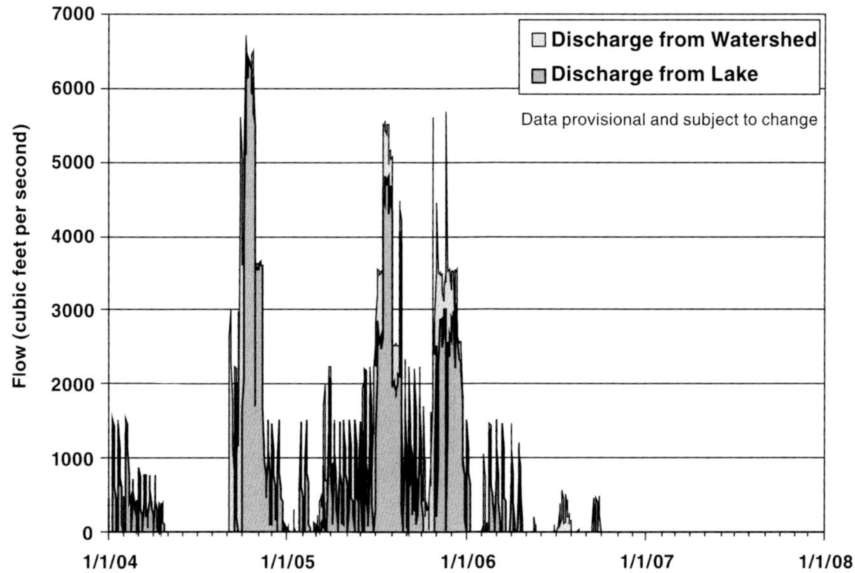

FIGURE 5-8 Total discharge to the St. Lucie Estuary from C-44 during Water Year 2006. The portion of the discharge accounted for by Lake Okeechobee releases is shown in blue and the portion from the St. Lucie river basin is shown in yellow.

SOURCE: P. Doering, SFWMD, personal communication (2008).

These hydrologic changes, particularly as a result of drainage from Lake Okeechobee and the northern Everglades, have resulted in large ecological changes and a reduction in productivity of living resources. These changes are generally caused by exaggerated variations in salinity within the estuaries. However, changes in estuarine circulation and density stratification, and increased loading by nutrients (including not only phosphorus, which receives most attention in Lake Okeechobee and the Everglades, but also nitrogen, which contributes to the eutrophication of estuarine environments), sediments, and toxic contaminants are also factors. As a consequence, in addition to salinity stress, estuarine organisms have to contend with low-oxygen conditions, harmful algal blooms (red tides and blue green algae), decreased light availability, toxins, and siltation (Abbott et al., 2007).

The deterioration of these ecosystems is especially evident in the loss of oyster reefs and beds of SAV, both of which provide important and productive habitats. Oysters have been particularly affected by stressful or lethal incursions of low-salinity and heavy siltation (Wilson et al., 2005). Marine vascular plants such as turtle grass (*Thalassia testudinum*) may be reduced or extirpated during low-salinity episodes, while freshwater vascular plants such as tape grass (*Vallisneria americana*) are stressed by high-salinity incursions up the Caloosahatchee Estuary (Doering et al., 1999; Doering and Chamberlain, 2000; Kraemer et al., 1999). For these reasons, increasing areal coverage of both oysters and SAV is an important restoration goal for the northern estuaries (Doering et al., 2002).

Water Quality Impacts South of Lake Okeechobee

The flora and fauna of the unimpacted areas of the southern part of the Everglades ecosystem are severely phosphorus-limited, and they are adapted to nutrient-poor conditions. As a result, any small addition of nutrients, especially phosphorus, can have a dramatic effect on the structure and productivity of this ecosystem (Childers et al., 2001; Gaiser at al., 2005).

At present, the nutrient-rich waters of Lake Okeechobee have minimal effects on the Everglades Protection Area because only 4 percent of the lake's outflow (on average) goes to the WCAs (see Box 5-1). Nutrients discharged from the Everglades Agricultural Area (EAA) and C-139 basins have been identified as the major sources impacting the downstream Everglades Protection Area. Source control strategies such as BMPs and STAs have been used to reduce phosphorus loads from these basins to the Everglades Protection Area (Adorisio et al., 2007).

Among the most prominent and widely documented effects of introducing water with high phosphorus concentrations into the greater Everglades region is the spread of cattails (*Typha* spp.) at the expense of sawgrass and other species typical of Everglades communities (Chiang et al., 2000; Craft et al., 1995; Noe et al., 2001; NRC, 2005; Scheidt and Kalla, 2007). Thus, sending water to the south that has high concentrations of phosphorus risks perturbing the vegetation community even further beyond its current disturbed state, which likely will affect other aspects of the biological community, including vertebrates.

To address these concerns, the state adopted a phosphorus criterion of 10 ppb (see Box 5-2) that is to be met during the first phase of its "Long Term Plan" (2003–2016). The phosphorus criterion is one part of a phosphorus standard, which also includes moderating provisions. These moderating provisions allow the use of best available phosphorus reduction technologies as a substitute for achieving the actual criterion. The assessment of compliance with the phosphorus criterion is based on a geometric mean from a network of monitoring

BOX 5-2
Judicial and Legislative Context for Lake Okeechobee Restoration Efforts

Water quality in Lake Okeechobee and the larger Everglades system has been a matter of considerable legislative, judicial, and administrative action since the 1980s. Rizzardi (2001) provides considerable insight into the complex legal issues that have accompanied those efforts. Litigation began in 1988 when the U.S. Attorney sued the SFWMD and the Florida Department of Environmental Regulation (DER) (now the Department of Environmental Protection [DEP]), alleging that those agencies were violating the state's water quality standards for phosphorus in the Everglades National Park (ENP) and the Loxahatchee National Wildlife Refuge (LNWR).[a] With the intervention of Governor Lawton Chiles in July 1991, the SFWMD, Florida DER, and U.S. Department of Justice entered into a settlement agreement that was subsequently adopted by the court as a consent decree.[b] The consent decree stipulated that state water quality standards would be met in ENP and LNWR by July 2002. The agreement also called for construction of a series of STAs and regulations requiring agricultural enterprises in the Everglades Agricultural Area (EAA) to implement best management practices BMPs.

The SFWMD initiated its surface water improvement and management (SWIM) planning process in 1991 to implement terms of the consent decree, and the Florida legislature passed the Everglades Protection Act to provide a statutory framework. When the SFWMD issued the SWIM plan in March 1992, it was subjected to numerous challenges under Florida's Administrative Procedure Act.[c] The legislature again stepped in and passed the Everglades Forever Act in 1994 (Ch. 373.4592, F.S.), which largely superseded the Everglades Protection Act and did the following (Rizzardi, 2001):

- in addition to the previously covered federal areas (ENP and LNWR), it extended jurisdiction to the entire Everglades;
- it changed time schedules for compliance from 2002 to 2006;
- it authorized the Everglades Construction Project that included six storm-water treatment areas (STAs);
- it authorized the use of *ad valorem* taxes to provide funding for those projects;
- it required rulemaking to establish a numeric criterion for phosphorus concentrations; and
- it authorized acquisition of agricultural lands.

Based on information gained from an enormous research effort, a phosphorus criterion of 10 ppb was established. During the 2006 Florida legislative session, moderating provisions were added to the Everglades Forever Act that effectively extended the water quality compliance deadline until 2016, at the earliest.

Special attention was given to water quality in Lake Okeechobee with passage of the Lake Okeechobee Protection Act (LOPA) by the Florida legislature in 2000. That legislation establishes authority for a comprehensive watershed program to reduce

continued

BOX 5-2 Continued

phosphorus loads to the lake based on a TMDL for total phosphorus (TP) developed by the Florida DEP. The legislation requires that the TMDL of 140 metric tons TP per year be met by 2015. The 2004 Lake Okeechobee Protection Plan (LOPP) developed under the act provided a phased, comprehensive approach to reduce TP loading to Lake Okeechobee. Efforts to clean up the lake were given an additional boost by the accelerate restoration of America's Everglades (Acceler8 program) and the Lake Okeechobee and Estuary Recovery (LOER) Plan initiated by the state in 2004 and 2005, which fast-tracked numerous CERP construction projects that affect the lake and the northern estuaries (e.g., C-43 and C-44 reservoirs; STAs north of the lake; Lake Okeechobee Watershed project). The LOPA was further expanded in 2007 with passage of the Northern Everglades and Estuaries Protection Program (described in more detail later in this chapter).

[a] *U.S. v. South Florida Water Management District,* Case No. 88-1886 CIV-HOEVELER (S.D. Fla.).

[b] *U.S. v. South Florida Water Management District*, 847 F. Supp. 1567 (S.D. Fla. 1992).

[c] *Sugar Cane Growers Cooperative of Florida; Roth Farms, Inc.; and Wedgworth Farms, Inc. v. SFWMD,* DOAH Case No. 92-3038 (petition filed 4/9/92); *Florida Sugar Cane League, Inc.; U.S. Sugar Corporation; and New Hope South, Inc. v. SFWMD,* DOAH Case No. 92-3039 (petition filed 4/27/92); *Florida Fruit and Vegetable Assn.; Lewis Pope Farms; W.E. Schlechter & Sons, Inc.; and Hundley Farms, Inc. v. SFWMD,* DOAH Case No. 92-3040 (petition filed 4/9/92).

stations established for this purpose within the Everglades Protection Area (see Box 1-1; SFWMD and FDEP, 2006, Appendix 2C-1). There are four components of the total phosphorus criterion for the Everglades, passed by the Florida legislature in 2003 and approved by the Environmental Protection Agency in 2005. All monitoring sites must:

- have a geometric mean of total phosphorus concentrations of less than or equal to 10 ppb in 3 of 5 years,
- have annual total phosphorus concentrations of less than or equal to 11 ppb across all stations,
- have total phosphorus concentrations less than or equal to 15 ppb annually at all monitoring stations, and
- have a five-year geometric mean of total phosphorus concentrations averaged across the network of less than or equal to 10 ppb.

Not later than December 31, 2008, the FDEP will evaluate the first 5 years of monitoring data and approve any incremental phosphorus reduction measures that are needed (373.4592(4)(e), F.S).

STEPS TOWARD REHABILITATION OF LAKE OKEECHOBEE AND AFFECTED DOWNSTREAM ECOSYSTEMS

Earlier sections of this chapter have described water quantity and quality problems in Lake Okeechobee and its watershed, as well as the effects of these problems on the lake biota and downstream ecosystems. The major challenges related to restoration of Lake Okeechobee are (1) excessive phosphorus loads, (2) large amounts of phosphorus stored in the lake sediments, (3) abnormally high and low water levels, and (4) rapid spread of exotic and nuisance plants. Improving water quality and the hydrologic regimes in and around Lake Okeechobee is critical to the long-term success of the Everglades restoration and to improving the condition of the northern estuaries. Thus, the committee agrees with this important premise of the CERP.

Improving Water Quality

Three major components need to be considered in water quality rehabilitation efforts: (1) water quality in Lake Okeechobee, with special attention to management of the large mass of phosphorus in the sediment; (2) phosphorus management in the Lake Okeechobee basin; and (3) effectiveness of the stormwater treatment to the south. The key questions that need to be addressed are: (1) Will Lake Okeechobee respond to phosphorus load reduction? (2) If so, how long will it take for the lake to recover and reach its background condition? (3) Are there any economically feasible management options to hasten the recovery process? Release of the internal load (a consequence of past excessive external loads) can extend the time required for the lake to reach its original condition. This lag time for recovery should be considered in developing management strategies for the lake. Existing, planned, and potential activities and their potential restoration benefits are discussed below.

Options for Managing Loads Internal to Lake Okeechobee

Several methods for managing sediment can address internal loading of contaminants; the most common ones include chemical treatment, oxidation, and dredging (Cooke et al., 1993). A feasibility study of alternatives was conducted to evaluate improvements in water quality by managing phosphorus released from sediment within the lake (SFWMD, 2003). The study considered approximately 30 possible actions, and ultimately, three options were evaluated in detail with respect to cost, effectiveness, and timeliness: (1) hydraulic dredging; (2) in-place chemical precipitation with aluminum compounds; and (3) no in-lake action.

Removal of mud sediments from Lake Okeechobee via dredging will require an order of magnitude greater effort than other lake restoration projects (Cooke et al., 1993). Removing 12 inches of sediments from the lake would take out approximately 94 years' worth of phosphorus accumulation. However, dredging of surface sediments alone will not reverse eutrophication unless external loads are also curtailed (Kleeberg and Kohl, 1999). Hydraulic dredging of deposits in the open lake was estimated to take over 15 years to accomplish and to cost roughly \$3 billion in 2002 dollars, although costs would vary with water level and depth of dredging. Despite these high costs, the dredging would leave behind a significant amount of phosphorus-enriched sediment, which would continue to release phosphorus into the water column for several decades. Several post-dredge sediment management options were identified, including beneficial reuse. However, the benefits of widespread dredging activities were not deemed to be cost-effective (SFWMD, 2003). Nevertheless, during the drought of 2006–2007, the SFWMD removed approximately 1,300 acre-feet (or 1.6 million cubic meters) of mud sediments along exposed shorelines in Lake Okeechobee (SFWMD and FDEP, 2008a). This large volume represents less than 1 percent of the 162,142 acre-feet of mud sediments estimated in the lake (Engstrom et al., 2006).

Chemical applications are intended to bind phosphorus; they usually include aluminum (alum), calcium (lime), or iron (ferric chloride; Cooke et al., 1993). In-lake treatments to control phosphorus concentrations have been used successfully elsewhere on a smaller scale (Cooke et al., 1993; Welch and Cooke, 1999). The SFWMD (2003) considered in-lake treatment using aluminum sulfate ("alum") and sodium aluminate to reduce dissolved and suspended phosphorus concentrations. Aluminum compounds could be particularly effective due to their dual mode of action for phosphorus removal. Concern has been expressed that application of aluminum sulfate compounds could have the unintended consequence of adding sulfate, which is linked to mercury methylation, to the ecosystem. The SFWMD predicted that aluminum compounds could inactivate existing phosphorus and much of the new phosphorus added to sediments for approximately 15 years at a cost of \$500 million in 2002 dollars. However, unless additional source controls are implemented to reduce phosphorus loads to the lake, the lake would progressively return to the original contaminated state, because the surface of aluminum oxy-hydroxides would become fouled and buried with sediments over time.

These actions were contrasted against a "no in-lake action" alternative, in which external source reductions would be emphasized. If the TMDL could be met by 2015, the algal bloom frequency would be reduced by 25 percent to less than 15 percent by 2015 and to below 10 percent by 2028. SFWMD (2003) estimated that the lake would reach a steady state in approximately 35 years with inflow at 140 mt of phosphorus per year. Based on this analysis, the SFWMD selected the no in-lake action alternative as the preferred option (SFWMD and FDEP, 2007).

The committee does not challenge the scientific basis for the decision to place priority on reducing phosphorus in the watershed over in-lake treatment. Nonetheless, the choice of the "no-action" alternative for in-lake treatment makes achievement of the target of 40 ppb concentration by 2015 very unlikely. Without in-lake treatment to substantially control resuspension of the sediment load, water-column concentrations are likely to remain in excess of the target for several decades. That reality may make achieving water objectives downstream of the lake much more difficult.

Managing External Phosphorus Loads

As noted in Box 5-2, over the past several decades a variety of federal and state agricultural programs have been used to reduce the flow of phosphorus from watersheds that empty into Lake Okeechobee. One of the most important was the Lake Okeechobee Protection Act (LOPA), enacted by the Florida legislature in 2000, which mandated preparation of a comprehensive plan to meet the TMDL of 140 mt per year of total phosphorus by 2015. The plan, known as the Lake Okeechobee Protection Plan (LOPP), published in 2004, relied on several on-going projects, expansion of cost-share programs to all agricultural activities, regional structural measures, and CERP reservoirs, STAs, wetlands restoration, and removal of phosphorus-rich sediment from tributaries.

More recently, the Northern Everglades and Estuaries Protection Program was established by action of the state of Florida in 2007 to strengthen protection of the Northern Everglades, including the estuaries, and to expand the use of the state's Save Our Everglades Trust Fund for use toward restoration of the Northern Everglades. In February 2008, the SFWMD released the *Lake Okeechobee Watershed Construction Project: Phase II Technical Plan* (LOWCP-II), a comprehensive plan to implement the Northern Everglades and Estuaries Protection Program (Box 5-3; SFWMD et al., 2008). The new plan expanded LOPP, with objectives to:

BOX 5-3
Assessment of the Lake Okeechobee Watershed Construction Project: Phase II Technical Plan (LOWCP-II)

Overall, the LOWCP II Technical Plan is a very useful document. It provides good background on earlier actions to improve water quality and ecological conditions of the Kissimmee River–Lake Okeechobee watershed. It also provides an extensive list and assessment of management measures relative to their state of design, likelihood of implementation, state of information about benefits, and costs. In addition, the plan includes an evaluation of four alternatives for management. Those who developed the plan should be applauded for assigning uncertainty levels (1–5) to the management measures (see Table 5-1). By assigning those levels, realistic expectations and challenges are presented. In doing so, weaknesses of the plan are also revealed. Many of the management measures are assigned Levels 3–5, meaning they are not more than a concept with considerable uncertainty about their benefits and costs and whether they will be implemented. In that sense, they share many of the same uncertainties as elements of the CERP.

Although the technical plan for the Northern Everglades is very good, there are several noteworthy issues. First, there may be an undue confidence placed on Level 1 and Level 2 BMPs regarding both the extent to which practices are likely to be implemented and their initial and continuing effectiveness. In general, BMPs include a number of practices that are neither intensively managed nor routinely inspected and frequently monitored. The Florida Ranchlands Environmental Services Project is an exception to the more general case of BMPs, but this program currently involves only 8,500 acre-feet of storage and just under 2,000 metric tons of phosphorus (SFWMD et al., 2008, pp. 9–11). Second, it would have been helpful to provide a listing in Chapter 9 of which management measures are included in the preferred plan along with their specific expected reductions in phosphorus and the uncertainty levels to which they were assigned. Only some of that information can be gleaned from the details of Table 7-9 and Table 8-2 in the technical plan. Estimated reductions are given only by the four alternatives and groups of management measures within the alternatives. Third, cost estimates are not associated with particular management measures. Without that relationship, readers are not able to judge for themselves the relative cost-effectiveness of the various measures. It is also not clear from the document as to whether streams of revenues coming from various sources identified in the plan will be sufficient to cover anticipated costs. Despite these issues, the committee commends the state on its Northern Everglades initiative. The state is making appropriate investments in improving water quality by initially focusing on source control in the Lake Okeechobee watersheds.

- Reduce phosphorus loads to meet the TMDL of 140 mt/yr for the lake;[2]
- Manage lake levels to keep them within ecologically desirable ranges;
- Manage releases from the lake to achieve salinity-related flow targets for the St. Lucie and the Caloosahatchee estuaries;
- Identify opportunities for alternative water supply sources in the watershed; and
- Maintain water supply capability for the Lake Okeechobee Service Area.

All watersheds that flow toward the lake are covered by the plan.

LOWCP-II lists and briefly discusses approximately 120 management measures that contribute to storage capacity and/or phosphorus load reduction (the water storage components are discussed in more detail later in this chapter). Each management measure is assigned to one of five levels based on confidence and certainty in its costs and benefits. The criteria (listed in Table 5-1) range from Level 1 projects—with well-established configuration, certain implementation, high benefits, and reliable cost estimates—to Level 5—with many uncertainties about their configuration, likelihood of implementation, benefits, and costs. The alternative identified as the "preferred plan" (see Table 5-2) was considered to be the most-efficient and most-effective combination of water storage capacity and phosphorus load reduction. Nevertheless, only 68 percent of the phosphorus load reduction in the plan was attributable to Level 1 and Level 2 management measures. Measures with moderate to low levels of implementation and projected benefits certainty (Levels 3 and 4) represented 15 and 14 percent, respectively, of the total planned phosphorus load reduction. Cost estimates for the first phase of implementation are $260–320 million for non-CERP projects and $1–1.4 billion for CERP projects (SFWMD et al., 2008). Given the current state budget and the recent allocation of only $50 million in state funds to all Everglades-related restoration efforts in 2009, progress could be much slower than had been originally anticipated when the project was launched with an anticipated Northern Everglades restoration budget of $100 million per year (complementing the $100 million per year pledged to CERP/Acceler8).

Water-Quality Treatment Downstream of Lake Okeechobee

Consideration of several factors leads the committee to conclude that there is a significant likelihood that the TMDL will not be achieved by 2015. First,

[2]SFWMD et al. (2008) estimate that the average annual loading of phosphorus on the lake is 514 mt per year. Considering an estimated 35 mt per year of the load is atmospheric deposition, the plan evaluates project alternatives against a target of 105 mt per year as the waterborne TMDL, or a reduction of 409 mt/yr.

TABLE 5-1 Levels Used to Classify the State of Maturity of Management Measures

Management Measure Level	Configuration Information	Implementation Certainty	Projected Benefits Certainty	Cost Estimate Confidence
1	Well Established	Certain	Very High	Very High
2	Known	High	High	High
3	Conceptual	Moderate	Moderate	Moderate
4	Limited	Low	Low	Low
5	Very Limited	Uncertain	Very Low	Very Low

SOURCE: SFWMD et al. (2008).

TABLE 5-2 Preferred Plan Features

	Management Measure Level[a]
Local Project Features	
Lake Okeechobee Watershed Phosphorus Source Control Programs	
SFWMD Phosphorus Control Programs	1 & 2
FDACS Agricultural BMP Programs	
Supplemental Nonagricultural BMP Programs	
Land Management Programs	
Comprehensive Planning/Land Development Regulations	3
Farm and Ranchland Protection Program Partnership	4
Florida Ranchlands Environmental Services Project	2
Alternative Water Storage Facilities	1, 3, & 4
Local Initiatives	1
Regional Features	
Storage	4
Stormwater Treatment Areas	1–5
Reservoir Assisted Stormwater Treatment Areas	4
Aquifer Storage and Recovery	(contingent on findings from test well)
Deep Injection Wells	4
Other Projects	
In-Lake Treatment	4
Innovative Nutrient Control Technologies	1,4, & 5
Wetland Restoration	3
Miscellaneous Projects	3 & 5

[a]See Table 5-1 for definition of management measures.
SOURCE: Information from SFWMD et al. (2008).

current trends on the loading of phosphorus to Lake Okeechobee (as shown in Figure 5-5) are upward, not downward. Second, a high degree of uncertainty about performance and financing is associated with many elements of the LOWCP-II. Third, implications of choosing the "no-action" option for in-lake treatment were discussed in the preceding section of this report. Failure to achieve or even approach the 40 ppb target in the lake may have serious consequences to current plans for restoring elements of the Everglades system downstream of Lake Okeechobee.

STAs are constructed wetlands used as buffers to retain nutrients and other contaminants. Most of the phosphorus in STAs is stored in soils and sediments via surface adsorption on minerals, precipitation, and immobilization in the cellular tissue of plants and microbes, which may ultimately be buried with refractory organic compounds. Wetland soils tend to accumulate organic matter due to suppressed rates of decomposition. Soil accretion rates for constructed wetlands can range between a few millimeters to more than one centimeter per year. Accretion rates in productive natural wetland systems such as the Everglades have been reported as high as one centimeter or more per year. The genesis of this new material is a relatively slow process, which may affect the nutrient retention characteristics of the wetland. With time, productive wetland systems will accrete organic matter (which ultimately forms peat) that has different physical and biological characteristics than the underlying soil.

STAs are usually managed to improve their overall performance and to maintain expected water quality. For example, small-scale wetlands can be managed effectively by altering hydraulic loading rates or integrating them with conventional treatment systems, while large-scale systems can be managed by controlling nutrient/contaminant loads. Management of newly accreted material by consolidation, hydrologic manipulation (water level drawdown), application of soil amendments and/or soil removal can improve the overall longevity of STAs to maintain water quality.

The first STA became operational in 1994 when just 3,800 acres of cells 1–4 of STA-1W were brought online (Figures 5-9 and 5-10). In 2007 the total effective area[3] of STAs was 34,276 acres. Addition of new units made possible a rapid increase in the volume of stormwater that could be treated, increasing from 183,000 acre-feet in 1996 to 1.44 MAF in 2006. The drought of 2007 caused a dramatic drop in the volume of water treated (SFWMD and FDEP, 2008a).

[3]The effective treatment area of an STA refers to the wetted or flooded area of the project where the vegetation grows, which water flows over; it does not include the levees, structures, etc. It is usually around 85 or 90 percent of the total project area (T. Piccone, SWFMD, personal communication, 2008).

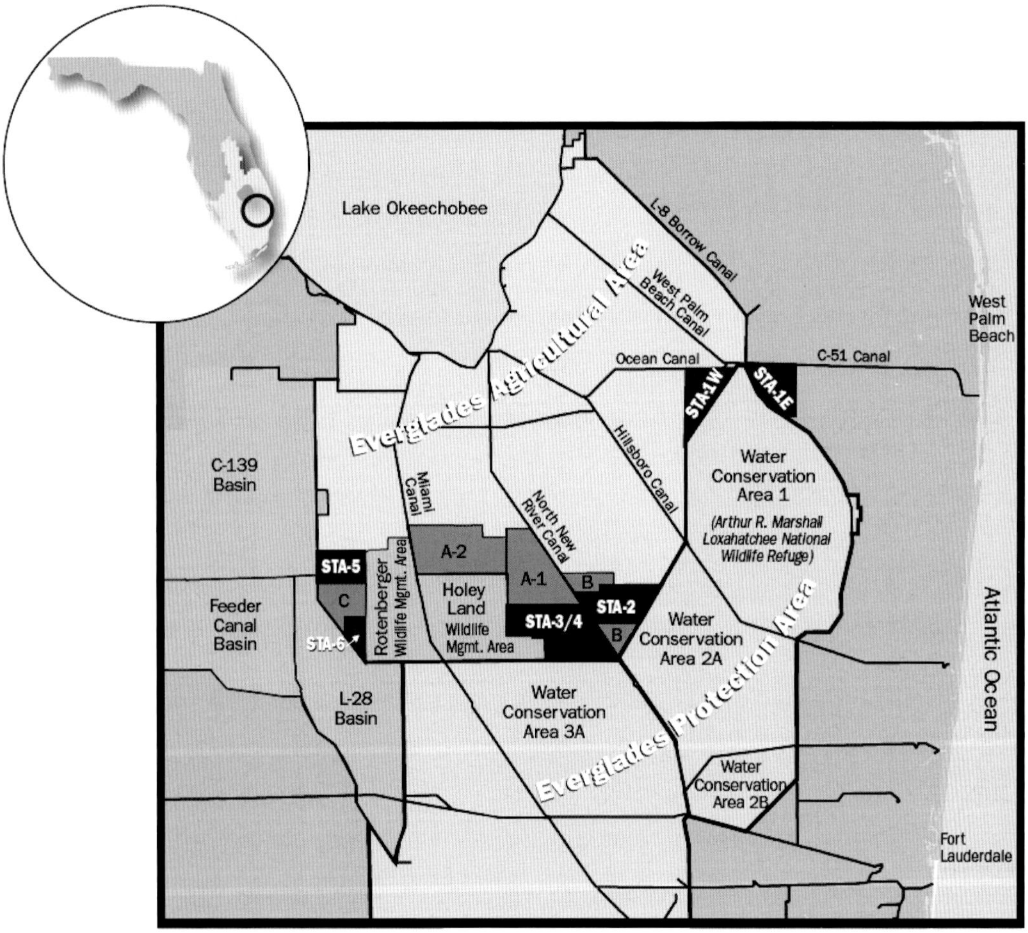

FIGURE 5-9 Location of stormwater treatment areas.

SOURCE: Pietro et al. (2008).

Performance of the STAs is measured primarily by two indicators: how efficiently they reduce the concentration of phosphorus in the inflow, and the total mass of phosphorus removed. A summary of efficiencies of STAs in removing phosphorus is given in Figure 5-11. A striking pattern in that figure is the performance of STA-1W. STA-1W performed at a high level of 74–82 percent reduction from 1996 through 2002. However, the performance has declined to only 47 percent in 2006. During the 5 years of operation, the performance of STA-2

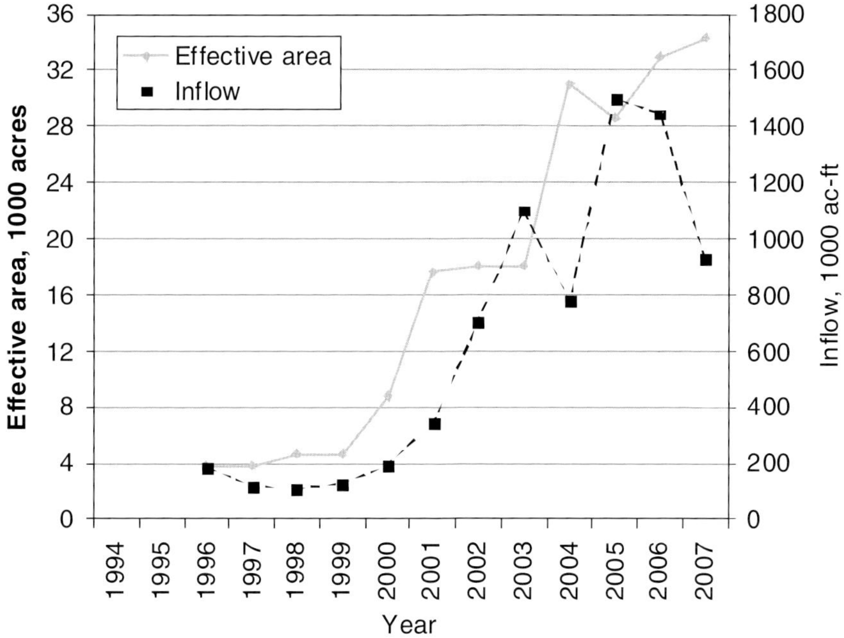

FIGURE 5-10 Effective areas and inflow treated by stormwater treatment areas 1994–2007. NOTE: Average effective treatment areas reflect treatment cells temporarily off-line for plant rehabilitation.

SOURCE: Data from Table 5-31 of SFWMD and FDEP (2008a).

initially improved but has shown recent declines. STA-5 is showing a decline in performance (Figures 5-11 and 5-12). No reliable trend can be determined for STA-3/4, which has been in operation for only 2 years.

The ongoing efforts to optimize performance of all the STAs and to rehabilitate STA1-W in particular are excellent examples of adaptive management. After design and construction of the STAs, their performance has been systematically monitored. Although general processes by which STAs remove phosphorus were well known at the design stage, their actual performance could be determined only through a rigorous monitoring program. Several studies have been undertaken to investigate why certain parts of the STAs were performing below expectations (Pietro et al., 2008). For example, in the original flow ways of STA-2, certain regions were found to be experiencing poor performance, and an investigation has been initiated to characterize phosphorus profiles as a

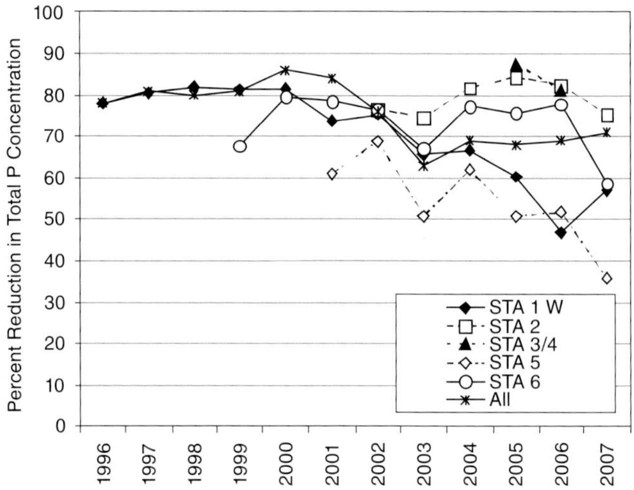

FIGURE 5-11 Time series of annual percent reductions in concentrations of total phosphorus for individual stormwater treatment areas.

SOURCE: Data from Table 5-31 of SFWMD and FDEP (2008a).

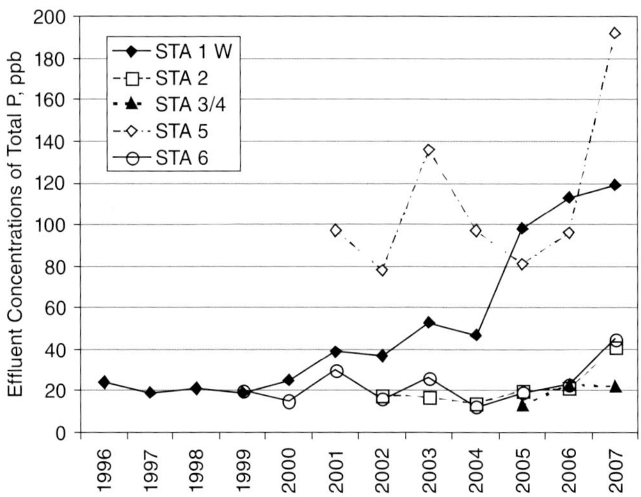

FIGURE 5-12 Effluent concentrations of total phosphorus of five stormwater treatment areas.

SOURCE: Data from SFWMD and FDEP (2008a).

function of the type of vegetation, time of year, and hydraulic and phosphorus loadings. Efforts are under way to rehabilitate STA-1W where reestablishment of vegetation was hindered by high turbidity. Since 1994, these STAs have removed approximately 800 mt of phosphorus from the agricultural drainage waters (Pietro et al., 2007).

It is difficult, but possible, to assess long-term trends in the phosphorus status of the Everglades. This difficulty is due, in part, to the spatial heterogeneity in phosphorus concentrations across the landscape as well as the year-to-year climatic variability. Data for the 2003–2007 water years showed that the interior portions of each WCA met the total phosphorus criterion, while the geometric means of total phosphorus concentrations of most of the individual sites in impacted areas of the WCAs exceeded both the 10 μg/L 5-year limit and the 15 μg/L annual site limit (Table 5-3; SFWMD and FDEP, 2008a).

The Environmental Protection Agency's Regional Environmental Monitoring and Assessment Program (R-EMAP) study (EPA, 2007) was conducted to evaluate the nutrient and mercury contamination status of the Everglades. The R-EMAP sampled the ecological condition of more than 750 miles of canals and more than 3,000 square miles of freshwater marsh extending from Lake Okeechobee southward to the mangrove along Florida Bay and from the eastern urbanized edge to Big Cypress National Preserve. The program utilizes a randomly located probability-based design, which included 199 separate canal locations and 990 marsh locations. With the large number of samples and probabilistic design, it is possible to rigorously evaluate spatial and temporal patterns in water and sediment quality.

The R-EMAP showed that during November 2005, 27.2 ± 7.5 percent of the Everglades Protection Area exceeded the 10 μg/L concentration of total phosphorus. This level of contamination was a substantial improvement from the survey conducted in September 1995 in which 57.8 ± 7.8 percent of the area exceeded total phosphorus concentration of 10 μg/L. While trends

TABLE 5-3 Geometric Mean of Total Phosphorus Concentrations for the Everglades Protection Area for Water Years 2005–2007

Everglades Areas	Inflow	Interior
Refuge	65.9	11.1
WCA-2A	26.2	14.8
WCA-3A	24.0	9.4
Park	9.8	5.8

NOTE: Concentrations are μg/L.

SOURCE: Data from SFWMD and FDEP (2008a).

in surface water concentrations are encouraging, soil concentrations of total phosphorus showed a contrasting pattern. The CERP has established a restoration goal of decreasing the areal extent of the Everglades with total phosphorus concentrations in soil > 500 μg/g, while maintaining or decreasing long-term average concentrations to 400 μg/g or less.[4] The 2005 R-EMAP showed that 24.5 ± 6.4 percent of the Everglades Protection Area exceeded 500 μg/g and 49.3 ± 7.1 percent exceeded 400 μg/g total phosphorus in soil. These values compare with the 1995 results, which showed that 16.3 ± 4.1 percent exceeded 500 μg/g and 33.7 ± 5.4 percent exceeded 400 μg/g total phosphorus in soil. This analysis suggests that although surface water concentrations of total phosphorus in the Everglades have improved markedly over the 10-year period 1995–2005, soils have experienced increased phosphorus contamination.

Data for 2007 indicate that the annual volume-weighted concentration of total phosphorus in the outflow for all STAs was 58 μg/L, with values ranging from 22 μg/L for STA-3/4 to 192 μg/L for STA-5 (Figure 5-12; SFWMD and FDEP, 2008a). This annual volume-weighted mean is above the 50 μg/L established in 1992 for the Interim Consent Decree for STA outflow and well above the 10 μg/L criterion of total phosphorus for the Everglades Protection Area. Given the phosphorus loading to the STAs and their long-term removal efficiency, it seems unlikely that the current configuration will allow for the 10 μg/L geometric mean criterion to be achieved. To decrease the loading of phosphorus into the Everglades and ultimately to achieve the total phosphorus criterion, the STA area needs to be expanded north of the Everglades Protection Area to allow for greater capacity for phosphorus removal, and improvements are needed in watershed management practices to decrease the inputs of phosphorus to the STAs. Additionally, improvements in the long-term phosphorus-removal efficiency of the STAs are needed.

The STAs were designed and are largely being optimized to decrease the transport of total phosphorus to the Everglades, but another waterborne contaminant of concern is sulfate because of its role in the methylation of mercury (Benoit et al., 2003). Gilmour et al. (2007) found that during high water, sulfur accumulates in the STAs under reducing conditions, and between 10 and 33 percent of inlet sulfate was retained. However, under drought conditions large quantities of sulfate can be released and subsequently transported to the Everglades. The drying and rewetting cycle of the STAs can also stimulate methylmercury production (Benoit et al., 2003). Thus, an objective for STA optimization is to increase the removal of sulfur, without net production of methylmercury to downstream drainage water (Rumbold and Fink, 2006).

---

[4]See *http://www.evergladesplan.org/pm/recover/eval_team_perf_measures_.aspx.*

Need for System-wide Accounting for Phosphorus and Other Contaminants

The current monitoring and assessment program for components of the Lake Okeechobee system is rich in detail and provides good indicators of performance (SFWMD and FDEP, 2008a); Chapters 3–5 and 10 of that report provide an impressive array of information about inflows, outflows, water quality, in-lake conditions, phosphorus-removal efficiencies in the STAs, and special investigations. Descriptions of the numerous monitoring and experimental studies provide considerable insights into successful operations and those that are experiencing less than desired outcomes. Although the performance and water quality of individual components are described in considerable detail, there is a need to better integrate and synthesize this information in the context of the entire system. For example, a comparison of reported phosphorus removals from the STAs (Pietro et al., 2008) with phosphorus loads to the EAA from water inflows (Van Horn et al., 2008) leads to the conclusion that the STAs have been removing a mass of phosphorus since 2001 that is approximately the same as the load on the EAA.

SFWMD and FDEP (2008a) report very high phosphorus loads from Lake Okeechobee in 2005 and 2006, but what is the fate of this phosphorus downstream? The mass of phosphorus leaving the STAs in 2006 and 2007 declined sharply from 86.9 mt in 2005 to 61.7 mt in 2007. Is real progress being made, or is the problem being either stored for future resuspension or simply discharged through the canals?

Despite the generous detail provided for various components of the system (e.g., lake water phosphorus concentrations and loads, STA outflow concentrations), the lack of integration of all the reports is a significant barrier to an evaluation of overall progress toward understanding and managing phosphorus loads to the Everglades Protection Area. By examining several components independently, it is possible to draw distinctly different conclusions about progress or the lack thereof: in terms of phosphorus, Lake Okeechobee conditions are deteriorating; in Everglades National Park, they might even be improving. This example illustrates why an annual system-wide accounting of water, phosphorus, and other contaminants (sulfur, mercury, and nitrogen) should be conducted. Such an integrated material balance would facilitate assessments of the role of the various water storage projects on phosphorus and nitrogen loading, as well as assessments of the subsequent impacts these management options would have on nutrient transport to the Everglades and the estuaries. Without a more complete accounting of inflows, outflows, storage, and extraction of water, phosphorus, and other contaminants for each component and how the flows from various components are related to each other, it is not clear that one can trace the mass of materials through the system. A comprehensive system-wide material balance analysis should be done for water and total phosphorus, as well as other critical contaminants that impact the Everglades ecosystem (i.e., sulfur, mercury, nitrogen). The transport and cycling of elements (e.g., phosphorus, carbon, sulfur) are generally closely coupled with one another (e.g., inputs of sulfur influence the transport and fate of phosphorus and mercury), and thus comprehensive material balances of contaminants of concern and associated major constituents will provide insight on ecosystem response to perturbations.

Increasing Water Storage

A fundamental premise of the CERP is that significantly increased water storage is needed to improve the condition of the South Florida ecosystem, including Lake Okeechobee, the estuaries, and the remnant Everglades ecosystem. As discussed previously in this chapter, modifications to the system (e.g., levees, canals, lake operations) have reduced the amount of water stored naturally in the ecosystem. As a result, some parts of the ecosystem are water starved while other parts are submerged, and the natural timing and amplitudes of highwater and drying events have been severely disrupted. Construction of storage for water in the Lake Okeechobee region is the single largest component of CERP and is proposed primarily in two forms: surface reservoirs and aquifer storage and recovery (ASR) wells. The Yellow Book plan proposed to provide approximately 5.5 MAF of new water storage, of which approximately 4 MAF can be attributed to ASR systems (assuming 30 percent injection loss) (NRC, 2005).

ASR pilot projects are currently under way to address technical feasibility issues associated with ASR (NRC, 2001, 2002a). Two ASR pilot project systems have been constructed and are about to begin cycle testing. "To date, no 'fatal flaws' have been uncovered...that might hinder the implementation of CERP ASR" (SFWMD and USACE, 2008), but the final technical ASR program assessment based on the operation of the pilot systems is not anticipated until 2012. The high costs of ASR, however, have caused SFWMD leaders to publicly question the scale of the proposed ASR effort (King, 2008). The ASR contingency plan still has not been completed, and therefore, discussions of alternative water storage options have been repeatedly postponed until this document is released. Meanwhile, some stakeholders question whether the CERP, even with ASR, provides sufficient storage to support rehabilitation of the estuaries and Lake Okeechobee, given uncertainties in future climate and precipitation patterns (Audubon of Florida, 2007).

Early CERP Storage Projects

A number of CERP projects that are designed to increase water storage and benefit Lake Okeechobee and the northern estuaries are under way.

- **The C-43 Basin Storage Reservoir (Figure 3-2, No. 1)** (170,000 acre-feet) is intended to improve the quantity and timing of freshwater flows to the Caloosahatchee Estuary by holding water to avoid excessive discharges and releasing water to maintain salinity gradients in the estuary.
- **The C-44 Basin Storage Reservoir (Figure 3-2, No. 8)** (50,600 acre-feet) **and STA** are components of the larger Indian River Lagoon-South restoration project and are designed to decrease and attenuate excess water flow and reduce the salinity impacts on the St. Lucie Estuary. A 6,300-acre STA will capture and treat some or all of the discharge from the reservoir before it enters the St. Lucie Canal and flows to the St. Lucie Estuary and Indian River Lagoon.
- **The Indian River Lagoon South project (Figure 3-2, No. 8)** will provide 195,000 acre-feet of water storage in four reservoirs and three natural storage areas to benefit the St. Lucie Estuary. The project also includes STAs and habitat restoration initiatives (e.g., muck removal in the estuary), described in Box 3-1.
- **The North Palm Beach County project (Figure 3-2, No. 13)** includes a water storage reservoir (48,000 acre-feet), intended to improve the timing and deliveries of flow to the Lake Worth Lagoon and Loxahatchee Estuary, enhance hydroperiods in the Loxahatchee Slough, and increase base flows to the Northwest Fork of the Loxahatchee River. This project also includes habitat restoration and water treatment components.
- **Site 1 Impoundment (Figure 3-2, No. 2)** (13,280 acre-feet) is intended to reduce water demands on Lake Okeechobee and the Arthur R. Marshall Loxahatchee National Wildlife Refuge.
- **The Everglades Agricultural Area Storage Reservoir, Phase 1 (Figure 3-2, No. 14)** (190,000 acre-feet) is expected to moderate high stages in Lake Okeechobee and discharges to the estuaries.

These projects are among those scheduled for early implementation in the CERP (see Figure 3-2 for project locations). The C-43 and C-44 reservoirs, the EAA Phase 1 Reservoir, and the Site 1 Impoundment projects were all included under the state of Florida's Acceler8 program. Also, the L-8 reservoir in the North Palm Beach County project was expedited by the state under a separate initiative. Construction is under way on the C-44 reservoir, the EAA Phase 1 Reservoir, and the L-8 reservoir (status reports for these projects are provided in Table 3-1). The

Indian River Lagoon-South and the Site 1 Impoundment projects were congressionally authorized in the 2007 Water Resources Development Act.

A number of issues have emerged as a result of technical review and stakeholder input with regard to the degree to which the CERP projects will achieve restoration goals. First, the storage capacity of the CERP plan, including ASR, is not large enough to detain peak freshwater discharges—such as resulted from the hurricanes of 2004 and 2005—sufficiently so as to prevent impacts from high flows into the estuaries. The impacts of excessive freshwater discharges to the estuaries could be further alleviated if more of the Lake Okeechobee outflow were allowed south through the EAA to the southern Everglades. However, as discussed previously, this is constrained by seepage management issues and water quality requirements in the Everglades, as well as flow capacity into Everglades National Park as facilitated by the Modified Water Deliveries to Everglades National Park (Mod Waters) project (see Chapter 4). Also, unmanaged flows from the Caloosahatchee watershed rather than from Lake Okeechobee may be significant (Figure 5-7) and may require additional efforts by landowners to store water.

Second, there may be conflicts between human and environmental demands on these water storage reservoirs once they are operational. During dry seasons and droughts, demands to deliver fresh water from the reservoirs for agricultural and municipal water uses may result in inadequate flows to the estuaries to maintain desired salinity gradients or inadequate lake levels to support the lake biota. Meeting those human demands may come at the expense of meeting environmental freshwater requirements (see Chapter 2). Also, several of the reservoirs are intended to support recreational uses, including boating and fishing. Once established, recreational users may oppose management actions, including water level fluctuations and drainage, required to achieve environmental benefits.

Finally, the water quality of this new water and potential adverse effects on the ecosystem need to be carefully considered. STAs are not included for the C-43 reservoir as they are for the Indian River Lagoon-South reservoirs. This deep-water reservoir is not expected to reduce phosphorus or nitrogen concentrations greatly because of its lack of macrovegetation, frequently limited residence time, and susceptibility to sediment resuspension; thus, nutrient loading to the Caloosahatchee Estuary may exceed that needed to achieve water quality objectives. A third draft of the EAA Reservoir Phase 1 project is now in development, primarily due to inadequate plans to address the water quality implications of the reservoir on the Everglades ecosystem.

Changes in Lake Operations

The USACE manages water levels in Lake Okeechobee through operation of a series of structures that can release water to the Caloosahatchee Estuary to the west, the St. Lucie Estuary to the east, and to outlet canals to the south of the lake. Safety concerns regarding the Herbert Hoover Dike, combined with high lake levels over the period 2003–2005, led to a number of damaging high water releases to the estuaries. Therefore, the USACE launched a review of the lake operating rules (called the Water Supply and Environmental [WSE] regulation schedule), which were adopted in 2000 and based on a historically drier time period.

Given the constraints on lake water discharges, there is a limited range of options for modifying the operating policy. Large releases to the Everglades ecosystem will be possible only with the appropriate conveyance and seepage management structures in place and reduced phosphorus concentrations, either from additional STAs or improvements to lake water quality. In the Lake Okeechobee Regulation Schedule Study, several alternatives were evaluated using the South Florida Water Management model, with 36 years' worth of historical records. The model was used to calculate stage duration curves (relative frequencies at which the lake is at or below stages varying from 8 to 18 feet) for a range of alternatives and compared to the WSE (No Action). At all relative frequencies, all of the proposed alternatives resulted in about a 1-foot decline in surface elevation compared to the WSE, thus reducing the total storage capacity of the lake by about 450,000 acre-feet. The alternatives had a positive effect on low flows to the estuaries, but they had no impact on very high flows (Table 5-4) (USACE, 2007b).

As a result of this study, a new regulation schedule was approved by the USACE in April 2008, and a new operation regime is now being implemented for an interim period. The new operation schedule consists of complex decision tree where releases are governed by hydrologic conditions at selected locations throughout the system. The new schedule incorporates improved climate fore-

TABLE 5-4 Comparison of the Frequencies of Flows into the Caloosahatchee Estuary under the Lake Okeechobee Regulation Schedule and the "No Action" Alternative for a 36-Year Simulation

	No. of Months in Flow Ranges	
Ranges of Flow (cfs)	No Action	Tentatively Selected Plan
< 450	198	131
450–2500	160	237
2500–4500	45	35
> 4500	29	29

SOURCE: *http://www.saj.usace.army.mil/cco/docs/lorss/LORSS_PM_Pres-6Aug07.pdf.*

casts as part of decision trees, and it allows longer and lower rates of release to the WCAs and the estuaries to reduce the impact of sudden pulse releases.

Water Storage in the Northern Everglades

The preferred plan for the Northern Everglades and Estuaries Protection Program includes water storage capacity ranging from 0.9 to 1.3 million acre-feet, including three reservoirs upstream of the lake, three large reservoir-assisted STAs, and a variety of other storage projects. The plan (SFWMD et al., 2008), however, does not provide full details on all the potential projects. Three non-CERP storage projects are described, with a combined volume of 441,000 acre-feet; three reservoir assisted STAs, with a capacity of 474,000 acre-feet, are listed in Table 5-4; and a variety of other specific projects have a combined capacity of 102,000 to 139,000 acre-feet. The balance of 250,000–280,000 acre-feet is not specified. Only 29 percent of the total storage capacity in the plan was represented by management measures with the highest level of implementation certainty (Level 1; see Table 5-1). Sixty-nine percent of the remaining planned storage capacity reflects Level 4 management measures with low implementation and projected benefits certainty.

The Northern Everglades Regional Simulation Model was used to examine the effects of the alternative plans on flows in the watershed compared to current base conditions and future base conditions without additional reservoirs. If these projects in the preferred plan are realized, there would be substantial improvement in all performance measures listed relative to current and future base conditions without those projects (Table 5-5).

Synergistic Opportunities from the Repair of the Herbert Hoover Dike

As discussed in the previous section, concerns about the structural integrity of the Herbert Hoover Dike currently limit the capacity to store water at higher stages in Lake Okeechobee, which creates more frequent high and damaging discharges to the northern estuaries. Optimal lake levels, however, are also determined by the desire to enhance conditions for lake biota and protect the lake's littoral zone. Nevertheless, rehabilitation of the Herbert Hoover Dike may offer synergistic opportunities for creating additional CERP storage and managing water levels for the benefit of the littoral zone, and the costs, benefits, and hydrologic and ecological viability of these options should be considered in any analysis of CERP storage alternatives.

Alternative Dike Configurations. Localized outward movement of the Herbert Hoover Dike was considered in the Yellow Book but not adopted. The concept

TABLE 5-5 Performance Measures for Water Quantity

	Target	Current Base	Future Base	Preferred Plan
Lake Okeechobee	(percent of the time)			
Extreme Low Lake Stage	100	91.7	94.1	99.1
Extreme High Lake Stage	100	83.6	87.9	95.4
Below Envelope—Weekly Average	100	61.0	66.0	89.2
Above Envelope—Weekly Average	100	58.0	65.9	80.3
Caloosahatchee Estuary	(number of months out of 432)			
Mean Monthly Flow > 2,800 cfs	3	82	55	51
Mean Monthly Flow > 4,500 cfs	0	38	25	18
Number of months Lake Okeechobee regulatory releases >2,800 cfs	0	21	13	9
Mean Monthly Flow < 450 cfs	0	190	32	18
St. Lucie Estuary				
Mean Monthly Flow > 2,000 cfs and < 3,000 cfs	0	37	38	33
Mean Monthly Flow > 3,000 cfs	0	28	21	18
Mean Monthly Flow < 350 cfs	31	134	26	15
Water Supply	(percent of target)			
Annual Percent Demand not Met (%)	0	4.7	4.2	2.4
Lake Okeechobee Service Area Mean Annual Percent Demand Not Met (%)	0	4.4	4.6	1.5

SOURCE: Data from SFWMD et al. (2008).

entails moving the dike outward in some locations so that the littoral zone can also move and support essential biotic functions in a different location of the lake. After all, the littoral zone did not exist at its current location historically; rather, it developed based on the management of the lake at lower water levels. An expanded dike configuration, if politically and societally feasible, could allow the lake to function at higher water levels once the dike has been rehabilitated. Although this committee makes no recommendations on this option, it echoes the advice of an earlier NRC committee (NRC, 2005) to keep an open mind about various water storage options, especially if current plans for storage are more expensive or less effective than expected.

Establishing a Flow Way. One option that has long been considered, and often rejected to date, is the concept of a flow way, a direct connection between Lake Okeechobee and WCA-3A to the south. With cost and feasibility of ASR still an issue, the flow way continues to appear among options for transporting water and has considerable advocacy among some environmental groups (e.g., A.R. Marshall Foundation, 2007). The concept involves "natural" flow of water south of Lake Okeechobee, through some areas of what is currently the EAA. Assuming a wetland environment, some water quality benefits might also accrue, although this would depend on the water depth. Some of its advantages and disadvantages as evaluated by this committee are outlined in Table 5-6. A major technical challenge is that subsidence and oxidation of peat have altered the historic land surface gradient between Lake Okeechobee and the WCAs, and some form of water management structures, including pumps, may be necessary to move the water through the region. If pumping is minimized, the storage areas themselves would need to be quite deep, functioning essentially like reservoirs.

Perhaps the primary objection raised by agency evaluations is that a hydraulically unmanaged flow from the lake to the WCAs "would not be present in dry or even normal years" (USACE and SFWMD, 1999). The Yellow Book conclusion is that "The need for flow ways would have to be justified for other reasons rather than hydrology alone." The A.R. Marshall Foundation (2007) argues that one such reason, apart from reestablishing the historic flow pathway, is that the flow way would be cost-effective relative to ASR by providing storage for large quantities of water in the subsided lands between the lake and the south. Of course, this assumes that a flow way, considering evapotranspiration losses, would provide storage equivalent to ASR.

The biggest and most-often cited impediment to a flow way was the sociopolitical task of obtaining land for this project that is currently in agricultural and urban use, but this hurdle might now have been greatly reduced with the announcement by the state of Florida that it is negotiating the potential acquisition of 187,000 acres of U.S. Sugar Corporation land in the EAA just south of Lake Okeechobee. CERP planners now have an opportunity to consider restoration alternatives that previously were unavailable (e.g., vastly increased STAs, additional surface storage, increasing flow from Lake Okeechobee to the Everglades ecosystem during wet periods).

These restoration opportunities would not be available if other kinds of development replaced agriculture as a primary land use in the EAA. Any reanalysis of the CERP should consider ways to optimize the restoration program and make it more cost-effective, while weighing the impacts of any associated trade-offs.

The high costs of rehabilitating the Herbert Hoover Dike have led to suggestions that a spillway at some as-yet-undetermined lake level could negate the need for extensive dike repair. Flows discharged out of such a spillway might logically enter a flow way, although it is unclear as to the degree to which such flows could be integrated into CERP storage and conveyance needs, given the current water quality issues in Lake Okeechobee. CERP agencies continue to

TABLE 5-6 Positive and Negative Characteristics of a Flow Way

Characteristic	Advantages	Disadvantages
Ecosystem processes	Creates hydrologic connection from the Northern to the Southern Everglades.	Transports low-quality water south into the Everglades.
	Increases water storage capacity of the natural system.	Reduces water quantity through additional ET from standing water in flow way.
	Mimics historic water flow path.	Requires land currently in agricultural production.
	May provide option for treatment as it flows, hence acting in part as an STA.	Water in flow way will likely be deep over most of its path if pumping costs are minimized. While sedimentation may be enhanced, wetlands vegetation growth may be inhibited.
	May increase habitat area available for certain species under wet and dry conditions.	Actual design is uncertain. May not be "miles wide," but rather more like a very wide canal.
	Contributes to "true" restoration of the Everglades. Good public perception.	
	Potentially high releases to the flow way would reduce or eliminate damaging releases to the estuaries.	
Hydrologic issues	Gravity feed from Lake O. to the WCAs.	Gravity feed hampered by subsidence in EAA. Greatly altered topography from early 1900s. Will likely require pumps to get water out.
	Water flows "naturally."	Current conveyance system with pumps and hydraulic structures offers flexibility in operations.
	Lake O. dike renovation may offer opportunity for synergistic connection with "the spillway," including opportunities for costsharing.	May require compartmentalization of flow way into "boxes" in order to reduce wind fetch and wave setup and resultant threat to levee integrity and freeboard.
		Water may seep out of constraining levees into remaining EAA.
Financial costs	Cost analysis not done, but might reduce need for some currently planned CERP storage and/or additional STA construction.	Cost of attaining additional EAA land is not documented.
	Alternative ASR life-cycle costs, including energy, may be more than capital and O&M costs of flow way. Flow way may have a smaller "carbon footprint."	Will require displacing communities and people at upstream end.
	Opportunities for cost sharing with HHD renovation.	

provide negative evaluations of the technical feasibility and cost-effectiveness of a flow way compared to current CERP plans, despite new potential cost benefits associated with the rehabilitation of the Herbert Hoover Dike (e.g., Strowd and Punnett, 2007).

CONCLUSIONS AND RECOMMENDATIONS

Lake Okeechobee is a critical linchpin of the South Florida ecosystem. However, the lake presently is plagued by both high and, more recently, very low water levels as well as poor water quality, especially phosphorus, that have affected its structure and functioning. The challenges of water quantity and quality in the lake have important ramifications for the entire ecosystem because the lake supports important elements of the region's biota, and it also has the potential to serve as a major source of water storage and water supply for downstream ecosystems. This potential will become more critical if other planned and proposed sources of water storage do not become available.

An integrated, system-wide view of water quality management is essential to the achievement of restoration goals for the South Florida ecosystem. Good data are available for study to understand the local dynamics of phosphorus and other contaminants, but a system-wide accounting is lacking for water and phosphorus as well as other important contaminants, such as sulfur, mercury, and nitrogen. An integrated system-wide accounting in various components of the basin (including soils, sediments, vegetation, and water) is needed to determine the mechanisms of contaminant transport throughout the ecosystem—from the Kissimmee River to Everglades National Park—to assess the implications of upstream ecosystem changes on downstream habitats, to determine appropriate management actions, and to evaluate system-wide progress to improve water quality. It also is crucial to determine to what degree the current status of the lake represents a changed condition that will resist restoration.

Recent water quality restoration initiatives in the Northern Everglades are not likely to achieve the stated water quality goals (40 ppb total phosphorus in the lake and 140 metric tons per year phosphorus input load) by the year 2015, and it might take decades for these goals to be met with current strategies. Using the "no-action alternative" to manage internal phosphorus loads in the lake is likely to delay achieving in-lake concentration goals by several decades, as concluded by the South Florida Water Management District. Also, although the Northern Everglades initiative's technical plan identifies numerous management measures to reduce phosphorus loads and appropriately assesses the challenges and uncertainties in the proposed plan, the strategies probably are not adequate to reduce external phosphorus loads sufficiently. The Northern

Everglades initiative is appropriately focused on reducing phosphorus inputs to the lake as an initial step, but given the uncertainties associated with the current management measures, this committee judges it unlikely that the current TMDL of 140 mt of phosphorus input to the lake will be met by the year 2015. More significant remediation strategies in the lake and its watershed will probably be needed to reduce the legacy phosphorus in the system and meet the TMDL goal.

Although the Northern Everglades plan represents a sizable effort, it will not be easy or inexpensive to reverse the lake's decline in water quality. One of the greatest challenges to this program may be securing the necessary funding to fully implement the initiative. The lake's importance in the ecosystem, however, justifies significant attention from researchers and planners and justifies the devotion of considerable resources to the lake.

In the near term, restoration planners should consider the consequences of the likely failure to achieve the phosphorus goals and develop alternative approaches. Alternatives may involve significant reallocation of priorities among restoration projects and/or significant changes to water quality criteria for downstream deliveries. One structural option is to increase the number and size of STAs. Given questions concerning the long-term effectiveness of STAs in phosphorus removal, the current phosphorus loadings to the STAs suggest that their current configuration will be insufficient to achieve the 10 µg/L phosphorus criterion in the Everglades Protection Area. Meanwhile, failure to achieve the water quality goals in Lake Okeechobee will affect the condition of the lake and the northern estuaries, and it will reduce the amount of additional water that can be delivered to the Everglades ecosystem. Alternative approaches to addressing these water quality issues may involve significant reallocation of priorities among restoration projects. Restoration planners should carefully consider the needs for additional STAs, considering the opportunities that may be made available by the state's potential land purchase in the Everglades Agricultural Area. In addition, methods of improving the long-term ability of STAs to remove phosphorus should be investigated. In-lake treatment of phosphorus may also be needed to expedite the rehabilitation of Lake Okeechobee as external loads are reduced.

Given concerns about the financial and technical feasibility of aquifer storage and recovery (ASR) at the large scale proposed in the CERP, additional opportunities for water storage should be investigated, and Lake Okeechobee may be an important component of those alternatives. Several important water storage projects are under development through the CERP and Acceler8, largely intended to modulate flows to the northern estuaries, and additional opportunities for water storage upstream of Lake Okeechobee are being considered within the Northern Everglades initiative. Nevertheless, alternative storage options should be considered as possible contingencies to ASR—the primary source of new water storage for the CERP, but for which there are concerns about financial and technical feasibility—including synergistic opportunities related to modifications of the Herbert Hoover Dike. This committee makes no specific recommendations as to the most appropriate storage options, but it encourages CERP planners to consider a wide array of alternatives and their costs and benefits.

Short-term and long-term trade-offs will be needed in the rehabilitation of Lake Okeechobee and northern estuaries. Moving appropriate volumes of water south into the Everglades and managing flows into the northern estuaries may pose conflicts with sustaining adequate water levels for the lake biota and other in-lake goals, and until the Herbert Hoover Dike is rehabilitated, the risk of its failure at high lake levels will constrain options. Given the current altered state of the whole system, goals for the lake, the northern estuaries, and other downstream interests might not be mutually compatible in all respects. As a result, trade-offs will have to be made. Modeling and adequate, reliable data will be needed to evaluate many of these trade-offs as discussed in NRC (2005) and Loucks (2006).

6

Building the Foundation for Adaptive Management

When the U.S. Congress approved the Comprehensive Everglades Restoration Plan (CERP) in the Water Resources Development Act of 2000 (WRDA 2000), there was clear recognition that the Central and South Florida Comprehensive Review Study (or Yellow Book; USACE and SFWMD, 1999) provided only a general outline for restoration of the Everglades and not a detailed restoration plan. Considering the need to move forward in the face of some uncertainties, the Yellow Book proposed an adaptive management plan founded on a comprehensive ecosystem monitoring program. This approach provides a mechanism for emerging scientific information to be incorporated into the plan and for unforeseen consequences of the restoration project to be addressed. Congress subsequently approved funding for an Adaptive Management and Monitoring program in WRDA 2000 and the Programmatic Regulations (33 CFR §385.31) directed the U.S. Army Corps of Engineers (USACE) to adopt an adaptive management approach.

In this chapter, progress toward building the necessary foundations for adaptive management is described, and major recent monitoring and assessment reports and associated issues are reviewed, building on previous National Research Council (NRC) reports (NRC, 2003b, 2007). The committee is specifically charged to review monitoring and assessment protocols and progress (see Chapter 1) because monitoring and assessment are essential for evaluating CERP progress. The chapter begins with an overview of the concept of adaptive management, and progress in creating and applying an adaptive management framework for the CERP is discussed. Next, progress in developing the monitoring and assessment plan is reviewed as a critical component of the adaptive management process. Specifically, the CERP System-wide Performance Measures report (RECOVER, 2007b) and the first full analysis of the status of the South Florida ecosystem (the 2007 System Status Report; RECOVER, 2007c) are discussed in detail. Previous NRC reports (NRC, 2003b, 2007) reviewed the monitoring plan titled Monitoring and Supporting Research (RECOVER, 2004).

Information management and the status of hydrologic and ecological models, which are essential to a well-functioning adaptive management process, are also discussed in this chapter.

ADAPTIVE MANAGEMENT

Adaptive management facilitates natural resource management or environmental restoration activities when uncertainty about the potential outcomes of management actions is present (NRC, 2007). It offers a means to proceed without a fixed design and to reduce uncertainty through the iterative refinement of management actions, ideally based on experimentation (Lee, 1999; Walters and Holling, 1990). Of the many applications of adaptive management, the most-effective ones have well-structured processes that include:

1. management objectives that are regularly revisited and accordingly revised,
2. a model or models of the managed system,
3. the monitoring and evaluation of outcomes,
4. mechanisms for incorporating what is learned into models guiding future decisions, and
5. a collaborative process for stakeholder participation and learning (NRC, 2004).

Through effective application of such a process, decision making moves from a trial-and-error process to one of experimentation based on continuous monitoring, assessment, and reevaluation.

Since its inception, the CERP has taken an adaptive management approach to address uncertainty in the restoration. This approach is mainly passive, entailing detailed scientific analysis, planning, monitoring, and assessment, combined with feedback to restoration design and operation (see Figure 6-1). In some cases, however, active (i.e., experimental) rather than passive adaptive management may better assist in achieving restoration goals, because substantial uncertainties remain about the degree to which a resilient, self-sustaining ecosystem can be restored under the dramatically changed environment of South Florida. Opportunities for active adaptive management experiments are numerous (NRC, 2007), and the CERP's Decompartmentalization (Decomp) project in particular is an example of a project that would likely benefit greatly from the active adaptive management approach described in Box 6-1. Opportunities also exist for incremental adaptive restoration, which was conceived by the previous committee (NRC, 2007) as a way to advance restoration in the face of contentious uncer-

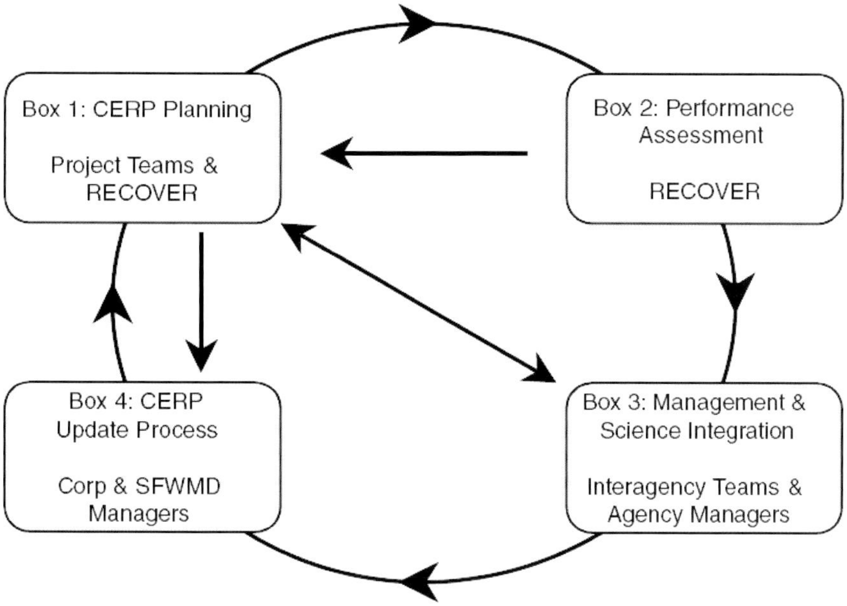

FIGURE 6-1 The CERP Adaptive Management Strategy.

SOURCE: Adapted from RECOVER (2006a).

tainties and sequencing constraints, while utilizing active adaptive management to resolve critical uncertainties and improve future project planning (see also Chapter 3).

The Restoration, Coordination, and Verification (RECOVER) team has fleshed out many dimensions of CERP adaptive management, including extensive work to create monitoring and assessment plans and protocols that are discussed later in this chapter. In 2006, RECOVER published the CERP adaptive management strategy, which outlines a framework for linking monitoring and assessment activities to management decisions and to plan updates and revisions at both system-wide and project levels (Figure 6-1; RECOVER, 2006a). As required by the Programmatic Regulations, the USACE and South Florida Water Management District (SFWMD) also have drafted a Guidance Memorandum for Assessment and Adaptive Management that recapitulates and elaborates upon the CERP adaptive management strategy (USACE and SFWMD, 2007a). The Guidance Memorandum lays out a framework for preparing CERP technical reports and for recommending changes or adaptive management actions as needed at

BOX 6-1
Decomp Adaptive Management Plan

The Water Conservation Area 3 Decompartmentalization and Sheet Flow Enhancement project (Decomp) has been described as "the heart of Everglades restoration" because of the tremendous ecological benefits the project provides (USACE and SFWMD, 2002). The objective of Decomp is to reestablish hydrologic connectivity between WCA-3A, WCA-3B, and Everglades National Park such that hydropatterns in these regions better approximate those experienced in these areas historically.

Decomp project planning is well behind schedule due to ongoing stakeholder conflicts (e.g., disagreements over the need to completely fill canals and thereby eliminate bass fishing habitat) and constraints in the project planning process (Light, 2006; NRC, 2007). To expedite progress the U.S. Army Corps of Engineers (USACE) and the South Florida Water Management District (SFWMD) are developing the Decomp project in three separate project implementation reports and a "Decomp physical model" (DPM), which combines features of a pilot study with an active adaptive management experiment. USACE preferred the term *physical model* over *experiment* because the latter implies research, which is not in the USACE purview. Because the specific plans are still under development, this section focuses on the general direction being taken with the Decomp adaptive management plan.

The plan to decompartmentalize the water conservation areas has raised a number of thorny scientific questions: What is the relationship between the extent of sheet-flow restoration and the rate and degree of ecosystem recovery? What are the effects of partial versus extensive backfilling of canals? How will levee degradation impact adjacent areas? How will tree islands respond to different water depths and hydroperiods? To date this scientific uncertainty has delayed restoration because interest groups have challenged the credibility of proposed modifications. Decomp adaptive management plan has been designed to reduce scientific uncertainty and to accommodate stakeholder concerns. The program combines gathering of baseline information, field trials of alternative decompartmentalization approaches (through the DPM), monitoring and assessment, and feedback to the agencies and stakeholders.

Currently the DPM consists of two proposed experiments that will be installed along the L-67 levee connecting WCA-3A to WCA-3B: the before-after control impact (BACI) flow way and the repeated measure flow way. The BACI flow way experiment would compare geomorphic and ecological effects associated with three canal-backfilling designs (complete, partial, and no backfill) installed in three contiguous 3,000-ft gaps in L-67C (see Figure 4-3). Water would be supplied by a 200-ft (61-m) gap located 2 miles upstream in L-67A. Although the experiment involves no replication, the BACI approach is intended to provide statistical validity to the results: outcomes will be compared to

project, subregional (or module), and system-wide levels of the restoration. The memorandum also provides guidance on reporting standards and expectations for peer review.

The draft *CERP Adaptive Management Implementation Guidance Manual* (RECOVER, 2007e) provides clear technical guidance to project delivery teams

trends in adjacent areas without levee breaches. Field studies are scheduled for 3 years. The repeated measure flow way consists of three 3,000-ft partially backfilled gaps in L-67C intended to measure local variation in the ecological effects of this restoration alternative (Sklar, 2007b).

The DPM designs entail field-scale engineering projects that may significantly reduce uncertainty while advancing Decomp design and implementation. Large-scale changes in the water management system supported by information from the DMP experiment provide opportunities for learning as well as providing some, albeit fairly small, ecological restoration benefits. This approach is consistent with the incremental adaptive restoration approach as suggested by a previous NRC committee (NRC, 2007).

The experimental design reflects and responds to operational, fiscal, and political realities. Concerns about the ability of the DPM to fully address scientific uncertainties associated with Decomp include:

1. The current design does not address the relationship between the extent of sheet-flow restoration and ecological restoration;
2. Management constraints may prevent moving enough water through the system to allow fair comparison of the ecological implications of complete versus partial canal backfilling;
3. The proposed design may not accommodate infrequent high flows that could have an important structuring role in the Ridge and Slough ecosystem; and
4. The 3-year assessment period may not allow sufficient time to distinguish treatment effects.

These concerns are well recognized by RECOVER scientists, and final plans for the DPM are evolving.

CERP scientists hope that the scientific uncertainties associated with Decomp can be greatly reduced by the DPM such that the contentious issues regarding the project design will be lessened or even resolved. Of much graver concern to implementing Decomp are the political uncertainties associated with completion of the Tamiami Trail component of the Modified Water Deliveries to Everglades National Park project (Mod Waters) (Chapter 4). Currently, it is not clear if Mod Waters ever will be completed as described in the authorizing legislation, or how the recommended Tamiami Trail modification plan will impact the ecological benefits that can be achieved by Decomp.

on whether and how to apply adaptive management at the project level. The manual also provides a brief discussion of system-wide implementation of adaptive management, promising a more detailed treatment of this topic in future versions. The manual sets out conditions for employing passive versus active management and advocates incremental adaptive management (NRC, 2007) as the preferred active adaptive management approach when there is stakeholder gridlock or political pressure for quick restoration benefits (RECOVER, 2007e).

Previous NRC reviews of CERP adaptive management concluded that the MAP is generally scientifically strong and that the adaptive management strategy provides a sound organizational basis for passive adaptive management (NRC, 2003b, 2007). Nevertheless, some remaining scientific and institutional challenges need to be squarely addressed, including improved analytical methods and ecological models for upscaling local monitoring data to subregional and system-wide evaluation, stronger coupling of hydrologic and ecological monitoring, and the need for a mechanism to document how science has been used in support of decision making. These are difficult scientific and technical challenges that face all large-scale wetland restoration efforts.

Social and institutional issues affecting the effectiveness of CERP adaptive management may ultimately prove less tractable than scientific issues. To date, the pace and scope of restoration have been driven more by special interest groups and the need to resolve stakeholder conflicts (for examples, see Box 4-1) than by the desire to learn (Light, 2006), and traditional command-and-control planning and governance structures have prevailed over more adaptive approaches (Gunderson and Light, 2006). Stakeholder and collaborative relationships appear to be fraying as the CERP suffers continued delays (NRC, 2007). Agency commitment to continued long-term funding of the monitoring and assessment plan is not assured. Moreover, as discussed in detail in Chapter 3, project-by-project implementation of the plan is laden with planning and review processes that, although well-intended, not only slow restoration but also may operate against effective adaptive management. For instance, analyses of various project alternatives as part of the project implementation report development process (see Chapter 3, Box 3-1) rely on traditional performance-based assessments that do not account for uncertainty or the potential for learning under alternative designs. This conventional "management for optimality" operates against adaptively managing for resilience and surprise (Gunderson and Light, 2006).

CERP MONITORING AND ASSESSMENT

The monitoring and assessment plan is the foundation of the CERP adaptive management program. The CERP monitoring and assessment plan provides the framework that the RECOVER teams use to measure and understand the ecosystem's responses to the CERP and to help determine how well the CERP is meeting its goals and objectives (RECOVER, 2004). Using a "performance assessment" process, the information generated from the monitoring and assessment plan provides information required to make informed decisions about the need to alter restoration plans through the adaptive management process (Figure 6-1). Thus, without effective, long-term monitoring and assessment, adaptive management cannot occur (NRC, 2004). Monitoring and assessment also can support CERP project planning, design, implementation, and operation.

Performance assessment (Box 2 in the adaptive management strategy; Figure 6-1) involves determining how the ecosystem is responding to restoration activities. Performance assessment consists of three activities: evaluation, monitoring, and assessment. *Evaluation* uses a variety of modeling approaches (discussed later in this chapter) and existing data and known relationships among variables to project potential ecosystem changes due to restoration activities. *Monitoring* is the gathering of data for one or more variables that can be used to generate quantitative indicators (or performance measures, described in more detail in the next section) that compress information about complex, related phenomena. *Assessment* in an adaptive management framework is not simply a report of the status of individual performance measures; it involves comparing the values of groups of linked performance indicators with known and expected values that are based on baseline data, modeling, or extensive literature review to determine the actual response of the system to manipulations, including CERP restoration actions.

The CERP monitoring and assessment plan is founded on conceptual ecological models that are an assembly of logical hypotheses that describe the relationships among societal actions, environmental stressors, and ecosystem characteristics and the linkages among the physical, chemical, and biological elements within the natural system. They identify key driving factors, processes, stressors, and functional relationships based on extensive reviews of scientific data (Ogden et al., 2005). The conceptual models cover 11 physiographic regions and the entire South Florida ecosystem (i.e., the total system conceptual model) (Figure 6-2a). For assessment purposes, these have been combined into four geographic modules (see Box 6-2 and Figure 6-2b) and two modules that relate to the total system (the South Florida Hydrology Module and the South Florida Mercury Module). These conceptual models are used to inform research, planning, operational management, and the performance assessment aspects of the adaptive management plan.

The monitoring aspects of performance assessment are described by three RECOVER documents (see Table 6-1):

1. *Monitoring and Supporting Research Plan* (MAP I) (RECOVER, 2004), which establishes the monitoring and assessment plan;

2. *Development and Application of Comprehensive Everglades Restora-*

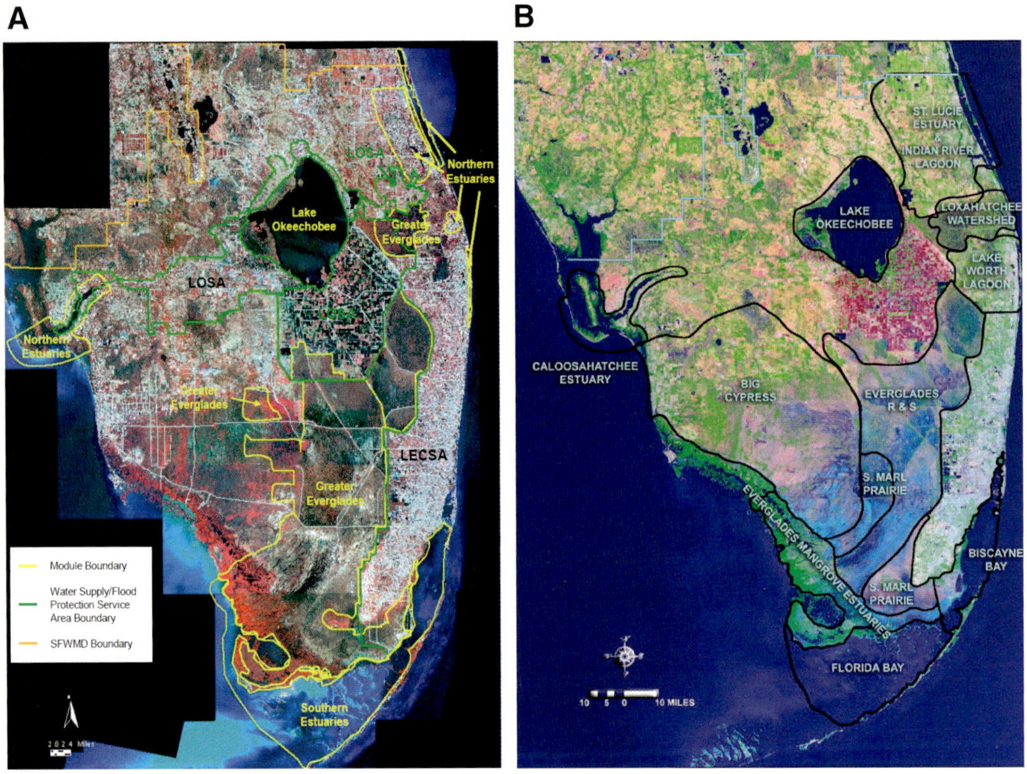

FIGURE 6-2 Boundaries of the 11 conceptual ecological models (A) and the four monitoring and assessment modules (B).

SOURCE: RECOVER (2004).

---

tion Plan System-wide Performance Measures (Performance Measures Report) (RECOVER, 2007b), an analysis and justification of monitored performance measures; and

3. *Draft Quality Assurance Systems Requirements* (QASR) (RECOVER, 2007d) containing quality assurance and data management protocols (see Box 6-3).

All aspects of monitoring including what is monitored and why (MAP I and Performance Measures Report), the spatial and temporal extent of the monitoring (MAP I), and the methods used to collect the data and assure data quality (QASR) are described in these documents. A framework for analyzing monitoring data and evaluating the restoration progress is described in a fourth

BOX 6-2
Monitoring and Assessment Plan Module Groups

The monitoring and assessment plan performance measures are grouped into four geographic modules (Figure 6-2b) and two additional modules for hydrology and mercury bioaccumulation monitoring. The modules are the organizing elements and research units of the monitoring and assessment plan. Each of the modules includes one or more of the conceptual ecological models briefly described in the text and in RECOVER documents (RECOVER, 2004, 2006f, 2007c) and in previous NRC reports (2004, 2007). The modules are designed to test the conceptual model working hypotheses as the CERP is implemented. The module groups are teams of scientists with technical expertise in ecology, hydrology, water quality, and human systems. The module groups oversee the monitoring and research programs in the four major regions of the Everglades and carry out the performance of assessment.

- **Northern Estuaries (NE) Module Group** includes St. Lucie/Indian River Lagoon-South, Caloosahatchee Estuary, Lake Worth Lagoon, and Loxahatchee River Estuary.
- **Greater Everglades Wetlands (GE) Module Group** includes the ridge and slough, southern marl prairies, mangrove estuaries, and Big Cypress swamp regions.
- **Southern Estuaries (SE) Module Group** includes Florida and Biscayne Bays, and the Southwest Florida Coast.
- **Lake Okeechobee (LO) Module Group.**
- **South Florida Hydrology Monitoring** module is designed to assist in evaluating water supply and protection for urban and agricultural areas.
- **Mercury Bioaccumulation.**

SOURCES: RECOVER (2004, 2006f).

TABLE 6-1 Activities and Components of the Monitoring and Assessment Plan

Component	Purpose	Status
Monitoring and Supporting Research (MAP I) (RECOVER, 2004)	Describes monitoring plan and research justifying plan. Performance measures are identified. Focus is the natural system. Implementation plan is included.	Final/January 2004; Revision of MAP I is planned for 2008
Assessing the Response of the Everglades Ecosystem to Implementation of the CERP (Assessment Report) (RECOVER, 2006e)	Describes the guidelines and procedures used to synthesize monitoring data to assess ecosystem response to the CERP.	Final/December 2006
Development and Application of Comprehensive Everglades Restoration Plan System-wide Performance Measures (Performance Measures Report) (RECOVER, 2007b)	Describes performance measures monitored; baseline information.	Final/October 2007
Draft Quality Assurance Systems Requirements (QASR) (RECOVER, 2007d)	Describes quality assurance protocols for all performance measures. Includes some information on data validation, management, and data archiving.	Final Draft (peer-review comments incorporated)/ October 2007

BOX 6-3
Quality Assurance Systems Requirements (QASR) Manual

The QASR document (RECOVER, 2007d) is an extensive manual developed and periodically updated by the CERP to ensure the quality of the data collected during both research and monitoring activities. It represents the mechanism by which the CERP can ensure that all data collected through the multiple agencies, projects, evaluation and assessment activities all adhere to the same standards and requirements of planning, performance, analysis and data archiving, and that data are comparable among components of the CERP. Separate sections of the report are devoted to detailed discussions of methods and quality assurance procedures for monitoring water quality, water quantity, hydraulics, soils and sediments, hydrometeorology, and biological entities. Other chapters specify quality assurance procedures and recommended analytic methods for laboratory chemical determinations and remotely sensed data. The document also addresses research and monitoring administration and lines of responsibility, requirements for data management and archiving, and data quality evaluation and assessment. Extensive appendices list specific EPA and other official sources of methods, and methods established in the peer-reviewed literature for a large number of analytes in water, soils, and sediments, for animal and plant groups (e.g., fish, amphibian, birds, macroinvertebrates, plants), for use of remote sensing data, and for methods of data analysis such as methods of measurement and calculations of flux rates. The set of documents constitutes perhaps the most comprehensive, thorough and wide-ranging summary of specific methods in existence, as it integrates over hydrology, meteorology, soil science, environmental chemistry, animal, plant, and microbial ecology, and remote sensing, as well as data management and assessment protocols. The QASR also includes a very detailed protocol for documenting sample acquisition, analysis methods, and QA/QC analysis. Data custody is assigned to both the agency collecting the data and the Information and Data Management Program of the CERP, which has created the CERPZone Infrastructure to foster data availability and sharing among CERP participants. If the standards and methods for data acquisition, verification and quality assurance, and data archiving and sharing can be fully implemented as described in this document, it will itself be a striking achievement of the CERP.

document—*Assessing the Response of the Everglades Ecosystem to Implementation of the CERP* (Assessment Report) (RECOVER, 2006e). Collectively, these four documents (see Table 6-1) constitute the monitoring and assessment plan. The first full implementation of the assessment strategy is described in the 2007 System Status Report (RECOVER, 2007c).

The following sections reflect the committee's evaluation of two major monitoring and assessment reports released between 2006 and 2008: the Performance Measures Report (RECOVER, 2007b) and the 2007 System Status Report (RECOVER, 2007c). The recently finalized Quality Assurance Systems Requirement Manual (Recover, 2007d) is also discussed in Box 6-3.

Performance Measures

Because of the complexity of ecosystems, there is no single, simple variable whose response can be used to answer questions about the status of ecosystems, their changes over time, and their responses to management. Instead, indicators are developed that have a suite of properties that support their use for management (Niemi and McDonald, 2004; NRC, 2000). The Performance Measures Report (RECOVER, 2007b) supports the monitoring plan by describing the set of indicators (referred to as performance measures; see Appendix F) used to determine the effects of CERP implementation. A performance measure may involve a single variable or, more commonly, is developed as a construct of the combined response of several readily measurable environmental attributes (e.g., population size, water nutrient concentration, water flow rates and durations). The choice of directly measured variables is justified by a review of the scientific literature demonstrating a causal relationship between the measured variables and the desired environmental outcome(s) of interest (restoration target). For example, mangrove forest production and soil accretion is an important performance measure of the food webs critical to fisheries production in Florida Bay and wading bird populations in the Everglades. To assess the condition of this indicator it is necessary to measure variables such as mangrove species composition, canopy density, root growth, decomposition of leaves and roots, sediment deposition, and wetland surface elevations.

Performance measures have been developed for both indicators of ecosystem condition and for critical stressors on the ecosystem (e.g., estuarine salinity, soil and water phosphorus concentrations, hydropatterns). This allows the evaluation and assessment processes to focus on the current understanding of cause-and-effect relationships. This is a great strength of the performance measure system, because the cause-and-effect view of ecosystem dynamics is crucial for implementing an adaptive management approach.

Performance measures play a key role in the restoration because they are used in the planning phase to evaluate project benefits at local and regional scales (termed "evaluation" in CERP documents), and they also are intended to be used to assess the effectiveness of projects once they are implemented (termed "assessment" in CERP documents). In project evaluation, hydrologic and ecological models are used to project the effects of various alternative project plans on performance measures. CERP performance assessment using the performance measures is described in more detail later in this chapter. Some performance measures might also serve as potential "report card" indicators. Report card indicators are key indicators used to inform the public about how the natural system is responding to restoration efforts (e.g., the abundance and nesting success of wading birds or the abundance of alligators, crocodiles, oysters, or pink shrimp). Such indicators have been used as a strategy to maintain public awareness and support for ecosystem restoration projects (e.g., Chesapeake Bay Program, 2002; Heinz Center, 2002; NRC, 2000). For some of the performance measures, interim goals have also been established (see Box 6-4; Appendix G).

BOX 6-4
Agreements on Interim Goals and Interim Targets

The 2003 Programmatic Regulations require the development of interim goals and interim targets as means for measuring the effectiveness of CERP in meeting objectives related to both restoration (interim goals) and societal needs for water supply and flood control (interim targets). The goals and targets are intended to provide an important basis for performance assessment within the adaptive management framework. Interim goals and targets are to be developed using the best available science, formally agreed to by the U.S. Army Corps of Engineers (USACE) and the state of Florida, used as a basis for reporting progress to the U.S. Congress on a 5-year basis, and revised as appropriate.

The First Biennial Review (NRC, 2007) noted the release in February 2005 of RECOVER's (2005b) recommendations for interim goals and interim targets for CERP. It observed that, in addition to their use in evaluating progress in meeting restoration objectives and water-related use needs, they also afforded a way to learn about the trajectories of system response and improve the understanding of ecosystem behavior. NRC (2007) also noted that RECOVER had to base its goals and targets on models that were in need of additional development and that were based on outdated sequencing assumptions.

Based on RECOVER's (2005b) recommendations, draft interim goals and interim targets were formally agreed to in 2007 (see Appendix G). The agreements (USACE et al., 2007; USACE and State of Florida, 2007) acknowledge the current limitations in the performance predictions on the interim goal and target performance predictions recommended by RECOVER because of uncertainties with the models and the science at the time they were developed. Consequently, almost all of the interim goals and targets included in the agreement are strictly qualitative (e.g., increase the coverage of oysters in northern estuaries, reduce high and low volume flows, reduce phosphorus concentrations in Lake Okeechobee, increase the spatial extent of natural habitat, increase water supplies, and maintain or improve level-of-service flood protection). While there are specific water storage and treatment capacities predicted, only the 10 microgram per liter Everglades phosphorus concentration goal is quantitative.

The agreements specify that the best available information and changes in assumptions since the Comprehensive Everglades Restoration Plan was first authorized and the current version of the Master Implementation Sequencing Schedule shall be utilized in developing incremental performance predictions for this initial suite of interim goals and targets. Until this is done, the qualitative nature of present goals and targets greatly limits their usefulness in assessing and reporting progress, other than in a directional sense.

Evaluation of the RECOVER Performance Measures Report

A variety of criteria were used to select the performance measures for the CERP (Box 6-5; see Appendix F for complete list of performance measures). These criteria provide a comprehensive and scientifically sound basis for evaluating and developing individual performance measures because they stress measures that reflect desired outcomes of the CERP and allow evaluation of system-wide responses. The criteria were used to reduce the number of performance measures from well over 200 in early versions of MAP I (NRC, 2003b) to 53 (47 for the natural system plus 6 for water supply and flood protection) in the current performance measures report (RECOVER, 2007b) (see Appendix F). This reduction of the number of performance measures is a significant accomplishment.

The current number of performance measures is not inherently problematic. However, RECOVER should continue to revise and prioritize the performance measures so that the total number of variables monitored is appropriate to their purposes for informing decisions and to the available funding for monitoring efforts. A weakness in the Performance Measures report and the monitoring plan is that no process for periodic review of the performance measures is described, although the need to revise and adapt the performance measures is recognized (RECOVER, 2007b). RECOVER clearly faces a challenge to maintain monitoring

BOX 6-5
Criteria for Establishing Performance Measures

- Changes "directly" in relation to CERP implementation
- Appears in a conceptual ecological model and/or is based in regulatory program(s)
- Includes strong indicator of ecosystem integrity or is a major cause of stress
- Includes indicators of important process, structure or environmental change
- Regional or system-wide scope (rather than project-level)
- Provides unique information (relative to other performance measures)
- Directly or indirectly measurable using indicators
- Strong degree of predictability and should distinguish CERP effects from other factors
- Availability of a mechanism to predict future performance
- Low measurement uncertainty
- Significance of species in species-based performance measures as listed taxa (threatened or endangered), as a keystone taxon, as having a high public profile (esthetics, public attention), as having high recreational or commercial value

SOURCE: RECOVER (2007b).

activities in support of the performance measures that are adequate in spatial and temporal resolution to detect change and minimize uncertainty.

A careful review of monitoring activities and data sources—Who is monitoring which variable? How accurate and precise are the data? How sustainable is the funding and organization to support continued data collection? for example—is crucial to ensure that the performance measures have adequate supporting monitoring data to be useful. RECOVER should also work to match the frequency of monitoring in support of the performance measures with the speed of change of the variables that are monitored and place increasing reliance on remotely sensed or automated data collection methods. The set of performance measures should be reviewed regularly to determine whether they adequately capture the crucial processes and stressors of the system, and whether adequate data collection for each could be sustained over the course of the restoration. This periodic review should also determine whether there is an appropriate balance between module-specific and whole-system measures. Yet, caution is urged before dropping a performance measure from the monitoring program without clear justification. The value of long-term data sets arises in unexpected ways when surprising ecosystem responses occur.

The performance measures report contains a standard list of information required for each performance measure. The performance measure documentation sheets contain information about the scientific basis for selection as an indicator; its relationship to the conceptual ecological models and adaptive management hypotheses as described in MAP I (RECOVER, 2004); the expected response of the indicator to implementation of the CERP; and the way in which it will be used to evaluate plans and assess restoration progress (Appendix H shows what a documentation data sheet should contain). The documentation sheets also provide an analysis of the uncertainties associated with each performance measure (although this information has not been developed for all performance measures) along with critical references and other relevant information. Thus, the report serves as a valuable resource for scientists and managers. Information that would make the documentation sheets of even greater value for assessment and evaluation include length of the data record, the time scale over which the performance measure is likely to respond to CERP-based and natural environmental changes, sampling frequencies, levels of accuracy and precision in environmental monitoring, and sources of monitoring data.

The Performance Measures report provides an excellent discussion of challenges associated with developing and applying performance measures. One specific challenge discussed is the potential for inappropriate claims of uncertainty to derail CERP projects. While the report suggests approaches for reducing uncertainty (as recommended by Breckling and Dong, 2000), these suggestions are quite general (e.g., "think nonlinearly"), and there is insufficient consideration of how to apply the approaches to specific performance measures in the development phase. For those performance measures with documentation about uncertainty, the analysis was based on some combination of formal uncertainty and sensitivity analysis based on quantitative modeling and consensus obtained during peer review of the performance measures. Once sufficient monitoring data have been gathered to generate baselines and confidence intervals for performance measures, additional analysis of the uncertainties associated with performance measure baselines should become part of the performance measure documentation sheet.

A major limitation to the application of performance measures is associated with the evaluation (planning) process. Only a few ecological performance measures—those with habitat suitability index (HSI) models based on the 2-mile by 2-mile South Florida Water Management Model (SFWMM)—can currently be used for evaluation (Tarboton et al., 2004). Performance measures with accepted HSIs include the Wetland Landscape Patterns performance measure for Ridge and Slough and Tree Islands, and the Wetland Trophic Relationships performance measure for Periphyton, Fish, Alligators, and Wading Birds. To date, development of evaluation performance measures has focused on the hydrologic instead of ecological performance measures. Greater effort is needed to develop and refine modeling tools for more of the ecological performance measures and to link those tools to those used to predict hydrologic patterns and water quality (see modeling later in this chapter).

Only some of the hydrologic performance measures are used for both evaluation and assessment of hydrologic performance measures, and in some instances the evaluation and assessment targets for a single hydrologic performance measure are not the same.[1] Some reconciliation of the evaluation and assessment performance measures seems appropriate. Furthermore, in the Greater Everglades module, there are no hydrologic or ecological performance measures that can be used for both evaluation and assessment (see Appendix F). In other cases, RECOVER is not yet able to use a performance measure for either evaluation or assessment purposes, but the indicator is considered to be so important in assessing restoration progress that it was included in the monitoring

[1]For example, the only performance measure that is used for both evaluation and assessment in the northern estuaries module is the estuarine salinity envelope: a two-part salinity target is used for planning and evaluation purposes (450–2,800 cfs at S-79 and 75 percent of the time the flow should be between 450–800 cfs), while a three-part target is used for assessment purposes (mean monthly flow > 300 cfs at S-79, greater frequency of flows at S-79 at approximately 500 cfs in dry years, and reduced numbers of mean monthly flows that exceed 2,800 cfs). Although both targets should result in the same information about the estuarine salinity envelope, they are different enough that comparisons between an evaluation and assessment may be difficult.

program while the performance measure is developed (e.g., Greater Everglades Module, Ridge and Slough Landscape Dynamics). It is important that RECOVER develop explicit methods of reconciling performance measure standards for evaluation and assessment and that all performance measures currently used in monitoring be completely developed.

2007 System Status Report

Assessment of restoration progress in support of adaptive management (e.g., as part of the performance assessment process described previously in this chapter; Figure 6-1, box 2), involves comparison of the status of the ecosystem against some baseline or reference condition.[2] System Status Reports are viewed by RECOVER as a way to provide a holistic description of the entire Everglades ecosystem. The first System Status Report contains an analysis of data collected through the CERP monitoring and assessment plan, historical data, and data from other sources (e.g., universities; federal, state, and local agencies) to provide a pre-CERP baseline of ecosystem conditions. Once CERP projects are constructed, these data will be essential to determining if changes occurring in the ecosystem are the result of implementing restoration projects. Until CERP projects are constructed, the reports will also be useful to document ecosystem trends. In the future, System Status Reports will also be used to determine if the CERP's interim goals are being met (Box 6-4 and Appendix G).

The 2007 System Status Report (RECOVER, 2007c) is the first full implementation of the CERP assessment strategy (RECOVER, 2006f). Therefore, the 2007 System Status Report also served to test the adequacy of the CERP monitoring and assessment plan for performance assessment. The 2007 System Status Report was developed based on experience gained from conducting a pilot assessment in 2006 (RECOVER, 2006d).

The CERP assessment strategy lists five steps that assessments must complete to determine CERP performance (RECOVER, 2006f):

- establish the ability to detect change for each performance measure,
- establish a baseline for each performance measure,

[2]Clarity is needed here because the terms *baseline* and *initial condition* are used in the RECOVER monitoring and assessment program to reflect two different standards of comparison for evaluating monitoring data. *Baseline* represents the historical condition as reflected in the most long-term data set available. A baseline reference permits evaluations of trends with respect to long-term patterns of variability. An *initial condition* or *reference condition*, in contrast, is based on data collected over a short period of time prior to an environmental change (e.g, a CERP project implementation); it permits evaluation of system response over time to a management action. See NRC (2003b) for additional discussion of this topic.

- measure change in response to the CERP,
- integrate and assess conceptual model hypothesis clusters, and
- scale up from the module level to the ecosystem level.

The first System Status Report primarily focused on the first two out of these five steps. The fact that no CERP projects have been fully implemented meant that it was not possible to complete a performance assessment and measure change in response to the CERP. As CERP projects are brought on-line, future assessments will complete each of the five steps. Although the status report does not assess CERP performance, it does report the initial condition of the ecosystem and can be used to gauge system response as CERP projects are implemented. For this reason, the first System Status Report is an extremely valuable document.

Although it is not clear yet how or if the 2007 System Status Report can be used to support adaptive management presently, there are clear advantages to having completed a full assessment of the ecosystem at this time. First, baselines and their variability have been established by the status report for many performance measures and interim goals. However, for some performance measures, the monitoring data set is only a year or two long (e.g., water quality and phytoplankton in Biscayne Bay), and thus what has been established is considered an initial or reference condition rather than a baseline. That some performance measures have been collected for only a short time is not unexpected because the CERP monitoring plan was not completed until 2004. The vast spatial extent and heterogeneous nature of the Everglades made the design and implementation of the monitoring plan exceptionally challenging. Additionally, limited resources (funding and manpower) meant that it was necessary to prioritize implementation of monitoring components so that data collection in support of some performance measures was delayed, making assessment difficult.

Second, the experience of completing a system assessment provided valuable information about the adequacy of the monitoring data (i.e., spatial and temporal extent, and types of data) and the data management system, the uncertainties associated with hypothesis clusters, the degree to which conceptual models incorporate correct linkages, the limitation of conceptual models for performance assessment, and the need for additional modeling capabilities. The lessons learned from completing the first System Status Report will be invaluable to refinement of the monitoring plan, the conceptual ecological models, and existing models and further prioritization of future monitoring and assessment efforts.

Approaches to Assessment

The 2007 System Status Report describes the activities and results of the assessment process for each of the four geographic modules (See Box 6-2 and Figure 6-2b). The monitoring and assessment plan module groups used three general approaches to assessment:

1. A collection of assessments of functionally different subregions (northern estuaries, Lake Okeechobee modules),
2. Module-wide statistical data analysis by performance measure and hypothesis cluster (southern estuaries module), and
3. Module-wide qualitative data description by performance measure and hypothesis cluster (greater Everglades module).

These different strategies for assessment complicate attempts to integrate assessment across the entire system. In addition, the types of monitoring data used for a single performance measure were not always consistent across the different modules. For example, fish abundance was quantified in multiple ways: as standing stock in g/m^2 in the Greater Everglades module, as total mass in kg for the Lake Okeechobee module, and as the number of fish per m^2 for the southern estuaries module. However, it appears that sufficient data might be collected so that a common metric could be calculated for all the modules. The diversity of subsystems within each module also contributed to the great variability in the metrics used in quantifying performance measures.

The northern estuaries module encompasses four very different estuaries and groups of scientists. Consequently, this group analyzed each estuary individually in the context of restoration. While the disturbances, performance measures, and restoration goals among the four northern estuaries are similar, differences in the magnitude of the stressors and in the physical characteristics of the estuaries (e.g., bathymetry, physical orientation, connection to the ocean) make establishing integrated baselines across the four estuaries a complex problem. The identity of the flora, fauna, and processes within the four estuaries also are similar, but the community structure and the interaction among the processes differ among estuaries such that different methods may be required to monitor the communities and/or process rates. Differences among investigators in approaches to monitoring and in the history of monitoring within the estuaries led to large differences in the degree of uncertainty associated with the individual northern estuaries module hypotheses from one estuary to another. The Lake Okeechobee module group faced similar problems—functionally dissimilar subregions and a long historical record for monitoring some performance measures by various investigators—and took an approach similar to that used by the northern estuaries module group. For the first full assessment, the approach used by these two groups is adequate because it emphasized the need for future assessments to establish common monitoring approaches among subregions and estuaries, to ensure that what is monitored is consistent with the interim goal performance measures, or to develop alternative methods that allow the monitoring results to be synthesized.

In contrast, the southern estuaries module team took an integrated approach to assessment that was facilitated by the similarity between Florida and Biscayne bays, the relatively homogeneous environments within the bays, and the long history of the team scientists working together as part of the Florida Bay program. Individual investigators pooled their data and reanalyzed the data from scratch. Their analyses were grouped by performance measure and hypothesis cluster rather than by geography or habitat type, which allowed them to address questions of the robustness of the monitoring plan and the ability to detect ecological response to the CERP to a greater extent than the other module teams. Such an approach that uses all available data to analyze the variance associated with a baseline should be a model for future assessments because it provides robust baselines as opposed to short snapshots, which can only establish initial conditions, and simplifies comparison of system response among modules.

The Greater Everglades module team integrated summaries of long-term data sets from all the scientists working in the region by performance measure and hypothesis cluster. The primary data sets were integrated by using visual overlays of the spatial data layers for performance measures, creating the first step toward more-robust statistical integration of the data and baseline characterization. Like the southern estuaries module group, the Greater Everglades group also has a strong history of working together and sharing data (e.g., Davis and Ogden, 1994) that facilitated the assessment process. Findings gleaned from each investigator were used to weave a story providing a snapshot of the entire central Everglades and how the period 2005–2006 differed from the long-term record for each performance measure where possible (e.g., the Aquatic Fauna Forage Base, Numbers of Wading Bird Nests, and Crocodilian Population Dynamics). For other Greater Everglades performance measures, the monitoring data sets were too short (2 years or less) to allow this type of analysis. For some performance measures (e.g., Ridge and Slough Landscape Dynamics and Water Depth Estimation System [EDEN]), monitoring remains in the planning or early implementation stages, and the 2007 System Status Report addressed progress developing these performance measures.

Assessment of the First System Status Report

Regardless of the approach taken and the length of the monitoring record within each module, the first System Status Report achieved its stated objectives within each module group. The module assessments established a baseline for some and the initial condition for most performance measures being measured, and in most cases they provided some discussion of the ability to detect performance measure changes or trends. The module assessments also identified gaps in monitoring and addressed the adequacy of the conceptual ecological models, considering their associated uncertainties. For those performance measures with interim goals, the module-level assessments compared the current condition of the performance measure to its interim goal. Analysis of the adequacy of the sampling design for each performance measure is missing from most assessments, although in several cases the available data are adequate to allow this type of analysis.

A major challenge in performance assessment is the ability to detect environmental change and attribute it to specific causes (CERP- or non-CERP-related). The amount and quality of data will have a large impact on this capacity, as will the methods of data analysis. There is no uniform approach to determine whether change has taken place, either among modules or performance measures.

The assessment also lacks a coordinated approach that allows comparisons to be made across the module groups. For example, each of the modules includes performance measures for fish, yet there is no clear way to make comparisons across the modules, or to allow the consideration of the effects of a restoration regime across modules. As an earlier NRC report (2005) described, there is a need for assessing trade-offs among modules when evaluating various restoration regimes because no restoration regime will benefit all modules equally. As a result, even if system-wide performance measures are not easily identified, a coordinated conceptual approach to assessment would at least facilitate such comparisons. Other measures examined within but not across modules were vegetation patterns and phosphorus. Because a defining characteristic of the historic Everglades was its vast spatial extent (SSG, 1993), evaluating the condition of fish or plants or phosphorus in a coordinated way *across the entire system* should be fundamental to the assessment process. Future assessments should include evaluation of similar ecological, as well as hydrologic, performance measures across module boundaries at the level of the entire ecosystem, as well as within artificial module boundaries. Thus, unified analysis of data from a variety of sources should be completed among the module groups to assess the status of performance measures in common among the modules/conceptual models.

The highly technical nature of much of the status report is a consequence of the focus on establishing baselines and change detection for the performance measures. As a result, the document is primarily of interest to scientists working on similar problems. Nonetheless, this type of analysis is critical to future assessments of changes in response to the CERP. For future system status reports with objectives that reach far beyond establishing baselines, this high degree of technical detail alone is unlikely to satisfy the needs of project managers and decision makers. Managers will need information relevant to the interim and ultimate restoration goals. To maximize the usefulness of future status reports for adaptive management, those reports should contain succinct summaries that clearly address whether the interim and longer-term goals are being met; if not, why; and what CERP operations or design changes are most likely to move ecosystem response closer to the interim goals.

Use of Performance Assessment to Improve the Monitoring and Assessment Plan

The System Status Report describes the possible outcomes of performance assessment and how each outcome would be applied to refine the monitoring plan (Figure 6-3). The procedure is based on Weinstein et al.'s (1997) decision framework for Delaware Bay salt marsh restoration. One possible outcome of performance assessment is the finding that there is insufficient monitoring either spatially or temporally to allow performance measures, conceptual models, and associated uncertainties to be examined in a rigorous way. This situation would necessitate changes in the monitoring plan, such as extending the spatial or temporal extent of monitoring to provide a more robust, sensitive baseline, or changing the monitoring methods. Another possible outcome of the performance assessment is the determination that monitoring results are inconsistent with or do not support the conceptual ecological models, hypotheses, or the interim goals. The response to this situation could require modification of the assessment tools (i.e., models), hypotheses, conceptual ecological models, or performance measures, or changes to the CERP or the monitoring protocols if it could be determined that the monitoring data were somehow inaccurate or inadequate.

A third potential outcome requires no action: that the performance measures are adequate to capture system behavior with the desired level of sensitivity, and the performance measure responses are consistent with the conceptual ecological models and hypotheses and support the interim goals. The System Status Report concludes that outcome 1 ("insufficient data and/or time available to establish pre-CERP conditions and identify trends") applies to most of the performance measures. Nevertheless, a clear methodology is needed to help scientists distinguish between the first and second potential outcomes (see Figure 6-3), because there is potential for wasting time and resources if the wrong outcome is followed. The current effort to revise the MAP (RECOVER, 2004) should include guidance in this regard.

When MAP I was released in 2004, RECOVER expected that it would need to adapt to new information learned from monitoring and assessment, surprises

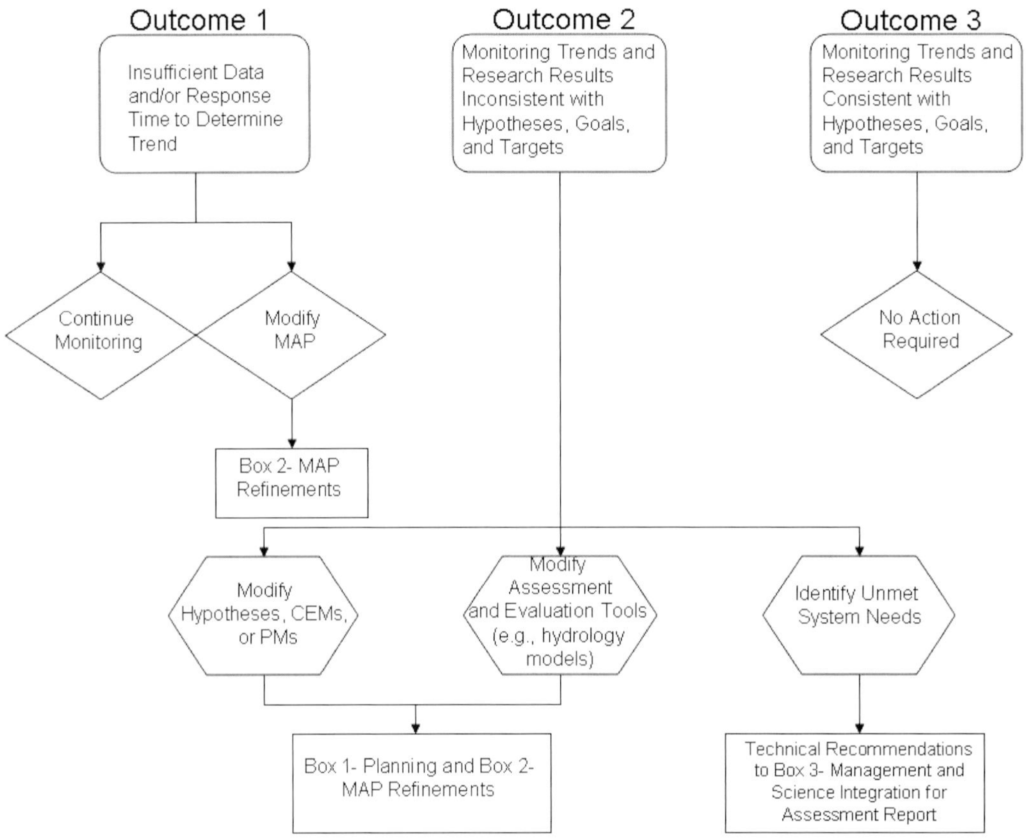

FIGURE 6-3 Application of System Status Report results to refinement of the monitoring plan.

SOURCE: Adapted from RECOVER (2007c).

in ecosystem response, changing implementation schedules, and changing management or societal priorities. With the completion of the first System Status Report and the uncertainties about funding for monitoring in the current fiscal climate (as discussed in the next section), RECOVER's efforts to revise MAP I are timely and clearly needed. Questions that should be addressed during the update might include:

- Based on analyses in the first System Status Report, what additional data sets might be needed for future assessments?

• Should the monitoring plan be based on modules and hypothesis clusters?
• Should the monitoring plan rely on complementary non-CERP monitoring programs?
• Can monitoring be accomplished more effectively by changing the spatial and temporal distribution of the sampling effort?
• Should performance measures for urban and agricultural systems be added to the monitoring plan?
• How can monitoring better support adaptive management?
• Does monitoring directly address the interim goals, and are the correct processes/drivers/stressors being measured so that when the interim goals are not met, the reasons for failing to achieve the goals can be determined?
• Should the conceptual models be redefined?

Financial Pressures on Monitoring Programs

At the current level of CERP funding ($10,000,000 per year), there are insufficient funds to support the system-wide monitoring plan as described in MAP I (RECOVER, 2004), even though this plan counts on leveraging project-level monitoring and monitoring being carried out by other federal, state, local, and private groups to supplement CERP-supported monitoring (RECOVER, 2004). Recent cuts to some of these complementary monitoring programs are of concern and emphasize that the CERP's monitoring program is vulnerable to changes in funding beyond its control (Sharfstein and Tipple, 2007). Within CERP, resolution of a dispute between project delivery teams and RECOVER over interpretation of CERP Guidance Memorandum No. 40[3] (USACE and SFWMD, 2008b) could add project-level ecological monitoring responsibilities to the MAP (E. Bush, USACE, personal communication, 2007), further stressing RECOVER's already limited monitoring budget. Monitoring is expensive, requiring investment in personnel, laboratory facilities and supplies, and access to field sites over an extremely large and often remote region. When costs rise, cutbacks in monitoring programs are often targeted to keep escalating costs down.

As NRC (2007) pointed out, the sustainability of the CERP monitoring plan over the long term would benefit from a reduced set of performance measures, but that committee also cautioned that there is danger in excluding too many measures, especially during the early stages of the CERP, until more is learned about which performance measures will be most useful. Thus, optimizing the

[3]CERP Guidance Memorandum No. 40 describes the requirements for incorporating monitoring and assessment activities and costs in planning, design, and implementation documents for CERP projects.

monitoring budget is challenging. If monitoring data are collected for too many performance measures, resources (money and manpower) are wasted for little gain in information. However, if the monitoring budget is cut too severely, critical information that is needed to guide adaptive management will be lost. Monitoring that is transferred from the individual projects to RECOVER will stretch an already thin monitoring budget even more. This will inevitably reduce the total number of performance measures, the frequency of measurement, and/or the spatial density of the measurements. If cuts to the monitoring program are sufficiently extensive, the information provided will be severely compromised, making it impossible to detect how the ecosystem is responding to restoration activities.

If monitoring and assessment information is going to be available to support adaptive management of the CERP, monitoring of ecosystem response to projects should be a priority. CERP monitoring responsibilities at all levels of project management across and within agencies should be clearly established. While monitoring in and of itself does not ensure restoration progress, without monitoring plans tailored to improve understanding of ecosystem response and the outcomes of project implementation from local to whole ecosystem scales, uninformed management decisions will be made with potentially undesirable ecosystem consequences. Investments in monitoring at the outset of restoration and during the entire restoration process are as important to the CERP as investments in construction projects.

INFORMATION AND DATA MANAGEMENT

Effective performance assessment in support of adaptive management relies on a well-designed information and data management system, which ensures that investments in monitoring are maximized and data are widely available to be utilized to the fullest extent. Primary responsibility for managing CERP data and information falls to the Information and Data Management (IDM) program, housed with the USACE in Jacksonville and the SFWMD in Ft. Lauderdale. As described in the *Program Management Plan: Information and Data Management* (USACE and SFWMD, 2007b), the CERP informatics strategy is based on a common information system (CERPZone) that enables Web-based sharing of information. CERPZone is not a centralized data repository for all CERP data. Instead, it is designed to facilitate search and retrieval of CERP data and documents over an "extranet" of participating agencies, universities, and other organizations. As of March 2008, 26 agencies and organizations were registered CERPZone users.

The CERPZone architecture is consistent with the principles laid out in the CERP Master Program Management Plan (USACE and SFWMD, 2000), which specifies general requirements such as tools for data and information sharing, networked servers for sharing documents, schedules, financial, scientific and geospatial data information, security, and content standards. The 2000 Master Program Management Plan also stipulates development of a document management and control system for tracking and documenting decisions affecting restoration design and implementation.

The IDM program currently operates with an annual budget of $6–7 million and is jointly funded by the SFWMD and the USACE. IDM activities are now closely coupled to those of the Interagency Modeling Center. In addition to developing and maintaining CERPZone, the IDM program, upon request and approval, provides tools and technical support to CERP projects. Examples of support activities include database administration, Web application development, GIS analysis and map production, and computing infrastructure support (e.g., server backups and data archiving). Thus, CERP activities such as RECOVER participate in CERPZone but are also "customers" that can request special services on a cost recovery basis (e.g., preparation of maps for the 2007 System Status Report).

CERPZone provides access to an impressive amount of information. Roughly 200,000 online documents and 1 Terabyte of data are catalogued in an electronic data catalogue (EDCat). Documents are stored and tracked using Documentum, an electronic document management system. A data access, storage, and retrieval (DASR) system and a GIS portal enable access to the data. A Model Management System has been developed to access model code and input and output data. Metadata management tools (e.g., Morpho[4]) are being used to standardize data documentation, and tools are being developed to support spatial queries (EDCat v. 2.0). Administrative tools are in place to manage CERPZone access, data storage requests, and technology project requests. A public Web site (*www.evergladesplan.org*) is maintained to share information with the general public. Future development plans through FY 2010 are listed in the April 2007 Program Management Plan (USACE and SFWMD, 2007b).

The CERP IDM program provides a sound structure and good functionality for collaborative CERP data and information management. Its effectiveness depends on user awareness of and competence in using available tools and applications, timely provision of research and monitoring data from primary data stewards to CERPZone, and community adherence to data standards and guidelines (these are articulated in a series of guidance memoranda).

Experiences of CERP staff to date indicate that the CERP information and data management system is working well for the documentation and retrieval of monitoring data. Most monitoring data and associated metadata collected under

[4] *http://knb.ecoinformatics.org/morphoportal.jsp.*

contracts executed with RECOVER are archived, discoverable, and accessible, and data gaps are mainly attributable to some individual principal investigators who have been slower to share their data. However, based on experience, assembling the 2007 System Status Report, integration of CERP monitoring and data, and data synthesis for ecosystem assessment are challenging and tools are needed to facilitate the production of standard assessments. Spatial query tools currently under development will help. CERP IDM program staff should work closely with RECOVER scientists to identify and prioritize products that should be routinely derived from CERP monitoring data or tools to help scientists produce such products.

MODELING IMPROVEMENTS IN SUPPORT OF ADAPTIVE MANAGEMENT

As discussed by NRC (2007), models are critical tools used in adaptive management to test the understanding of and to predict the ecological and hydrologic consequences of management alternatives and ecosystem drivers (e.g., rainfall, sea-level rise, climate change). The CERP was developed using simulation models to evaluate expected outcomes of various restoration scenarios. Both monitoring and modeling support the adaptive management process by providing information to allow informed alterations to the CERP during its implementation. The monitoring program will measure ecosystem response to restoration, and the modeling program provides a system-level context for integrating the responses. As abstract representations and simplifications of the complex real world, models are useful tools for integrating and updating current knowledge of a system and for identifying and prioritizing critical uncertainties.

South Florida restoration activities are supported by an enormous and impressive multiagency modeling effort. Numerous models have been or are being developed by researchers from agencies such as the SFWMD, USACE, other federal agencies, independent consultants, and academic institutions in the United States and elsewhere. The models vary in stage of development and application; some have been widely applied for evaluation and planning of CERP projects, while others are still being developed, calibrated, verified, or reviewed.

The following sections review the current state of restoration modeling of hydrologic and ecological systems and evaluate the status against modeling needs for effective adaptive management.

Hydrologic Modeling

The current system model for the Everglades restoration continues to be the SFWMM, developed and maintained by the SFWMD. Operating on a

2-mile-by-2-mile grid (hence "2 × 2 model"), the SFWMM simulates the vertical water balance of rain, evapotranspiration, and groundwater seepage as well as overland flow and flow at hydraulic structures. Generally operating on a daily time step, the model simulated a 31-year historic meteorological period (1965 to 1995) for the original CERP planning in the 1990s, whereas recent runs have extended the period of record by 5 years (1965 to 2000), with ongoing efforts to extend the record further forward in time. Extending the simulation period to a starting date prior to 1965 is complicated by fewer rain gages and other monitors, although conceivably this could be done through extrapolation and time series techniques. The physics of the 2 × 2 model are well documented, and the SFWMM has undergone extensive peer review (e.g., Bras et al., 2005).

However, the SFWMM consists of more than just a simulation of rainfall, runoff, and flow; the operations of the massive and complex SFWMD system are also incorporated into the model, as are operating rules for all CERP components. These operational algorithms are embedded in the Fortran code and are difficult to change and understand by all but the most familiar users. That is, if changes in operations of CERP components are to be simulated, the process involves hard-coding the changes into the SFWMM, rather than in the form of external model input. The SFWMD is well aware that this restriction limits the flexibility of the model with regard to straightforward evaluation of alternative plans and is working for a more flexible means to input operating rules in the next generation of CERP water simulation, namely, the Regional Simulation Model, or RSM (J. Obeysekera, SFWMD, personal communication, 2008).

The issue is exacerbated in that system response is more sensitive to operating rule changes than to most changes in the physics of the model. Because of the intricate way in which operations are coded into the SFWMM, it is difficult to tell what change in rules causes what change in output, as well as the relative impact of different operating rules within the system and the relative impact of operating rules versus, say, climate change. This again points to the urgency of differentiating policy from physics in the crucial CERP modeling tools, as is under way in the development of the RSM.

The RSM is intended to address this issue by incorporating two simulation engines: one for hydrology and hydraulics (the Hydrologic Simulation Engine, or HSE) that itself has substantial improvements in the physics, and one for management (the Management Simulation Engine, or MSE), for easy variation of operations. Two external panels (Bales et al., 2007; Chin et al., 2005) have recently reviewed the latest version of RSM (and its application to the natural system, the NSRSM). The RSM is approximately 2 or 3 years away from replacing the SFWMM for routine CERP analysis (e.g., first use for CERP as a whole may be in 2010), but it is being tested now in more self-contained basins north of

Lake Okeechobee (Van Zee, 2007) and is being applied successfully to a variety of local problems (e.g., the water conservation areas, Everglades National Park, and lower east coast [the so-called "Glades model"]; C-111; and Biscayne Bay coastal wetlands).

Further improvements in numerical modeling have been identified in a number of earlier reports (e.g., the strategic modeling plan of 2003 [Plato Consulting, Inc., 2003]; NRC, 2007; RECOVER 2005c). Included in these reports are statements that the need to model the "fate and transport of nutrients, sediments and nonpoint source pollution is imperative" (RECOVER, 2005c, p. 40). The SFWMD intends to eventually add nutrient and sediment transport models linked to the RSM. Another limitation of the RSM is that it is a two-dimensional model, where only the average flow in the vertical direction is considered. An eventual linkage of the RSM with local project-scale models to analyze fine-scale problems (e.g., sediment transport for tree island or seepage management) is in the planning stage. Limited staff and funding resources mean that these additional envisioned improvements to the RSM will not be accomplished for many years. Hydrologic and hydraulic applications for site-specific CERP project analysis will continue to be handled by local-scale models (e.g., hydraulic models for channels), often with the SFWMM or the RSM used to provide boundary conditions.

As recommended in the Strategic Modeling Plan (Plato Consulting, Inc., 2003), RSM development and application are using software engineering and project management processes based upon the Capability Maturity Model Integrated (CMMI)[5] developed by the Software Engineering Institute at Carnegie Mellon. CMMI is a process improvement model consisting of industry best practices that organizations use to improve their business, project management, and software and engineering processes in order to increase product quality, work efficiencies, and customer satisfaction. SFWMD modelers report that the implementation of CMMI processes within the Hydrologic and Environmental Systems Modeling Department of the SFWMD has brought tremendous value and improvements to the RSM, and future progress depends upon continuing these efforts. Similarly, the Interagency Modeling Center has been instrumental in achieving consistency of modeling efforts across agency boundaries. The design coordination team, consisting of SFWMD and USACE personnel, set the modeling priorities. The priority modeling tasks are then executed by the Interagency Modeling Center (J. Obeysekera, SFWMD, personal communication, 2008).

The pace of model development is limited by the staff resources available—a problem typical of all agencies in the CERP. Staff time for model development is also impacted by the need for production runs of the SFWMM and/or the RSM.

[5]Additional information on CMMI can be found online at *http://www.sei.cmu.edu/cmmi/*.

Possible reductions in SFWMD funding will not help this situation. One way in which agencies have tried to adapt to limitations on resources is to use the simplest model that will address specific needs. If the SFWMM is not needed to resolve a local question, then a simpler or more site-specific hydrologic or hydraulic model may suffice. One example of "simpler" is to use the SFWMM to develop regional boundary conditions surrounding a smaller application area, within which the more sophisticated RSM is used to provide small-scale, detailed modeling (J. Obeysekera, SFWMD, personal communication, March 2008). Another example of "simpler" is the use of a regional spreadsheet model in the analysis of Tamiami Trail modifications for the Modified Water Deliveries to Everglades National Park project (USACE and SFWMD, 2008a). The model uses measured flows as inputs to a 25-year simulation of flows entering Shark River Slough.

The Natural System Regional Simulation Model (NSRSM), the application of RSM to the predevelopment Everglades but with adjustments for topography and landscape constraints, is hampered (just as it has been for the NSM) by lack of predevelopment data with which to calibrate and verify simulation results. Ongoing SFWMD efforts to resolve predevelopment topography will aid in this effort.

Regarding application of models such as RSM to climate change scenarios, natural variability in rainfall patterns is partially accounted for by the current 36-year input rainfall time series. Current efforts to develop an input time series dating back to 1914 will certainly be an improvement in developing an understanding of long-term climate fluctuations and trends. In addition, for climate-change analysis, scenarios of future rainfall time series are needed, as well as changing sea-level boundary conditions.

The CERP could not have been developed without models like the SFWMM and cannot continue to progress without models like the RSM. It is clear that agency staff and managers recognize the value of modeling, and modeling is performed in every case in which it is needed. The principal impact of limited resources is to slow model development. Use of the SFWMM instead of the RSM because the RSM is not yet ready for application to the full CERP leads to evaluations that are likely less accurate and certainly more cumbersome. Technical challenges in model development continue to be addressed by a capable model development team.

Ecological Modeling

Ecological models linking hydrology and water quality to species and ecosystem dynamics have the potential to significantly improve the effectiveness of

CERP monitoring and assessment and to accelerate learning through adaptive management (DeAngelis et al., 2003; NRC, 2003b, 2007). A spectrum of models, ranging from qualitative conceptual models to complex dynamic simulation models, has been developed to link water management to ecological outcomes. While each has value, spatially explicit models that are logically coupled to the water management models and support landscape-level planning and decision making have been especially important to CERP (Sklar et al., 2001) Two ecological models in particular, the Across Trophic Level System Simulation (ATLSS) model and Everglades Landscape Model (ELM), were developed in large part to support restoration design, monitoring, and evaluation (DeAngelis et al., 1998; Sklar et al., 2001).

ATLSS is a set of spatially structured models operating at 0.01—1 km^2 resolution. Coarse (2 × 2) output from the SFWMM is downscaled to finer-scale hydrologic grids (currently the rate-limiting step in model operation) and then used in process models for lower trophic levels such as zooplankton and phytoplankton, structured population models for fish and macroinvertebrates, and individual-based models for large vertebrates such as Cape Sable seaside sparrows (CSSSs), wood storks, alligators, and white-tailed deer. ELM is a spatially structured "patch" model that couples hydrology, energy, and nutrients to species-specific plant growth and vegetation pattern across the landscape (Sklar et al., 2001).

Between 1997 and 2001 ATLSS was used intensively for evaluating CERP alternatives (*http://www.atlss.org/*). Researchers have continued to extend ATLSS capabilities, adding fire and vegetation succession modules and refining and testing spatially explicit species index models and population viability analyses for target species such as the snail kite, CSSS, and American alligator. Software has been developed to support model visualization and decision making, and ATLSS code has been revised to allow the model to be operated using parallel computing to improve scalability and reduce run time (Wang et al., 2006).

Despite the large investment in ATLSS models for restoration planning, they have not been used in CERP implementation and are not integrated with MAP monitoring efforts for model validation and refinement, integrated assessment, or for forecasting alternative management scenarios. This lack of integration may be due in part to technical complexity and computing requirements of the ATLSS models. Until recently, the expertise and infrastructure to operate ATLSS models existed only at the University of Tennessee, hindering integration with the water management models. Several modules (e.g., periphyton, crayfish, snail kite, and alligator modules) could be readily applied at the 2-mile-by-2-mile scales of current water management models (although the variable grid sizes of the RSM [0.1 to 2 miles] now under development would be much better for linking to the ecological models). ATLSS has recently been installed and is being operated at Everglades National Park, and the Department of the Interior (DOI) intends to locate one or two ecological modelers at the Interagency Modeling Center to operate ATLSS and other ecological models. The latter would be a positive step, as limited DOI involvement has precluded coupled hydrologic-ecological modeling at the Interagency Modeling Center. Given DOI's relatively low level of investment in modeling staff, it will be probably at least 2 years before ATLSS will be operational at the Interagency Modeling Center.

Recent development of ELM (version 2.5) has focused on linking water management scenarios to regional water quality outcomes, specifically total phosphorus concentrations in surface water and net phosphorus accumulation in the Greater Everglades ecosystem. An external peer review of ELM 2.5 concluded that the model formulation was fundamentally sound and relatively unbiased. The peer-review panel argued for close integration of ELM with CERP monitoring and adaptive management and stated that ELM "may be the primary method that SFWMD should use to guide any monitoring changes in the future, not just for ELM itself, but to monitor progress in restoring the Everglades" (Mitsch et al., 2007). Nevertheless, ELM was not integrated into system status monitoring and reporting for 2006 and 2007.

In summary, ecological models have seen diminishing use in CERP since the development of the CERP during the mid-1990s (the Restudy). The trend is not due to technical problems with the models and is not a hiatus while next-generation models are developed; rather, it reflects reduced staffing for model application and development. The committee can only echo previous NRC reports in stating that integrated hydro-ecological modeling has an important role in project planning, monitoring, assessment, and adaptive management. To improve the application of ecological models for the CERP planning and management, the DOI needs to invest more attention and resources in ecological modeling and data management activities at the Interagency Modeling Center.

CONCLUSIONS AND RECOMMENDATIONS

To facilitate restoration progress despite some scientific and engineering uncertainty, Congress mandated an adaptive management approach for the CERP. Adaptive management requires the support of effective monitoring and assessment protocols and adequate hydrologic and ecological models. In this chapter, recent progress and major issues with respect to CERP science to support the adaptive management process were reviewed. The major findings are highlighted below.

The RECOVER team has now produced nearly all of the elements needed to implement a decision-making framework using adaptive management to assess scientific uncertainty. Documents describing the adaptive management process (RECOVER, 2006a; 2007e), and all aspects of performance assessment (i.e., the monitoring plan [RECOVER, 2004], an assessment plan [RECOVER, 2006f], performance measures [RECOVER, 2007b], and quality assurance requirements [RECOVER, 2007d]) are completed. Conceptual ecological models that are the foundation of the monitoring and assessment documents have been peer-reviewed and published. The information management and data management system and the Interagency Modeling Center are actively developing tools to support the assessment and planning aspects of decision making and assisted in production of the 2007 System Status Report, the first in a series of assessment reports that documents the ecosystem response to implementation of CERP projects. The System Status Reports are a critical component of the adaptive management strategy; they are the vehicle used to transmit new scientific information to restoration managers.

These are significant accomplishments, and their importance should not be underestimated. However, the CERP adaptive management scheme could be improved by addressing several major issues, which are summarized below.

In order for monitoring and assessment information to adequately support CERP adaptive management, a robust program of ecological monitoring should remain a priority. While monitoring in and of itself does not ensure restoration progress, without monitoring to understand ecosystem response to project implementation from local to whole ecosystem scales, uninformed management decisions will be made with potentially undesirable ecosystem consequences. A well-justified and documented set of performance measures has been developed, and a scientifically robust process for updating, refining, and adding to the set of performance measures is in place. RECOVER should continue to move forward to fully develop those performance measures that are currently monitored and to reconcile performance measure standards for evaluation and assessment. The periodic review of performance measures should consider ways to make sure that the total number of variables monitored is appropriate to their purpose for informing decisions and to the funding available for monitoring efforts. It also is important to match the frequency of monitoring with the speed of change of the variables that are monitored and to increase reliance on remotely sensed data collection methods. Revisions of the monitoring and assessment system should be firmly grounded in the use of the data for planning and management decision making.

The 2007 System Status Report achieved its stated objectives to test the monitoring and assessment plan and establish as long a baseline as possible to capture the natural variance of CERP performance measures. In most cases, the System Status Report also provided some discussion of the ability to detect performance measure change or trends. In doing so, the first System Status Report serves as the reference that will be used to gauge system response as CERP projects are implemented, and it is extremely valuable. Insights learned during the production of the report should be incorporated into the revision of the Monitoring and Assessment Plan (MAP I) and the conceptual ecological models, as needed, and for reprioritization of the performance measures.

To maximize the usefulness of System Status Reports for adaptive management, RECOVER should develop succinct summaries in future reports that clearly address whether the interim and longer-term goals are being met; if not, why not; and what CERP operations or design changes are most likely to move ecosystem response closer to the interim goals. RECOVER should also aim for a more coordinated approach in future assessments that allows for additional consideration of restoration effects across modules.

The CERP Information and Data Management program provides a sound structure and good functionality for collaborative CERP data and information management. Performance assessment depends on a well-designed information and data management system so that monitoring data can be widely available and utilized to the fullest extent. The effectiveness of the Information and Data Management program depends on user awareness of and competence in using available tools and applications, timely provision of research and monitoring data, and community adherence to the articulated data standards and guidelines. Based on experience assembling the 2007 System Status Report, it appears that timely integration of CERP monitoring data remains challenging, and efforts to develop tools to facilitate the production of standard assessments should continue.

Integrated hydrologic, ecological, and water quality modeling tools are needed for science to have a fully developed role in CERP decision making and ecosystem management. CERP planning and assessment of performance indicators are dependent on the modeling tools; as model development and implementation lag, so does access to more accurate and functional tools. Models are needed for each ecological indicator (performance measures) to compare predicted and monitored indicator responses to the CERP for effective adaptive management decision making. This will only occur when:

- ecological modeling and data management activities are fully incorporated and funded in the CERP's Interagency Modeling Center,
- Water quality and sediment transport models become routinely available and integrated with the new Regional Simulation Model (RSM), and

- These physical-chemical models can be readily linked to ecological models.

Linking of models (the third point above) continues to be a particularly slow endeavor. Impediments to achieving the synergy between planning, monitoring, and assessment needed to support effective adaptive management are related to shrinking resources, loss of staff, and time. Shrinking CERP resources mean that the trade-off between use of staff for model development versus for model production runs for CERP planning favors the latter. Moreover, if staff numbers are reduced, the knowledge and training of departing professionals go with them. This committee recognizes that resources are limited but notes that model development is a long-term proposition and should continue with as much support as possible so the tools required to restore and manage the ecosystem are available in the future.

7

Synthesis of CERP Progress

The Comprehensive Everglades Restoration Program (CERP) is a project of grand vision and great ambition to restore the Everglades, but the project is bogged down in budgeting, planning, and procedural matters and is making only scant progress toward achieving its goals. Meanwhile, the ecosystems that it is intended to save are in peril.

In this chapter, the committee synthesizes its evaluation of the CERP by standing back from the details and outlining in the broadest terms the reason for these critical conclusions. The chapter begins with science and engineering issues, followed by a summary of the management and policy issues that affect restoration progress. It concludes with the committee's assessment of what is needed in the near future to reverse unfavorable trends and achieve effective restoration.

This chapter has three major points: (1) the condition of the Everglades ecosystem is declining; (2) the CERP is entangled in procedural matters involving federal approval of projects and lacks consistent infusions of financial support from the federal government; and (3) without rapid implementation of the projects with the greatest potential for Everglades restoration, the opportunity for meaningful restoration may be permanently lost.

The South Florida ecosystem at certain times and places has too much water, and at others it has too little. From its inception, the primary goal of the CERP has been to "get the water right," under the presumption that if water controls could adequately replicate pre-drainage hydrology in the remnant Everglades, a sustainable and functioning ecosystem would follow. Getting the water right, however, is not as simple as it sounds. There are formidable water-related issues that remain unresolved in the restoration effort, including where and how some of that water should be stored, transported, and delivered, and when such deliveries should occur. The magnitude, duration, frequency, and timing of flows are all keys to a functioning, sustainable restored ecosystem, but managers are still experimenting with how these hydrologic characteristics might best lead to restoration. Getting the water right also has important social dimensions (e.g., water supply, flood control) that may or may not work in concert with improving environmental conditions.

In addition to water quantity, water quality is also a continuing issue in Everglades restoration. Excess phosphorus is the contaminant of greatest concern, with continuing inputs from some agricultural areas in the Everglades watershed. Although these contributions of phosphorus are less than they would have been without the tremendous effort on the part of the state of Florida to create vast stormwater treatment areas (STAs), phosphorus concentrations in the waters of the South Florida ecosystem remain at unhealthy levels for its native plant communities. High phosphorus concentrations in Lake Okeechobee also limit the volume of water that can be moved south into the Everglades ecosystem, and phosphorus-laden sediments are likely to be a continuing source of contamination for several decades.

Successful restoration of the Everglades depends upon consideration of the many interconnected parts of this extensive and integrated watershed. If water quantity or quality is altered in the Kissimmee River, the effects of the change are transmitted to Lake Okeechobee, the Caloosahatchee and St. Lucie rivers and their estuaries on the east and west coasts, and to the remnant Everglades ecosystem, including Everglades National Park far to the south. Similarly, decisions about the planning and funding of individual projects in the CERP affect not only the individual projects but also distant components of the system. Given current fiscal constraints, funds spent in one location may imply less funding for other locations. These circumstances necessitate clear restoration priorities and a system-wide vision for restoration. Continuation of the current piecemeal approach to planning, authorizing, and funding CERP projects will make successful restoration unlikely.

CONTINUING DETERIORATION OF THE NATURAL SYSTEM

The creation of the CERP plan in the late 1990s was a response to a broad recognition among the public, interest groups, governmental managers, and state and national legislators that the Everglades ecosystem was in serious decline. Today, as described in Chapter 2, the decline continues, so that failure to press forward with restoration results in additional deterioration to the natural system. To do nothing is, in fact, to do harm. The nation risks losing some populations of iconic wildlife associated with the Everglades. Populations of some bird species, including the Cape Sable seaside sparrow and the snail kite, are at risk. Wading birds have redistributed themselves to new locations outside their former ranges that included Everglades National Park, one of the jewels of the national park system. Native species struggle to compete with a daunting array of exotic invasive species, and releases of new exotic and invasive species into the Everglades continue unabated. Even basic functional components of the Everglades landscape are in jeopardy because dikes, levees, and roadway embankments have eliminated sheet flow from many critical areas of the Everglades: the tree islands that dot the River of Grass have declined in number and area during the 20th century, and the ridge and slough topography is gradually degrading. Resolving these issues will require major restoration actions that attempt to reverse the impacts of development and decades of hydrologic alterations and water management that did not focus primarily on the Everglades ecosystem.

A sense of urgency in the need to move forward quickly with the CERP is partly a product of concern about the declining ecosystem and its lack of resiliency, but two other forces further reinforce the need for prompt action: escalating costs and development. Costs of construction materials (particularly cement), labor, and land are all escalating in South Florida, and rising fuel costs also force increases in construction costs. Population increase in South Florida continues to produce urban sprawl, with its attending conversion of natural and agricultural landscapes into urban and suburban development. The density of settlement in the region results in increased water demands that make CERP more difficult to accomplish.

The CERP was launched in 2000 in an effort to get the water right in the South Florida ecosystem and for Everglades National Park, but this project is founded upon numerous restoration projects that, although they are not formally part of the CERP, are nonetheless central to the success of the CERP in achieving its restoration goals. For example, the effort to provide more water passing under Tamiami Trail and rehydrate the northeastern portion of Everglades National Park (Mod Waters, see Chapter 4) is essential to restoring appropriate flow volumes, velocities, and water distribution in Everglades National Park and to remedying several endangered species issues (see Chapter 2). At present, the Mod Waters project remains unfinished nearly 20 years after authorization. The recently recommended Tamiami Trail alternative represents only a small step in the right direction. If the restoration goals are to be achieved, the project should be designed and perceived only as a step toward ultimate construction of a plan that provides more environmental benefits in conjunction with CERP implementation or an alternative mechanism.

Progress of the CERP primarily has been made in planning and establishing the administrative framework for the restoration and its adaptive management, and a tremendous amount of effort has been put forth to date. However, natural system restoration benefits from the CERP itself are extremely limited, because no CERP construction projects have been completed. Nevertheless, in a broader context, efforts to restore the Everglades include some significant accomplishments. The non-CERP Kissimmee River restoration project in the Northern Everglades has re-created some of the basic terrain elements of a meandering river connected to its flood plain through occasional flooding, and the removal of a dam in its lowest reaches has improved hydrologic connectivity within the system. STAs have had success in reducing phosphorus transport into the Everglades ecosystem, and treatment areas such as these will be essential to successful restoration. Land purchases by the state of Florida have made good progress toward obtaining the necessary spaces for project components. The state's Acceler8 program has attempted to move some projects forward. The CERP faces many challenges in achieving its objectives, including climate change, population growth, urban sprawl, and water quality. But the committee is optimistic that, given the political will, these challenges can largely be addressed by regional planning and by embracing adaptive management and scientifically informed decision making. The results of the restoration may not be exactly those that were envisioned, and some trade-offs may be necessary, but the committee expects that CERP efforts, if implemented under an effective adaptive management framework and—above all—if undertaken expeditiously, will create more resilient ecosystems that should fare better in facing future environmental stresses.

SCIENTIFIC KNOWLEDGE AND CERP PROGRESS

Scientific knowledge is a critical foundation for decisions about the planning, design, and adaptive management of the CERP. Over the past two decades, knowledge about the Everglades ecosystem has expanded enormously, and CERP planners, engineers, and scientists now have significant data to guide the restoration. Recent research has produced substantial new knowledge, such as a more sophisticated understanding of water flow and its importance in the Everglades landscape, and this new knowledge is being employed to inform the planning process. Although much of the knowledge about Everglades functioning includes some uncertainty, this NRC committee concurs with the findings of its predecessor NRC committee: There are no gaps in knowledge or ranges of uncertainties that are large enough that they should impede CERP progress. In many cases, scientific knowledge will improve and uncertainty will decrease as project construction moves forward in an adaptive management framework.

When stakeholders and agencies use uncertainties as an excuse to halt progress in contentious projects, an incremental adaptive restoration (IAR) approach, as described in NRC (2007), may help advance restoration progress while fostering learning that can help resolve project-planning conflicts and improve future project planning and design. In an IAR approach, large complex projects (or a combination of multiple interrelated projects) are undertaken in increments, with each step providing substantial, measurable restoration and new learning benefits. These increments are small enough to allow completion of the step in a relatively short period of time (about 5 years or less), but large enough to provide significant restoration benefits and to supply new knowledge to inform future project planning. CERP planners are considering some applications of IAR, tailoring it to their particular needs, and their preliminary assessment is that the concept is a helpful management tool.

FUNDING AND IMPLEMENTATION PROBLEMS THAT LIMIT RESTORATION PROGRESS

As reported in NRC (2007), one of the major political advantages enjoyed by the CERP as it was developed was the strong and united support of South Florida stakeholders. These groups—including nongovernmental organizations, Native American tribes, and state and federal agencies—have not always agreed with one another, but they were able to reach accord on major issues concerning the direction of the CERP. In the past, this agreement could be reached because the CERP, in effect, offered something for everyone. Now the South Florida coalition is experiencing new strains. Presentations by representatives of these groups to this committee indicate that slow-to-develop funding from the federal government and cost escalations have led to a well-grounded fear that the CERP may not be completed as envisioned, which is leading to new tensions among stakeholders. Recognition is dawning that some CERP projects and outcomes may be left out and that some stakeholder groups might not get what they wanted from the project. This is already the case with the Regional Water Availability Rule, which places increased responsibility on cities to find new water supplies to meet growing needs and to protect water for the environment. Mod Waters (see Chapter 4) is another example of a project with objectives sharply compromised by cost escalations and the imposition of a budget constraint. The maintenance of the stakeholder coalition is a key to the achievement of CERP's restoration goals, and CERP planners will need to invest more effort in maintaining common goals for the overall project.

Although lack of timely restoration progress by the CERP, to date, has been largely due to the complex federal planning process and the need to resolve conflicts among stakeholders, future progress is likely to be limited by the availability of funding and an authorization and funding mechanism that was not designed for a project of this magnitude and complexity and seems ill suited for it (see Chapter 3). When the CERP was initiated in 2000, restoration planners anticipated that the U.S. Congress would provide project authorization through

Water Resource Development Acts (WRDAs) every 2 years, with funding to follow in U.S. Army Corps of Engineers (USACE) appropriations bills. In fact, WRDAs emerged only in 2000 and 2007. Another 7-year hiatus until the next WRDA bill would be potentially devastating to restoration progress. Only three CERP projects (Picayune Strand, Site 1 Impoundment, and Indian River Lagoon) have been authorized to date (beyond those projects conditionally authorized in WRDA 1999 and 2000), and only authorized CERP projects with approved project implementation reports (PIRs) can receive federal appropriations.

In periods of restricted funding and limited capability to move forward on many fronts for restoration, the ability to set priorities and implement them is critical. Yet, restoration priorities have not been established. The latest project schedule is already out of date given current project progress and availability of funding, although a revised implementation schedule is in development. In addition, the present project authorization and funding system lacks consideration of the restoration effort as a whole. Three projects have survived the rigorous project approval and authorization process, and each of these three projects will provide restoration benefits. However, the committee has seen no systematic analysis showing that they provide the greatest restoration benefits for the natural system or that they deserve the highest priority for funding compared to other components. Instead, they are minimally contentious projects with the strongest stakeholder support. As a result, in the years ahead, CERP planners may be forced to choose between using limited available federal funds on authorized projects and conserving federal funds for higher-priority projects that may not be authorized for years to come.

Meanwhile, Florida has aggressively funded CERP projects, so that what was envisioned as a state-federal partnership has become highly unequal. Because Florida is providing funding through land acquisition, Acceler8, and its Northern Everglades initiative, state objectives (particularly improving the condition of Lake Okeechobee and the northern estuaries) are likely to be achieved faster than restoration to benefit areas with stronger federal interests, such as Everglades National Park. And as delays in construction continue, costs increase, which potentially leave the federal government with an even greater funding burden.

Program management processes should support an efficient process of designing, planning, and implementing individual components of the CERP, but sometimes these processes are instead impediments to progress. Congressionally mandated procedures for USACE were designed for administering individual projects, but the CERP is a conglomeration of more than 60 components, all designed to operate within an integrated system. As a result, project teams plan and implement components one at a time without following the logical dictates that derive from recognition of the interconnectedness of these projects with each other or with an overall watershed or ecosystem. Meaningful assessments of project benefits in such a complex system (i.e., next added increment) cannot be done one piece at a time. As a result, new program management, decision-making, and funding approaches are needed to support more timely completion of the CERP (see Chapter 3).

There is considerable frustration about the administrative process among managers, decision makers, and researchers in South Florida. To many of them, it appears that planning rather than doing, reporting rather than constructing, and administering rather than restoring are consuming their talents and time. Improved federal and state procedures that recognize the interconnected nature of the components of the CERP and allow meaningful priorities to be implemented (see Chapter 3) will reinvigorate the restoration mission and its most important personnel.

THE CERP AND THE PUBLIC

The general public is cognizant of the poor condition of Everglades National Park, and many understand that degraded habitats exist far upstream of the park, throughout South Florida. Regardless of citizens' specific expectations, it is essential to demonstrate some restoration progress to sustain the public's support. Restoration is expensive and time-consuming, and the scale of Everglades restoration is so large and complex that it will require decades for completion even under the most-efficient circumstances. Demonstrable restoration progress has been lacking, and without such progress, regional and national support for the initiative may falter.

Strong political leadership is essential to support and maintain restoration progress. Elected officials and agency leaders can strengthen public support for this important mission. Every participating party will not obtain its desired objectives completely, but strong restoration leadership could harness the collective support for restoration progress to work through the difficult decisions that have to be made. Strong restoration leadership can also identify impediments to effective restoration, such as unclear priorities and episodic funding, and develop programmatic solutions to improve restoration progress. Without such leadership, the CERP will face substantial additional delays, risking the loss of public confidence and support and leaving a treasured ecosystem to continue its perilous decline.

CONCLUSIONS

If the sweeping vision of environmental restoration of the Everglades is to be realized, demonstrable progress must come soon. Heretofore, management of the Everglades has resulted in its diminution, and the CERP has not, to date, been effective in halting the decline of the South Florida ecosystem. If the CERP continues on its present course at its current pace, the ecosystem will continue to lose its resiliency, which could lead to rapid and deleterious changes that might be very difficult or impossible to reverse, and more importantly, the restoration effort will lose the support of the public at large. Clear funding priorities, modifications to the project planning, authorization, and funding process to support system-wide restoration goals, and strong leadership are needed to move the restoration forward and begin to reverse the decades of decline. To do nothing is to do harm. To find ways to press forward with the CERP is a statement by this generation to future generations that we accept responsibility for the restoration of the Everglades as one of the nation's priceless ecological treasures.

References

1000 Friends of Florida. 2006. Florida in 2060, Not a Pretty Picture: An Executive Summary. Available online at *http://www.1000friendsofflorida.org/PUBS/2060/2060-executive-summary-Final.pdf*. Accessed January 11, 2008.

Abbott, L., T. Barnes, R. Bennett, R. Chamberlain, T. Coley, T. Conboy, C. Conrad, K. Cunniff, P. Doering, M. Gostel, D. Haunert, K. Haunert, M. Hedgepeth, G. Hu, M. Hunt, S. Kelly, C. Madden, A. McDonald, R. Robbins, D. Rudnick, E. Shornick, P. Walker, and Y. Wan. 2007. Chapter 12: Management and Restoration of Coastal Ecosystems. In 2007 South Florida Environmental Report. Available online at *https://my.sfwmd.gov/pls/portal/docs/PAGE/PG_GRP_SFWMD_SFER/PORTLET_PREVREPORT/VOLUME1/vol1_table_of_contents.html*. Accessed July 19, 2008.

Abtew, W., R. S. Huebner, C. Pathak, and V. Ciuca. 2007. Appendix 2-2: Stage-Storage Relationship of Lakes and Impoundments. In 2007 South Florida Environmental Report. Available online at *https://my.sfwmd.gov/pls/portal/docs/PAGE/PG_GRP_SFWMD_ SFER/PORTLET_PREVREPORT/VOLUME1/appendices/v1_app_2-2.pdf*. Accessed July 19, 2008.

Achenbach, J. 2008. Florida Moves to Restore Wetlands: Sugar Corp. Purchase Would Aid Everglades. Washington Post. June 25, 2008, p. A1.

Ad Hoc Senior Scientists. 2007. Draft Recommendations and Conclusions from an Ad-hoc Senior Scientists Workshop on Comprehensive Everglades Restoration Plan (CERP) "Restoration Priorities." September 14, 2007. Miami, FL: Florida Atlantic University.

Adorisio, C., C. Bedregal, J. Gomez, J. Madden, C. Miessau, D. Pescatore, P. Sievers, S. Van Horn, T. van Veen, and J. Vega. 2007. Phosphorus source controls for the basins tributary to the Everglades Protection Area. In 2007 South Florida Environmental Report. Available online at *http://www.sfwmd.gov/sfer/SFER_2007/index_draft_07.html*. Accessed August 21, 2008.

Alexander, R. B., E. W. Boyer, R. A. Smith, G. E. Schwarz, and R. B. Moore. 2007. The role of headwater streams in downstream water quality. Journal of the American Water Resources Association 43: 41–59.

Alvarado, M. A., and S. Bass. 2007. Regional water bird abundance: Everglades National Park area. Pages 32–37 in M. I. Cook and H. K. Herring, eds., South Florida Wading Bird Report. Volume 13. October 2007. West Palm Beach, FL: South Florida Water Management District.

Alvarez, J., G. D. Lynne, T. H. Spreen, and R. A. Solove. 1994. The economic importance of the EAA and water quality management. In A. B. Buttcher and F. T. Izuno, eds., Everglades Agricultural Area (EAA): Water, Soil, Crop, and Environmental Management. Gainesville, FL: University Press of Florida.

Alverez, C., M. Eng, and A. Mayes. 2002. Assessment of Opportunities for Multi-Stakeholder Collaboration on the Environmental Impact Statement for the Combined Structural and Operational Plan for Modified Water Deliveries to Everglades National Park and C-111 Canal Projects. Available online at *http://www.ecr.gov/pdf/everglades_final_report.pdf*. Accessed August 22, 2008.

Appelbaum, S. 2004. Power Point: Comprehensive Everglades Restoration Plan—Briefing for CISRERP. Jacksonville, FL: U.S. Army Corps of Engineers.

Appelbaum, S. 2008a. South Florida Ecosystem Restoration Integrated Delivery Schedule. Presentation to the South Florida Ecosystem Restoration Task Force, May 21, 2008. Available online at *http://www.sfrestore.org/tf/minutes/2008_meetings/21,22may08/3a_IDS%20Presentation%20 5May08.pdf.* Accessed May 8, 2008.

Appelbaum, S. 2008b. Modified Water Deliveries to Everglades National Park: Tamiami Trail Modifications. Limited Reevaluation Report Working Group Briefing. Presentation to the South Florida Ecosystem Restoration Working Group and Science Coordination Group, April 28, 2008. Available online at *http://www.sfrestore.org/wg/wgminutes/2008meetings/28,29apr2008/ 2_MWD-TTM_wg_april_2008.pdf.* Accessed August 20, 2008.

A.R. Marshall Foundation. 2007. Risk Assessment of the Storage Approach to Restoration in the Northern Everglades Watershed (EAA Region), Summary of White Paper presented to CISRERP, August 29, 2007. Palm Brach, FL: A.R. Marshall Foundation.

Armentano, T. V., et al. 2002. Vegetation pattern and process in tree islands of the southern Everglades and adjacent areas. Pages 225–282 in F. H. Sklar and A. van der Valk, eds., Tree Islands of the Everglades. Boston, MA: Kluwer Academic.

Audubon of Florida. 2007. Lake Okeechobee: Everything in Harmony, Restoration Needs. Available online at *http://www.audubonofflorida.org/pubs_OkeechobeeReport.html.* Accessed May 22, 2008.

Baiser, B., and J. L. Lockwood. 2006. The influence of water level on the breeding success of Cape Sable Seaside Sparrows (*Ammodramus maritimus mirabilis*): Getting the water right for the recovery of an endangered species in the Everglades. Pages 47–48 in J.L Lockwood, B. Baiser, R. Bolton, and M. Davis. Detailed Study of Cape Sable seaside sparrow nest success and causes of nest failure, 2006 Annual Report.

Balbin, M. 2008. Water Use Efficiency Planning with Conserve Florida Guide. Available online at *http://www.sfwmd.gov/pls/portal/docs/PAGE/PG_GRP_SFWMD_NEWS/PORTLET_ CONSERVESUMMIT/TAB19809840/CONSERVEFLORIDAGUIDE_3_28_0 8_MARIBELBALBIN. PDF.* Accessed July 30, 2008.

Bales, J. D., R. L. Bras, W. Graham, L. Gunderson, and P. Stone. 2007. Scientific Peer Review of the Natural System Regional Simulation Model (NSRSM) v2.0, Panel Final Report, July 9, 2007. Report to South Florida Water Management District, West Palm Beach, FL. Available online at *https://my.sfwmd.gov/pls/portal/docs/PAGE/PG_GRP_SFWMD_HESM/PORTLET_RSM_ PEERREVIEW/TAB2564291/NSRSMFINALV3.PDF.* Accessed June 15, 2008.

Bean, M. J. 2006. Second generation approaches. Pages 274–285 in D. Goble, M. J. Scott, and F. W. Davis, eds., The Endangered Species Act at Thirty: Renewing the Conservation Promise. Washington, DC: Island Press.

Beisner, B. E., D. T. Haydon, and K. Cuddington. 2003. Alternative stable states in ecology. Frontiers in Ecology and the Environment 1: 376–382.

Beissinger, S. R. 1986. Demography, environmental uncertainty, and the evolution of mate desertion in the Snail Kite. Ecology 67: 1445–1459.

Beissinger, S. R. 1990. Alternative foods of a diet specialist, the Snail Kite. Auk 107: 327–333.

Beissinger, S. R. 1995. Modeling extinction in periodic environments: Everglades water levels and Snail Kite population viability. Ecological Applications 5: 618–631.

Beissinger, S. R., and N. F. R. Snyder. 2002. Water levels affect nest success of the snail kite in Florida: AIC and the omission of relevant candidate models. Condor 104: 208–215.

Benoit, J. M., C. C. Gilmour, A. Heyes, R. P. Mason, and C. L. Miller. 2003. Geochemical and biological controls over methylmercury production and degradation in aquatic ecosystems. Pages 262–297 in Y. Cai and O. C. Braids, eds., Biogeochemistry of Environmentally Important Trace Elements. Washington, DC: American Chemical Society.

Blake, N. 1980. Land into Water—Water into Land: A History of Water Management in Florida. Tallahassee, FL: University Presses of Florida.

Brandt, L. A., J. E. Silveira, and W. M. Kitchens. 2002. Tree islands of the Arthur R. Marshall Loxahatchee National Wildlife refuge. Pages 311–335 in F. H. Sklar and A. van der Valk, eds., Tree Islands of the Everglades. Dordrecht, The Netherlands: Kluwer Academic Publishers.

Bras, R. L., A. S. Donigian, W. Graham, V. Singh, and J. Stedinger. 2005. The South Florida Water Management Model, Version 5.5, Review of the SFWMM Adequacy as a Tool for Addressing Water Resources Issues, Final Panel Report to South Florida Water Management District, West Palm Beach, Fla. Available online at *https://my.sfwmd.gov/pls/portal/docs/PAGE/PG_GRP_SFWMD_HESM/PORTLET_SFWMM/TAB2632161/FINAL_PEER_REVIEW_REPORT_1DEC05.PDF*. Accessed June 15, 2008.

Breckling, B., and Q. Dong. 2000. Uncertainty in ecology and ecological modeling. In S. E. Joergensen and F. Muller, Handbook of Ecosystem Theories and Management. Boca Raton, FL: CRC Press.

Brezonik, P. L., and D. R. Engstrom. 1998. Modern and historic accumulation rates of phosphorus in Lake Okeechobee, Florida. Journal of Paleolimnology 20: 31–46.

Brooks, H. K. 1974. Lake Okeechobee. Pages 256–286 in P. J. Gleason, ed., Environments of South Florida: Past and Present. Miami, FL: Miami Geological Survey.

Burnham, W., T. J. Cade, A. Lieberman, J. P. Jenny, and W. R. Heinrich. 2006. Hands-on restoration. In D. D. Goble, J. M. Scott and F. W. Davis, eds., The Endangered Species Act at Thirty, Volume 1, Renewing the Conservation Promise. Washington, DC: Island Press.

Carpenter, S. R. 2003. Regime Shifts in Lake Ecosystems: Pattern and Variation. Oldendorf/Luhe, Germany: Inter-Research.

Carpenter, S. R., and L. H. Gunderson. 2001. Coping with collapse: Ecological and social dynamics in ecosystem management. Bioscience 51: 451–457.

Cassey, P., J. L. Lockwood, and K. H. Fenn. 2007. Using long-term occupancy information to inform the management of Cape Sable seaside sparrows in the Everglades. Biological Conservation 139: 139–149.

Cave, D. 2008. A dance of environment and economics in the Everglades. New York Times, July 31, 2008, pp. A14, A16.

Chamberlain, R., and D. Hayward. 1996. Evaluation of water quality and monitoring in the St. Lucie Estuary, Florida. Journal of the American Water Resources Association 32: 681–696.

Chesapeake Bay Program. 2002. The State of the Chesapeake Bay: A Report to the Citizens of the Bay Region. Annapolis, MD: Chesapeake Bay Program.

Chiang, C., C. B. Craft, D. W. Rogers, and C. J. Richardson. 2000. Effects of four years of nitrogen and phosphorus additions on Everglades plant communities. Aquatic Botany 68: 61–78.

Childers, D. L., R. D. Jones, J. C. Trexler, C. Buzzelli, J. Boyer, A. L. Edwards, E. E. Gaiser, K. Jayachandaran, D. Lee, J. F. Meeder, J. Pechmann, J. H. Richards, and L. J. Scinto. 2001. Quantifying the effects of low level phosphorus enrichment on unimpacted Everglades wetlands with in situ flumes and phosphorus dosing. In J. Porter and K. Porter, eds., *The Everglades, Florida Bay, and Coral Reefs of the Florida Keys*. Boca Raton, FL: CRC Press.

Chin, D. A., J. A. Dracup, N. L. Jones, V. M. Ponce, R. W. Schaffranek, and R. Therrien. 2005. Peer Review of the Regional Simulation Model (RSM). Available online at *https://my.sfwmd.gov/pls/portal/docs/PAGE/PG_GRP_SFWMD_HESM/PORTLET_RSM_PEERREVIEW/TAB2564283/FINAL_REPORT_9_23_05.PDF*. Accessed June 15, 2008.

Cook, M. I., and H. K. Herring, Editors. 2007. South Florida Wading Bird Report. Volume 13. West Palm Beach, FL: South Florida Water Management District.

Cooke, G. D., E. B. Welch, A. B. Martin, D. G. Fulmer, J. B. Hyde, and G. D. Schrieve. 1993. Effectiveness of Al, Ca, and Fe salts to control the internal loading in shallow and deep lakes. Hydrobiologia 253, 323–35.

Craft, C. B., J. Vymazal, and C. J. Richardson. 1995. Response of Everglades plant communities to nitrogen and phosphorus additions. Wetlands 15: 258–271.

Crozier, G., and M. Cook, eds. 2004. South Florida Wading Bird Report, Volume 10. Available online at *http://www.sfwmd.gov/pls/portal/docs/PAGE/PG_GRP_SFWMD_WATERSHED/PORTLET%20-%20EVERGLADES%20FLORIDA%20BAY/TAB1832037/WADINGBIRDREPORT04.PDF.* Accessed August 15, 2008.

Curnutt, J. L, J. Comiskey, M. P. Nott, and L. J. Gross. 2000. Landscape-based spatially explicit species index models for Everglades restoration. Ecological Applications 10: 1849–1860.

Darby, P. 2007. Task 3: Examination of interactions between apple snails and habitat structure and hydrology throughout the range of the snail kite, including a review of apple snail abundance and availability as prey items for the kite. Unpublished status report to the U.S. Fish and Wildlife Service, Vero Beach, FL. June 22, 2006.

Darby, P. C., L. B. Karunaratne, and R. E. Bennetts. 2005. Technical Report: The influence of hydrology and associated habitat structure on spatial and temporal patterns of apple snail abundance and recruitment. Pensacola, FL: University of West Florida/U.S. Geological Survey.

Darby, P. C., R. E. Bennetts, and L. B. Karunaratne. 2006. Apple snail densities in habitats used by foraging snail kites. Florida Field Naturalist 34: 37–47.

Darby, P. C., R.E. Bennetts, and H. F. Percival. 2008. Dry down impacts on apple snail (*Pomacea paludosa*) demography: Implications for wetland water management. Wetlands 28: 204–214.

David, P. G. 1994a. Wading bird nesting at Lake Okeechobee, Florida: A historic perspective. Colonial Waterbirds 17: 69–77.

David, P. G. 1994b. Wading bird use of Lake Okeechobee relative to fluctuating water levels. Wilson Bulletin 106: 719–732.

Davis, S. M. and J. C. Ogden (ed.). 1994. Everglades: The ecosystem and its restoration. Delray Beach, FL: St. Lucie Press.

Dean, C. 2008. The Preservation Predicament. The New York Times. Environment Section. January 29, 2008. Available online at *http://www.nytimes.com/2008/01/29/science/earth/29habi.html?_r=1&ref=science&oref=slogin.* Accessed August 14, 2008.

DeAngelis, D. L., L. J. Gross, M. A. Huston, W. F. Wolff, D. M. Fleming, E. J. Comiskey, and S. M. Sylvester. 1998. Landscape modeling for Everglades ecosystem restoration. Ecosystems 1: 64–75.

DeAngelis, D. L., L. J. Gross, E. J. Comiskey, W. M. Mooij, M. P. Nott, and S. Bellmund. 2003. Chapter 6: The Use of Models for a Multiscaled Ecological Monitoring System. In D. Busch and J. Trexler, eds., Ecological Monitoring of Ecoregional Initiatives: Interdisciplinary Approaches for Determining Status and Trends of Ecosystems. Washington, DC: Island Press.

Department of the Army. 2008. Fiscal Year 2009 Civil Works Budget for the U.S. Army Corps of Engineers. Office, Assistant Secretary of the Army (Civil Works). Available online at *http://www.usace.army.mil/cw/cecwb/budget/budget.pdf.* Accessed June 11, 2008.

Diekmann, J. E., and K. B. Thrush. 1986. Project Control in Design Engineering. A Report to the Construction Industry Institute. Austin, TX: University of Texas.

Doering, P. H. 1996. Temporal variability of water quality in the St. Lucie Estuary, South Florida. Journal of the American Water Resources Association 32: 1293–1306.

Doering, P. H. 2007. Power Point: Ecological Conditions in the Northern Estuaries. West Palm Beach, FL: South Florida Water Management District.

Doering, P. H., and R. H. Chamberlain. 1999. Water quality and source of freshwater discharge to the Caloosahatchee Estuary, Florida. Journal of the American Water Resources Association 35: 793–806.

Doering, P. H., and R. H. Chamberlain. 2000. Experimental studies on the salinity tolerance of turtle grass *Thalassia testudinum.* Pages 81–97 in S. A. Bortone, ed., Seagrass: Monitoring Ecology, Physiology, and Management. Boca Raton, FL: CRC Press.

Doering, P. H., R. H. Chamberlain, K. M. Donohue, and A. D. Steinman. 1999. Effect of salinity on the growth of *Vallisneria americana Michx.* from the Caloosahatchee Estuary, Florida. Florida Scientist 62: 89–105.

Doering, P. H., R. H. Chamberlain, and D. E. Haunert. 2002. Using submerged aquatic vegetation to establish minimum and maximum freshwater inflows to the Caloosahatchee Estuary, Florida. Estuaries 25: 1343–1354

DOI (Department of the Interior). 2006. Modified Water Deliveries to Everglades National Park: Audit Report No. C-IN-MOA-0006-2005. Washington, DC: Office of Inspector General.

DOI and USACE (U.S. Army Corps of Engineers). 2005. Central and Southern Florida Project Comprehensive Everglades Restoration Plan: 2005 Report to Congress. Available online at *http://www.evergladesplan.org/pm/program_docscerp_report_congress_2005.cfm.* Accessed June 6, 2008.

Douglas, M. 1947. The Everglades: River of Grass. New York: Rhinehart.

Doyle, T. W. 2003. Predicting Future Mangrove Forest Migration in the Everglades under Rising Sea Level. USGS Fact Sheet FS-030-03. Available online at *http://www.nwrc.usgs.gov/factshts/030-03.pdf.* Accessed August 21, 2008.

Durrenberger, T., D. Denslow, and J. Dewy. 2007. Soaring House Prices and Wages of Local Government Employees. Florida Focus, June 2007.

Emanuel, K., R. Sundararajan, and J. Williams. 2008. Hurricanes and global warming: Results from downscaling IPCC AR4 simulations. Bulletin of the American Meteorological Society 89(3): 347–367.

Engstrom, D. R., S. P. Schottler, P. R. Leavitt, and K. E. Havens. 2006. A reevaluation of the cultural eutrophication of Lake Okeechobee using multiproxy sediment records. Ecological Applications 16: 1194–1206.

ENP (Everglades National Park). 2008. A Review of Discharge into Everglades National Park, Technical Memorandum. Homestead, FL: South Florida Natural Resources Center.

EPA (Environmental Protection Agency). 2007. Everglades Ecosystem Assessment: Water Management and Quality, Eutrophication, Mercury Contamination, Soils and Habitat. Monitoring for Adaptive Management: A R-EMAP Status Report. Available online at *http://www.epa.gov/region4/sesd/reports/epa904r07001/epa904r07001.pdf.* Accessed August 13, 2008.

FDEP (Florida Department of Environmental Protection). 2000. The State of Florida's draft total phosphorus total maximum daily load (TMDL) for Lake Okeechobee. Tallahassee, FL: FDEP.

FDEP. 2007. Emergency Authorization to Discharge Water from the L-8 Reservoir Project. Memorandum #DEP07-0393. Available online at *www.dep.state.fl.us/legal/ Final_Orders/2007/dep07-0393.doc.* Accessed May 24, 2008.

FDOT (Florida Department of Transportation). 2007a. Update on Highway Construction Cost Trends in Florida. Available online at *http://www.dot.state.fl.us/planning/ policy/costs/Update-0407.pdf.* Accessed May 27, 2008.

FDOT. 2007b. Transportation Cost Reports: Inflation Factors. Available online at *http://www.dot.state.fl.us/planning/policy/costs/inflation-073007.pdf.* Accessed May 27, 2008.

Feldman, D. L. 2008. Barriers to adaptive management: Lessons from the Apalachicola-Chattahoochee-Flint compact. Society & Natural Resources 21: 512–525.

Ferriter, A., B. Doren, R. Winston, D. Thayer, B. Miller, B. Thomas, M. Barrett, T. Pernas, S. Hardin, J. Lane, M. Kobza, D. Schmitz, M. Bodle, L. Toth, L. Rodgers, P. Pratt, S. Snow, and C. Goodyear. 2008. Chapter 9: The Status of Nonindigenous Species in the South Florida Environment. In 2008 South Florida Environmental Report. Available online at *https://my.sfwmd.gov/pls/portal/docs/PAGE/PG_GRP_SFWMD_SFER/PORTLET_SFER/TAB2236041/VOLUME1/chapters/v1_ch_9.pdf.* Accessed July 19, 2008.

Fisher, M. M, K. R. Reddy, and R. T. James. 2001. Long-term changes in the sediment chemistry of large subtropical lake. Lake and Reservoir Management 17: 217–232.

Fisher, M. M., K. R. Reddy, and R. T. James. 2005. Internal nutrient loads from sediments in a shallow, subtropical lake. Lake and Reservoir Management 21: 338–349.

Flaig, E. G., and K. R. Reddy. 1995. Fate of phosphorus in the Lake Okeechobee watershed, Florida, USA: Overview and recommendations. Ecological Engineering Volume 5(2–3): 127–142.

Florida Association of Realtors. 2007. UF: Florida population growth slows but still remains high. Available online at *http://www.andyweiser.com/articles/2008_Florida_ Population_Continues_Growth.htm*. Accessed August 15, 2008.

Florida Department of Community Affairs. 2007. Rural Land Stewardship Area Program: 2007 Annual Report to the Legislature, December 31, 2007. Available online at *http://www.dca.state.fl.us/fdcp/dcp/RuralLandStewardship/RLSA2007ReportLegislature.pdf*. Accessed August 15, 2008.

Florida Office of Economic and Demographic Research. 2007. Florida Population: Components of Change. Tallahassee, FL: Florida Office of Economic and Demographic Research.

Folke, C., S. Carpenter, B. Walker, M. Scheffer, T. Elmqvist, L. Gunderson, and C. S. Holling. 2004. Regime shifts, resilience, and biodiversity in ecosystem management. Annual Review of Ecology and Systematics 35: 557–581.

Frederick, P. C., and M. G. Spalding. 1994. Factors affecting reproductive success of wading birds (*Ciconiiformes*) in the Everglades ecosystem. Pages 659–691 in S. M. Davis and J. C. Ogden, eds., Everglades: The ecosystem and its restoration. Boca Raton, FL: St. Lucie Press.

Gaiser, E. E., J. C. Trexler, J. H. Richards, D. L. Childers, D. Lee, A. L. Edwards, L. J. Scinto, K. Jayachandran, G. B. Noe, and R. D. Jones. 2005. Cascading ecological effects of low-level phosphorus enrichment in the Florida Everglades. Journal of Environmental Quality 34: 717–723.

GAO (Government Accountability Office). 2007. South Florida Ecosystem: Restoration Is Moving Forward but Is Facing Significant Delays, Implementation Challenges, and Rising Costs. GAO-07-520. Washington, DC: GAO.

Gawlik, D. E., and D. A. Rocque. 1998. Avian communities in bayheads, willowheads, and sawgrass marshes of the central Everglades. Wilson Bulletin 110: 45–55.

Gilmour, C., W. Orem, D. Krabbenhoft, and I. Mendelssohn. 2007. Appendix 3B-3: Preliminary Assessment of Sulfur Sources, Trends and Effects in the Everglades in 2007 South Florida Environmental Report. Available online at *https://my.sfwmd.gov/pls/portal/docs/PAGE/PG_GRP_SFWMD_SFER/PORTLET_PREVREPORT/VOLUME1/appendices/v1_app_3b-3.pdf*. Accessed August 22, 2008.

Goble, D. D. 2006. Evolution of at-risk protection. Pages 6–23 in M. J. Scott, D. Goble, and F. W. Davis, eds., The Endangered Species Act at Thirty: Conserving Biodiversity in Human-Dominated Landscapes. Washington, DC: Island Press.

Groffman, P., J. Baron, T. Blett, A. Gold, I. Goodman, L. Gunderson, B. Levinson, M. Palmer, H. Paerl, G. Peterson, N. Poff, D. Rejeski, J. Reynolds, M. Turner, K. Weathers, and J. Wiens. 2006. Ecological thresholds: The key to successful environmental management or an important concept with no practical application? Ecosystems 9: 1–13.

Grosskruger, P. L. 2007. Saving the Everglades One Small Project at a Time. Miami Herald. December 8, 2007.

Grunwald, M. 2006. The Swamp: The Everglades, Florida, and the Politics of Paradise. New York: Simon and Schuster, Inc.

Gulf Engineers and Consultants, and Taylor Engineering. 2003. Municipal and Industrial (M&I) Water Use Forecast, Initial Comprehensive Everglades Restoration Plan (CERP) Update. Contract No. DACW17-01-D-0012 GEC Project No. 22423101. Available online at *http://www.evergladesplan.org/pm/recover/recover_docs/icu/muni_ind_water_report_final.pdf*. Accessed August 22, 2008.

Gunderson, L. H. 2000. Ecological resilience—in theory and application. Annual Review of Ecology and Systematics 31: 425–439.

Gunderson, L. H., and A. R. Light. 2006. Adaptive management and adaptive governance in the Everglades ecosystem. Policy Science 39: 323–334.

Gurnett, K. 2001. With Snowbirds Comes the Sprawl. Times Union, February 25, 2001.

Hallac, D., Bass, S., and M. Alvarado. 2007. Interior Everglades National Park and the Cape Sable Seaside Sparrow: Population trends, risk, and considerations for restoration. Presentation to the Avian Ecology Workshop Panel, Florida International University, August 13, 2007. Homestead, FL: South Florida Natural Resources Center, Everglades National Park.

Harris, W. G., M. M. Fisher, X. Cao, T. Osborne, and L. Ellis. 2007. Magnesium-rich minerals in sediment and suspended particulates of South Florida water bodies: Implications for turbidity. Journal of Environmental Quality 36(6): 1670–1677.

Havens, K. E. 2003. Submerged aquatic vegetation correlations with depth and light attenuating materials in a shallow subtropical lake. Hydrobiologia 493: 173–186.

Havens, K. E., and D. E. Gawlick. 2005. Lake Okeechobee Conceptual Ecological Model. Wetlands 25: 908–925.

Havens, K. E., and W. W. Walker. 2002. Development of a total phosphorus concentration goal in the TMDL process for Lake Okeechobee, Florida (USA). Lake and Reservoir Management 18: 227–238.

Havens, K. E., D. Fox, S. Gornak, and C. Hanlon. 2005. Aquatic vegetation and largemouth bass population responses to water-level variations in Lake Okeechobee, Florida (USA). Hydrobiologia 539: 225–237.

Havens, K. E., K.-R. Jin, N. Iricanin, and R. T. James. 2007. Phosphorus dynamics at multiple time scales in the pelagic zone of a large shallow lake in Florida, USA. Hydrobiologia 581: 25–42.

Heinz Center. 2002. The State of the Nation's Ecosystems: Measuring the Lands, Waters and Living Resources of the United States. Washington, DC: The H. John Heinz Center for Science.

Hoffman, W., G. T. Bancroft, and R. J. Sawicki. 1994. Foraging habitat of wading birds in the water conservation areas of the Everglades. Pages 585–614 in S. M. Davis and J. C. Ogden, eds., Everglades: The Ecosystem and Its Restoration. Boca Raton, FL: St. Lucie Press.

Hofmockel, K. 1999. Effects of hydrologic management decisions on marsh structure in Water Conservation Area 2A of the Everglades, Florida. Durham, NC: Duke University.

Ibelings, B. W., R. Portielje, E. H. R. R. Lammens, R. Noordhuis, M. S. van den Berg, W. Joosee, and M. L. Meijer. 2007. Resilience of alternative stable states during the recovery of shallow lakes from eutrophication: Lake Veluwe as a case study. Ecosystems 10: 4–16.

IPCC (Intergovernmental Panel on Climate Change). 2007. Climate Change 2007: The Physical Science Basis. Cambridge, UK: Cambridge University Press.

James, R. T., and K. E. Havens. 2005. Outcomes of extreme water levels on water quality of offshore and nearshore regions in a large shallow subtropical lake. Archiv für Hydrobiologie 163: 225–239.

James, R. T., and J. Zhang. 2008. Chapter 10: Lake Okeechobee Protection Program—State of the Lake and Watershed. In 2008 South Florida Environmental Report. Available online at *https://my.sfwmd.gov/pls/portal/docs/PAGE/PG_GRP_SFWMD_SFER/PORTLET_SFER/TAB2236041/VOLUME1/chapters/v1_ch_10.pdf.* Accessed July 19, 2008.

James, R. T., J. Zhang, S. Gornak, S. Gray, G. Ritter, and B. Sharfstein. 2006. Chapter 10: Lake Okeechobee Protection Program State of the Lake and Watershed. In 2006 South Florida Environmental Report. Available online at *https://my.sfwmd.gov/pls/portal/docs/PAGE/PG_GRP_SFWMD_SFER/PORTLET_PREVREPORT/2006_SFER/VOLUME1/chapters/v1_ch_10.pdf.* Accessed August 21, 2008.

Johnson, K. G., M. S. Allen, and K. E. Havens. 2007. A review of littoral vegetation, fisheries, and wildlife responses to hydrologic variation at Lake Okeechobee. Wetlands 27: 110– 126.

Jones, L.A . 1948. Soils, Geology, and Water Control in the Evergaldes Region. Bulletin 442. Gainesville, FL: University of Florida Agricultural Experiment Station.

Jurenas, R. 1992. Sugar Policy Issues. Congressional Research Service, Library of Congress, Washington, DC, 16 pages.

Kam, N. 2008. Negotiations Salvage Some Everglades Funds. Palm Beach Post. April 28, 2008.

Karl, T. R., G. A. Meehl, C. D. Miller, S. J. Hassol, A. M. Waple, and W. L. Murray (eds.). 2008. Weather and Climate Extremes in a Changing Climate. Regions of Focus: North America, Hawaii, Caribbean, and U.S. Pacific Islands. Washington, DC: National Oceanic and Atmospheric Administration, National Climate Data Center.

Karunaratne, L. B., P. C. Darby, and R. E. Bennetts. 2006. The effects of wetland habitat structure on Florida apple snail density. Wetlands 26: 1143–1150.

Kiker, C. F., J. W. Milon, and A. W. Hodges. 2001. Adaptive learning for science-based policy: The Everglades restoration. Ecological Economics 37: 403–416.

King, R. 2008. Costs erode water district's support for storage wells. Palm Beach Post. January 28, 2008.

Kleeberg, A., and J. G. Kohl. 1999. Assessment of the long-term effectiveness of sediment dredging to reduce benthic phosphorus release in shallow Lake Müggelsee (Germany). Hydrobiologia 394: 153–161.

Knutson, T. R., J. J. Sirutis, S. T. Garner, G. A. Vecchi, and I. M. Held. 2008. Simulated reduction in Atlantic hurricane frequency under twenty-first-century warming conditions. Nature Geoscience 1: 359–364.

Kraemer, G. P., R. H. Chamberlain, P. H. Doering, A. D. Steinman, and M. D. Hanisak. 1999. Physiological responses of *Vallisneria americana* transplants along a salinity gradient in the Caloosahatchee Estuary (SW Florida). Estuaries 22: 138–148.

Kushlan, J. A., O. L. Bass Jr., L. L. Loope, W. B. Robertson Jr., P. C. Rosendahl, and D. L. Taylor. 1982. Cape Sable Sparrow management plan. South Florida Research Center Report M-660.

Lee, K. 1999. Appraising adaptive management. Conservation Ecology 3(2): 3. Available online at *http://www.ecologyandsociety.org/vol3/iss2/art3/*. Accessed December 13, 2005.

Light, A. 2006. Tales of the Tamiami Trail: Implementing adaptive management in Everglades restoration. Journal of Land Use 22: 59–99.

Light, S. S., and J. W. Dineen. 1994. Water control in the Everglades: A historical perspective. Pages 47–84 in S. M. Davis and J. C. Ogden, eds., Everglades: The Ecosystem and Its Restoration. Boca Raton, FL: St. Lucie Press.

Lockwood, J. L., D. A. LaPuma, B. Baiser, M. Boulton, and M. J. Davis. 2006. Detailed study of Cape Sable seaside sparrow nest success and causes of nest failure. 2006 annual report. Vero Beach, FL: U.S. Fish and Wildlife Service.

Lodge, T. E., 2005. The Everglades Handbook: Understanding the Ecosystem. 2nd edition. Boca Raton, FL: CRC Press.

Lord, L. 1993. Guide to Florida Environmental Issues and Information. Winter Park, FL: Florida Conservation Foundation.

Loucks, DP. 2006. Modeling and managing the interactions between hydrology, ecology and economics. Journal of Hydrology 328 (3–4). Special Issue: SI: Coupling of engineering and biological models for ecosystem analysis. Pp. 408–416.

Marshall, C., Jr., R. Pielke Sr., L. Steyaert, and D. Willard. 2004. The impact of anthropogenic land cover change on the Florida peninsula sea breezes and warm season sensible weather. Monthly Weather Review 132: 28–52.

Martin, J., W. M. Kitchens, and J. E. Hines. 2007a. Importance of well-designed monitoring programs for the conservation of endangered species: Case study of the snail kite. Conservation Biology 21: 472–481.

Martin, J., W. Kitchens, C. Cattau, A. Bowling, S. Stocco, E. Powers, C. Zweig, A. Hottaling, Z. Welch, H. Waddle, and A. Paredes. 2007b. Snail kite demography draft annual progress report 2006. Gainesville, FL: USGS Florida Cooperative Fish and Wildlife Research Unit University of Florida.

Marx, D. E., and D. E. Gawlik. 2007. Wading Bird Colony Timing, Location, and Size at Lake Okeechobee, 2005–2007. Final report to National Park Service. Homestead, FL: Everglades National Park.

Mayer, A. L., and M. Rietkerk. 2004. The dynamic regime concept for ecosystem management and restoration. Bioscience 54: 1013–1020.
McCabe, G. J., M. A. Palecki, and J. L. Betancourt. 2004. Pacific and Atlantic Ocean influences on multidecadal drought frequency in the United States. Proceedings of the National Academy of Sciences 101: 4136–4141.
McDonald, M. V. 1988. Status survey of two Florida Seaside Sparrows and taxonomic review of the seaside sparrow assemblage. USFWS Florida Cooperative Fish and Wildlife Research Unit. Technical Report 32.
McPherson, B. F., and R. Halley. 1996. The South Florida Environment: A Region under Stress. USGS Circular 1134.Washington, DC: United States Printing Office.
Meier, M. F., M. B. Dyurgerov, U. K. Rick, S. O'Neel, W. T. Pfeffer, R. S. Anderson, S. P. Anderson, and A.F. Glazovsky. 2007. Glaciers dominate eustatic sea-level rise in the 21st Century. Science 317: (5841): 1064–1067. DOI: 10.1126/science.1143906.
Miami-Dade County. 2008. Comprehensive Development Master Plan, Adopted 2015–2025. Available online at *http://www.miamidade.gov/planzone/Library/Maps/CDMP2015.pdf*. Accessed July 30, 2008.
Milly, P., J. Betancourt, M. Falkenmark, R. Hirsch, Z. Kundzewicz, D. Lettenmaier, and R. Stouffer. 2008. Climate change: Stationarity is dead: Whither water management? Science 319: 573–574.
Mitsch, W .J., L .E. Band, and C. F. Cerco. 2007. Everglades Landscape Model (ELM), Version 2.5: Peer Review Panel Report. Available online at *https://my.sfwmd.gov/pls/portal/docs/PAGE/PG_GRP_SFWMD_HESM/PORTLET_ELM/PORTLET_SUBTAB_ELM_PEERREVIEW/TAB2644097/ELM_PEER_REVIEW_REPORT_FINAL_010307.PDF*. Accessed June 15, 2008.
Mooney, C. 2007. Storm World: Hurricanes, Politics, and the Battle over Global Warming. Orlando, FL: Harcourt.
Morgan, J., and S. Allen. 2007. Melaleuca Eradication and Other Exotic Plants—Implement Biological controls. Available online at *http://www.sfrestore.org/wg/wgminutes/2007meetings/18,19oct2007/Melaleuca%20Eradication%20Working%20Group%2010182007.pdf*.
Nicholson, E., and H. P. Possingham. 2007. Making conservation decisions under uncertainty for the persistence of multiple species. Ecological Applications 17(1): 251–265.
Niemi, G. J., and M. E. McDonald. 2004. Application of ecological indicators. Annual Review of Ecology and Systematics 35: 89–111.
NOAA. 2006. FAQ/State of the Science: Atlantic Hurricanes and Climate. Available online at *http://hurricanes.noaa.gov/pdf/hurricanes-and-climate-change-09-2006.pdf*. Accessed February 26, 2008.
Noe, G. B., D. L. Childers, and R. D. Jones. 2001. Phosphorus biogeochemistry and the impacts of phosphorus enrichment: Why is the Everglades so unique? Ecosystems 4: 603–624.
Nott, N. P., O .L. Bass Jr., D. M. Fleming, S. E. Killeffer, N. Fraley, L. Manne, J. L. Curnutt, T. M. Brooks, R. Powell, and S. L. Pimm. 1998. Water levels, rapid vegetational changes, and the endangered Cape Sable Seaside Sparrow. Animal Conservation 1: 23–32.
NPG (Negative Population Growth). 2002. Focus on Florida: Population, Resources, and Quality of Life. Available online at *http://www.npg.org/specialreports/FL/fl_report.html*. Accessed December 23, 2007.
NRC (National Research Council). 1996. Upstream. Washington, DC: National Academy Press.
NRC. 1999. New Strategies for America's Watersheds. Washington, DC: National Academy Press.
NRC. 2000. Ecological Indicators for the Nation. Washington, DC: National Academy Press.
NRC. 2001. Aquifer Storage and Recovery in the Comprehensive Everglades Restoration Plan: A Critique of the Pilot Projects and Related Plans for ASR in the Lake Okeechobee and Western Hillsboro Areas. Washington, DC: National Academy Press.
NRC. 2002a. Regional Issues in Aquifer Storage and Recovery for Everglades Restoration. Washington, DC: National Academy Press.

NRC. 2002b. Florida Bay Research Programs and Their Relation to the Comprehensive Everglades Restoration Plan. Washington, DC: National Academy Press.

NRC. 2003a. Does Water Flow Influence Everglades Landscape Patterns? Washington, DC: The National Academies Press.

NRC. 2003b. Adaptive Monitoring and Assessment for the Comprehensive Everglades Restoration Plan. Washington, DC: The National Academies Press.

NRC. 2003c. Science and the Greater Everglades Ecosystem Restoration: An Assessment of the Critical Ecosystems Initiative. Washington, DC: The National Academies Press.

NRC. 2004. River Basins and Coastal Systems Planning Within the U.S. Army Corps of Engineers. Washington, DC: The National Academies Press.

NRC. 2005. Re-Engineering Water Storage in the Everglades: Risks and Opportunities. Washington, DC: The National Academies Press.

NRC. 2007. Progress Toward Restoring the Everglades: The First Biennial Review–2006. Washington, DC: The National Academies Press.

NRC. 2008. Desalination: A National Perspective. Washington, DC: The National Academies Press.

Ogden, J. C., S. M. Davis, K. J. Jacobs, T. Barnes, and H. E. Fling. 2005. The use of conceptual ecological models to guide ecosystem restoration in South Florida. Wetlands 25: 795–809.

Parker, G. G. 1974. Hydrology of the pre-drainage system of the Everglades in southern Florida. Pages 18–27 in P. J. Gleason, ed., Environments of South Florida: Present and Past. Memoir 2. Miami, FL: Miami Geological Society.

Parker, G. G., G. E. Ferguson, S. K. Love, et al. 1955. Water resources of southeastern Florida, with special reference to the geology and ground water of the Miami area. Wat. Sup. Pap. 1255. Tallahassee, FL: U.S. Geological Survey.

Perry, W. 2004. Elements of South Florida's comprehensive Everglades restoration plan. Ecotoxicology 13: 185–193.

Pietro, K., R. Bearzotti, M. Chimney, G. Germain, N. Iricanin, and T. Piccone. 2007. Chapter 5: STA Performance, Compliance and Optimization. In 2007 South Florida Environmental Report. Available online at *https://my.sfwmd.gov/pls/portal/docs/PAGE/PG_GRP_SFWMD_SFER/PORTLET_PREVREPORT/VOLUME1/chapters/v1_ch_5.pdf*. Accessed August 11, 2008.

Pietro, K., R. Bearzotti, G. Germain, and N. Iricanin. 2008. Chapter 5: STA Performance, Compliance and Optimization. In 2008 South Florida Environmental Report. Available online at *https://my.sfwmd.gov/pls/portal/docs/PAGE/PG_GRP_SFWMD_SFER/PORTLET_SFER/TAB2236041/VOLUME1/chapters/v1_ch_5.pdf*. Accessed June 15, 2008.

Pimm, S. L., J. L. Lockwood, C. N. Jenkins, J. L. Curnutt, M. P. Nott, R. D. Powell, and O. L. Bass Jr. 2002. Sparrow in the grass: a report on the first ten years of research on the Cape Sable seaside sparrow (*Ammodramus maritimus mirabilis*). Report to the National Park Service, Everglades National Park, Fla. Available online at *http://www.nicholas.duke.edu/people/faculty/pimm/cssp/CSSSReportLarge.pdf*. Accessed August 6, 2008.

Plato Consulting, Inc. 2003. South Florida Water Management District Strategic Modeling Plan. Report to SFWMD. West Palm Beach, FL: Plato Consulting, Inc.

Rahmstorf, S. 2007. A semi-empirical approach to projecting future sea-level rise. Science 315 (5810): 368–370.

RECOVER (Restoration, Coordination, and Verification Program). 2004. CERP Monitoring and Assessment Plan: Part 1, Monitoring and Supporting Research. Available online at *http://www.evergladesplan.org/pm/recover/recover_map.aspx*. Accessed August 12, 2008.

RECOVER. 2005a. 2005 Assessment Strategy for the Monitoring and Assessment Plan. West Palm Beach, FL: RECOVER.

RECOVER. 2005b. The Recover Team's Recommendations for Interim Goals and Interim Targets for the Comprehensive Everglades Restoration Plan. Available online at *http://www.evergladesplan.org/pm/recover/recover_docs/igit/igit_mar_2005_report/ig_it_rpt_main_report.pdf*. Accessed June 15, 2008.

RECOVER. 2005c. Report on Evaluation Tools, Models, Work Plans and Budgets. Available online at *http://www.evergladesplan.org/pm/recover/recover_docs/sys_mdl/rec_mod_ndsrpt_mainreport.pdf.* Accessed June 15, 2008.

RECOVER. 2006a. Comprehensive Everglades Restoration Plan Adaptive Management Strategy. Available online at *http://www.evergladesplan.org/pm/recover/recover_docs/am/rec_am_stategy_brochure.pdf.* Accessed June 15, 2008.

RECOVER. 2006b. Comprehensive Everglades Restoration Plan System-wide Performance Measures. Available online at *http://www.evergladesplan.org/pm/recover/eval_team_ perf_measures.cfm.* Accessed June 26, 2006.

RECOVER. 2006c. Quality Assurance Systems Requirements (QASR) Manual for the Comprehensive Everglades Restoration Plan. Available online at *http://www.evergladesplan.org/pm/program_docs/qasr.cfm.* Accessed August 8, 2006.

RECOVER. 2006d. System Status Report: Pilot Assessment Reports for the Monitoring and Assessment Modules. West Palm Beach, FL: RECOVER.

RECOVER. 2006e. Assessing the Response of the Everglades Ecosystem to Implementation of the CERP (MAP II). West Palm Beach, FL: RECOVER.

RECOVER. 2006f. Monitoring and Assessment Plan (MAP), Part 2 2006 Assessment Strategy for the MAP. Available online at *http://www.evergladesplan.org/pm/recover/recover_map_ part2.aspx.* Accessed June 15, 2008.

RECOVER. 2007a. CERP System-wide Performance Measure Documentation Sheet. Lake Okeechobee Performance Measure: Vegetation Mosaic. Available online at *http://www.evergladesplan.org/pm/recover/recover_docs/et/lo_pm_vegetationmosaic.pdf.* Accessed June 14, 2008.

RECOVER. 2007b. Development and Application of Comprehensive Everglades Restoration Plan System-wide Performance Measures. Available online at *http://www.evergladesplan.org/pm/recover/perf_systemwide.aspx.* Accessed June 15, 2008.

RECOVER. 2007c. Final 2007 System Status Report. Available online at *http://www.evergladesplan.org/pm/recover/assess_team_ssr_2007.aspx.* Accessed August 22, 2008.

RECOVER. 2007d. Draft Quality Assurance Systems Requirements. Available online at *http://www.evergladesplan.org/pm/program_docs/qasr.aspx.* Accessed June 15, 2008.

RECOVER. 2007e. Draft CERP Adaptive Management Implementation Guidance Manual (Version 1.2). December 2007. Jacksonville and West Palm Beach, FL: USACE and SFWMD.

Reddy, K. R., E. G. Flaig, and D. A. Graetz. 1996. Phosphorus storage capacity of uplands, wetlands and streams of the Lake Okeechobee basin. Agriculture, Environment and Ecosystems 59: 203–216.

Reddy, K. R., G. A. O'Connor, and C. L. Schelske. 1999. Phosphorus Biogeochemistry in Subtropical Ecosystems: Florida as a Case Example. Boca Raton, FL: CRC/Lewis Publishers.

Reddy, K. R., M.M. Fisher, Y.Wang, J.R.White, and R. Thomas James. 2007. Potential effects of sediment dredging on internal phosphorus loading in a shallow, sub-tropical lake. Lake and Reservoir Management 23: 27–38.

Reynolds, J. E. 2006. Strong Non-agricultural Demand Keeps Agricultural Land Values Increasing. Gainesville, FL: University of Florida Institute of Food and Agricultural Sciences Extension.

Rizzardi, K. W. 2001. Alligators and litigators: A recent history of Everglades regulation and litigation. The Florida Bar Journal LXXV (3): 18.

Rizzardi, K. W. 2007. Statement [on November 16, 2007] of Keith W. Rizzardi (on the ESA) to the National Research Council's Committee on Independent Scientific Review of Everglades Restoration Progress. White Paper. West Palm Beach, FL: South Florida Water Management District.

Rodgers, J. A. 2007. Breeding success of *Rostrhamus sociabilis* (Snail Kites) at two Florida lakes. Southeastern Naturalist 6: 35–46.

Rogers, M. W., and M. S. Allen. 2008. Hurricane impacts to Lake Okeechobee: Altered hydrology creates difficult management trade offs. Fisheries 33(1): 11–17.

Rosenthal, E. 2008. New Trend in Biofuels Has New Risks. The New York Times, May 21, 2008.

Ross, M. S., and D. T. Jones, eds. 2004. Tree Islands in the Shark Slough Landscape: Interactions of vegetation, hydrology and soils. Final Report. Miami, FL: Southeast Environmental Research Center, Florida International University.

Ross, M. S., S. Mitchell-Bruker, J. P. Sah, S. Stothoff, P. L. Ruiz, D. L. Reed, K. Jayachandran, and C. L. Coultas. 2006. Interaction of hydrology and nutrient limitation in the ridge and slough landscape of the southern Everglades. Hydrobiologia 569: 37–59.

Rumbold, D. G., and L. E. Fink. 2006. Extreme spatial variability and unprecedented methylmercury concentrations within a constructed wetland. Environmental Monitoring and Assessment 112: 115–135.

Scheffer, M., and S. R. Carpenter. 2003. Catastrophic regime shifts in ecosystems: linking theory to observation. Trends in Ecology & Evolution 18: 648–656.

Scheffer, M., and E. H. van Nes. 2004. Mechanisms for marine shifts: can we use lakes as microcosms for oceans? Progress in Oceanography 60: 303–319.

Scheffer, M., S. Carpenter, J. A. Foley, C. Folke, and B. Walker. 2001. Catastrophic shifts in ecosystems. Nature 413: 591–596.

Scheidt, D. J., and P. I. Kalla. 2007. Everglades Ecosystem Assessment: Water Management and Quality, Eutrophication, Mercury Contamination, Soils and Habitat: Monitoring for Adaptive Management: A R-EMAP Status Report. USEPA Region 4, Athens, Ga. EPA 904-R-07-001. Available online at *http://www.epa.gov/region4/sesd/reports/epa904r07001.html*. Accessed May 23, 2008.

Schortemeyer, J. L. 1980. An evaluation of water management practices for optimum wildlife benefits in Conservation Area 3A. Tallahassee, FL: Florida Game and Fresh Water Fish Commission.

Science and Technology Committee of the Miami-Dade County Climate Change Task Force. 2007. Statement on Sea Level in the Coming Century. Available online at *http://www.alachuacounty.us/assets/uploads/images/epd/documents/ECSC/Statement.pdf*. Accessed July 22, 2008.

Scott, T., L. Fernandez, and M. F. Allen. 2006. Land use planning. Pages 206–217 in M. J. Scott, D. Goble, and F. W. Davis, eds., The Endangered Species Act at Thirty: Conserving Biodiversity in Human-Dominated Landscapes. Washington, DC: Island Press.

SCT (Science Coordination Team). 2003. The Role of Flow in the Everglades Ridge and Slough Landscape. Available online at *http://www.sfrestore.org/sct/docs/SCT%20Flow%20Paper%20-%20Final*. Accessed January 30, 2006.

Secretary of Interior. 1994. The Impact of Federal Programs on Westlands. Report to Congress. Chapter 7: Florida Everglades. Available online at *http://www.doi.gov/oepc/wetlands2/v2ch7.html*.

SEI (Sustainable Ecosystems Institute). 2007. Everglades Multi-Species Avian Ecology and Restoration Review Final Report, November 2007. Portland, Ore.

SFERTF (South Florida Ecosystem Restoration Task Force). 2000. Coordinating Success: Strategy for Restoration of the South Florida Ecosystem. Available online at *http://www.sfrestore.org/documents/isp/sfweb/sfindex.htm*. Accessed January 30, 2006.

SFERTF (South Florida Ecosystem Restoration Task Force). 2006. Coordinating Success: Strategy for Restoration of the South Florida Ecosystem and Tracking Success: Biennial Report for FY 2004–2006 of the South Florida Ecosystem Restoration Task Force Integrated Financial Plan. Volume 1 of 2. Available online at *http://www.sfrestore.org/documents/2004_2006_strategic_plan_volume%20I.pdf*. Accessed May 19, 2008.

SFERTF. 2007a. Tracking Success: 2007 Integrated Financial Plan for the South Florida Ecosystem Restoration Task Force. Available online at *http://www.sfrestore.org/documents/2007_ifp.pdf*. Accessed May 19, 2008.

SFERTF. 2007b. South Florida Ecosystem Restoration Land Acquisition Strategy. Available online at *http://www.sfrestore.org/issueteams/latt/documents/LAS2007final-ac.pdf*. Accessed May 8, 2008.

SFWMD (South Florida Water Management District). 2003. Summary of Evaluation of Alternatives for the Lake Okeechobee Sediment Management Feasibility Study. Final Report to the South Florida Water Management District prepared by Blasland, Bouck, and Lee, Inc., Boca Raton, Fla. Presented at public outreach meeting January 14, 2003.

SFWMD. 2007. Quick Facts on Everglades Restoration Progress. West Palm Beach, FL: South Florida Water Management District.

SFWMD and FDEP (Florida Department of Environmental Protection). 2005. Comprehensive Everglades Restoration Plan Annual Report. Available online at *http://www.everglades plan.org/pm/pm_docs/cerp_annualreport/2004_cerp_ann_rpt.pdf.* Accessed June 9, 2008.

SFWMD and FDEP. 2006. South Florida Environmental Report: Volume I: The South Florida Environment. Available online at *https://my.sfwmd.gov/pls/portal/docs/PAGE/PG_ GRP_SFWMD_SFER/PORTLET_PREVREPORT/2006_SFER/VOLUME1/VOL1_TABLE_OF_CONTENTS.HTML.* Accessed June 15, 2008.

SFWMD and FDEP. 2007. South Florida Environmental Report: Volume I: The South Florida Environment. Available online at *https://my.sfwmd.gov/pls/portal/docs/PAGE/PG_ GRP_SFWMD_SFER/PORTLET_PREVREPORT/VOLUME1/vol1_table_of_contents.html.* Accessed June 15, 2008.

SFWMD and FDEP. 2008a. South Florida Environmental Report: Volume I: The South Florida Environment. Available online at *https://my.sfwmd.gov/pls/portal/docs/PAGE/PG_GRP_SF-WMD_SFER/PORTLET_SFER/TAB2236041/VOLUME1/vol1_table_of_contents.html.* Accessed June 15, 2008.

SFWMD and FDEP. 2008b. South Florida Environmental Report: Volume II: District Annual Plans and Reports. Available online at *https://my.sfwmd.gov/pls/portal/docs/PAGE/PG_GRP_SFWMD_SFER/PORTLET_SFER/TAB2236045/volume2/vol2_table_of_conte nts.html.* Accessed May 8, 2008.

SFWMD, FDEP, and FDACS (Florida Department of Agriculture and Consumer Services). 2008. Lake Okeechobee Watershed Construction Project Phase II Technical Plan. Available online at *https://my.sfwmd.gov/pls/portal/docs/PAGE/PG_GRP_SFWMD_KOE/PORTLET_NORTHERNEVERGLADES/TAB2302093/LAKEO_WATERSHED_CONSTRUCTION_PROJ_PHASE_II_TECH_PLAN.PDF.* Accessed June 15, 2008.

SFWMD and USACE. 2008. Aquifer Storage & Recovery Program: Interim Report 2008. Available online at *http://www.evergladesplan.org/pm/projects/project_docs/pdp_asr_combined/052808_asr_report/052808_asr_interim_rpt.pdf.* Accessed August 1, 2008.

Sharfstein, B., and D. Tipple. 2007. Power Point: The RECOVER Budget, Status and Challenges. West Palm Beach, FL: RECOVER.

Sheikh, P. A. 2005. CRS Report for Congress: Everglades Restoration: Modified Water Deliveries Project. Washington, DC: Congressional Research Service.

Sheikh, P. A., and N. T. Carter. 2006. CRS Report for Congress: Everglades Restoration—The Federal Role in Funding. Washington, DC: Congressional Research Service.

Sklar, F. H. 2007a. Power Point Presentation: Presentation to the NRC Committee on Independent Scientific Review of Everglades Restoration Progress. Miami, Florida. November 15, 2007. West Palm Beach, FL: South Florida Water Management District.

Sklar, F. H. 2007b. Power Point Presentation: The Decomp Physical Model (DMP). West Palm Beach, FL: South Florida Water Management District.

Sklar, F., and A. van der Valk (eds.). 2002. Tree Islands of the Everglades. Boston, MA: Kluwer Academic Publishers.

Sklar F. H., H. C. Fitz, Y. Wu, et al. 2001. The design of ecological landscape models for Everglades restoration. Ecological Economics 37(3): 379–401.

Snyder, G. H. 2004. Everglades Agricultural Area Soil Subsidence and Land Use Projections. Available online at *http://www.evergladesplan.org/pm/projects/project_docs/pdp_08_eaa_store/pdp_08_sub_land_use_report.pdf.* Accessed September 2, 2008.

Snyder, G. H., and J. M. Davidson. 1994. Everglades agriculture: Past, present, and future. Pages 85–116 in Everglades: The Ecosystem and its Restoration. Boca Raton, FL: St. Lucie Press.

Snyder, N. F. R., and H. A. Snyder. 1969. A comparative study of mollusc predation by Limpkins, Everglade Kites, and Boat-tailed Grackles. Living Bird 8: 177–223.

Snyder, N. F. R., S. R. Beissinger, and R. E. Chandler. 1989. Reproduction and demography of the Florida Everglade (Snail) Kite. Condor 91: 300–316.

Society for Ecological Restoration International Science & Policy Working Group. 2004. The SER International Primer on Ecological Restoration. Tucson, AZ: Society for Ecological Restoration International.

Søndergaard, M., E. Jeppesen, T. L. Laruidsen, C. Skov, E. H. Van Nes, R. Roijackers, E. Lammens, and R. Portielje. 2007. Lake restoration: Successes, failures and long-term effects. Journal of Applied Ecology 44: 1095–1105.

SSG (Science Sub-Group). 1993. Federal Objectives for the South Florida Restoration by the Science Sub-Group of the South Florida Management and Coordination Working Group. Available online at *http://www.sfrestore.org/sct/docs/subgrouprpt/index.htm*. Accessed January 30, 2006.

State of Florida. 2007. Governor Crist Signs Executive Orders to Reduce Greenhouse Gases, July 16, 2007. Available online at *www.floridaenergyproducers.com/GOVERN~4.DOC*. Accessed May 25, 2008.

Steinman, A. D., K. E. Havens, H. J. Carrick, and R. VanZee. 2002a. The past, present, and future hydrology and ecology of Lake Okeechobee and its watersheds. Pages 19–37 in J. Porter and K. Porter, eds., The Everglades, Florida Bay, and Coral Reefs of the Florida Keys. An Ecosystem Handbook. Boca Raton, FL: CRC Press.

Steinman, A., K. Havens, and L. Hornung. 2002b. The managed recession of Lake Okeechobee, Florida: Integrating science and natural resource management. Conservation Ecology 6: 17.

Stieglitz, W. O., and R. L. Thompson. 1967. Status and Life History of the Everglade Kite in the United States, 1st edition. Washington, DC: U.S. Government Printing Office.

Stokstad, E. 2008. Big land purchase triggers review of plans to restore Everglades. Science 321 (5885): 22; DOI: 10.1126/science.321.5885.22.

Stone, P. A., P. J. Gleason, and G. L. Chmura. 2002. Bayhead tree islands on deep peats of the northeastern Everglades. Pages 71–115 in F. Sklar and A. van der Valk eds., Tree Islands of the Everglades. Dordrecht, The Netherlands: Kluwer Academic Publishers,

Strowd, T. B., and R. Punnett. 2007. Regional Storage/Treatment and Conveyance Issues. Power Point Presentation to 10-County Coalition Meeting, June 7, 2007. Available online at *https://my.sfwmd.gov/pls/portal/docs/PAGE/PG_GRP_SFWMD_WRAC/PORTLET_SUBTAB_WRAC_ARC_RPT_LOK/TAB20383267/10%20COUNTY%20COALITION%20REGIONAL%20STORAGE%206-7-07%20TBS-REP.PDF*. Accessed August 13, 2008.

Swain, H. 2006. A reality check from Florida. In D. D. Goble, J. M. Scott, and F. W. Davis, eds., The Endangered Species Act at Thirty, Volume 1, Renewing the Conservation Promise. Washington, DC: Island Press.

SWET, Inc. 2008a. Task 2 Report—Evaluation of Existing Information. For project entitled "Technical Assistance in Review and Analysis of Existing Data for Evaluation of Legacy Phosphorus in the Lake Okeechobee Watershed." West Palm Beach, FL: SWET, Inc.

SWET, Inc. 2008b. Draft Task 3 Report—Legacy P Abatement Plan. For project entitled "Technical Assistance in Review and Analysis of Existing Data for Evaluation of Legacy Phosphorus in the Lake Okeechobee Watershed." Submitted to the South Florida Water Management District. West Palm Beach, FL: SWET, Inc.

Sykes, P. W., Jr. 1979. Status of the Everglade Kite in Florida—1968–1978. Wilson Bulletin 91(4): 495–511.

Sykes, P. W., J. A. Rodgers, Jr., and R. E. Bennetts. 1995. Snail kite. Birds of North America 171: 1–32.

TAME Melaleuca (The Areawide Management and Evaluation of Melaleuca). 2004. A Century of Melaleuca in South Florida. Available online at *http://tame.ifas.ufl.edu/pdfs/publications/CenturyofMelaleuca_000.pdf*. Accessed July 25, 2008.

Tarboton, K. C., M. M. Irizarry-Ortiz, D. P. Loucks, S. M. Davis, and J. T. Obeysekera. 2004. Habitat Sutiability Indices for Evaluating Water Management Alternatives. Technical Report. South Florida Water Management District.

Trimble, P. J., E. R. Santee, and C. J. Neidrauer. 1998. Preliminary estimate of impacts of sea-level rise on the regional water resources of southern Florida. Journal of Coastal Research 26: 252–255.

Trulock, S. 2007. Power Point: Authorization Process. February 12, 2007. Jacksonville, FL: USACE.

Union of Concerned Scientists. 2006. Climate Change in the U.S. Northeast: A Report of the Northeast Climate Impacts Assessment. Available online at *http://www.climatechoices.org/assets/documents/climatechoices/NECIA_climate_report_final.pdf*. Accessed August 20, 2008.

USACE (U.S. Army Corps of Engineers). 1992. General Design Memorandum and Environmental Impact Statement: Modified Water Deliveries to Everglades National Park. Atlanta, GA: USACE.

USACE. 1999. Herbert Hoover Dike Major Rehabilitation Evaluation Report. Available online at *http://www.saj.usace.army.mil/Divisions/Everglades/Branches/HHDProject/DOCS/HHD/HHD-abridgedMRR.pdf*. Accessed June 20, 2008.

USACE. 2003. Central and Southern Florida Project, Modified Water Deliveries to Everglades National Park, GRR/SEIS for the Tamiami Trail, December 2003. Jacksonville, FL: USACE.

USACE. 2005. Final Revised General Reevaluation Report/Second Supplemental Environmental Impact Statement (RGRR/SEIS) for the Tamiami Trail Modifications: Modified Water Deliveries to Everglades National Park. Jacksonville District. Available online at *http://www.saj.usace.army.mil/dp/mwdenp-c111/tt_LRR-PRP.htm*. Accessed July 19, 2008.

USACE. 2006. Lake Okeechobee and the Herbert Hoover Dike: Frequently Asked Questions. Jacksonville, FL: USACE.

USACE. 2007a. Civil Works Construction Cost Index System. 30 September 2007. EM 1110-2-1304. Available online at *http://www.usace.army.mil/publications/eng-manuals/em1110-2-1304/entire.pdf*. Accessed July 21, 2008.

USACE. 2007b. Revised Draft Supplemental Environmental Impact Statement—Lake Okeechobee Regulation Schedule. June 2007. USACE, Jacksonville, Fla.

USACE. 2007c. Design Modifications for the Canal 111 (C-111) Project Miami-Dade County, Florida: Draft Environmental Assessment. Available online at *http://planning.saj.usace.army.mil/envdocs_M_P/Miami-Dade/C-111/EA_and_BA-C-111SD_200706_2.pdf*. Accessed May 19, 2008.

USACE. 2007d. Memorandum for the Assistant Secretary of the Army (Civil Works) on Comprehensive Everglades Restoration Plan—Water Quality Improvements. May 25, 2007.

USACE and DOI. 2008. Modified Water Deliveries to Everglades National Park Tamiami Trail Modification Limited Reevaluation Report and Environmental Assessment. April 2008 Draft. Available online at *http://www.saj.usace.army.mil/dp/mwdenp-c111/*. Accessed July 21, 2008.

USACE and SFWMD. 1999. Central and Southern Florida Comprehensive Review Study Final Integrated Feasibility Report and Programmatic Environmental Impact Statement. Available online at *http://www.evergladesplan.org/pub/restudy_eis.cfm#mainreport*. Accessed January 30, 2006.

USACE and SFWMD. 2000. Master Program Management Plan Volume I: Management Processes. Comprehensive Everglades Restoration Plan. Available online at *http://www.evergladesplan.org/pm/pm_docs/mpmp/mpmp_final_000818.pdf*. Accessed June 15, 2008.

USACE and SFWMD. 2002. Central and Southern Florida Project Comprehensive Everglades Restoration Plan Project Management Plan: WCA-3 Decompartmentalization and Sheet Flow Enhancement Project Part 1. Available online at *http://www.evergladesplan.org/pm/pmp/pmp_docs/pmp_12_wca/decomp_main_apr_2002.pdf*. Accessed March 30, 2006.

USACE and SFWMD. 2004a. CERP Guidance Memorandum 016.00: Sea Level Rise Considerations for Formulation and Evaluation of CERP Projects. USACE and SFWMD, Jacksonville, FL.

USACE and SFWMD. 2004b. Central and Southern Florida Project Indian River Lagoon-South. Final Integrated Project Implementation Report and Environmental Impact Statement. Jacksonville, FL: USACE.

USACE and SFWMD. 2005a. Programmatic Regulations: Master Implementation Sequencing Plan 1.0 (April 6, 2005). Available online at *http://www.evergladesplan.org/pm/pm_docs/misp/040605misp_report_1.0.pdf.* Accessed January 30, 2006.

USACE and SFWMD. 2005b. North Palm Beach County Phase 1 Project Management Plan. May 2005. Available online at *http://www.evergladesplan.org/pm/projects/_proj_17_npbc_1.aspx.* Accessed July 19, 2008.

USACE and SFWMD. 2005c. Central and Southern Florida Project: Picayune Strand (Formerly Southern Golden Gate Estates Ecosystem Restoration) Final Integrated Project Implementation Report/Environmental Impact Statement. Available online at *http://www.evergladesplan.org/pm/projects/docs_30_sgge_pir_final.cfm.* Accessed March 22, 2006.

USACE and SFWMD. 2006. Site 1 Impoundment: Project Final Integrated Project Implementation Report and Environmental Assessment. Revised Final August 2006. Available online at *http://www.evergladesplan.org/pm/projects/docs_40_site_1_pir.aspx.* Accessed May 27, 2008.

USACE and SFWMD. 2007a. Comprehensive Everglades Restoration Plan: Programmatic Regulations: Six Program-wide Guidance Memoranda. Revised Final Draft. Available online at *http://www.evergladesplan.org/pm/pm_docs/prog_regulations/072707_prog_regs_rev_final_dft_gm.pdf.* Accessed May 7, 2008.

USACE and SFWMD. 2007b. Program Management Plan: Information and Data Management Available online at *http://www.evergladesplan.org/pm/pm_docs/data_mgmt/info_data_pmp_april_2007_final.pdf.* Accessed June 15, 2008.

USACE and SFWMD. 2008a. Preliminary Draft Incremental Adaptive Restoration Proposal for the Southern Everglades Restoration Projects. [White paper]. January 24, 2008.

USACE and SFWMD. 2008b. CERP Guidance Memorandum CGM Number-Revision: 040.01. Project-level Water Quality and Hydrometeorologic Monitoring and Assessment. Available online at *http://www.cerpzone.org/documents/cgm/CGM_040-01_Final_5-20- 08.pdf.* Accessed August 12, 2008.

USACE and the State of Florida. 2007. Intergovernmental Agreement among the United States Department of the Army and the State of Florida Establishing Interim Targets for the Comprehensive Everglades Restoration Plan. Available online at *http://www.evergladesplan.org/pm/pm_docs/prog_regulations/081607_int_targets.pdf.* Accessed August 19, 2008.

USACE, DOI, and the State of Florida. 2007. Intergovernmental Agreement Among the United States Department of the Army, the United States Department of the Interior, and the State of Florida Establishing Interim Restoration Goals for the Comprehensive Everglades Restoration Plan. Available online at *http://www.evergladesplan.org/pm/pm_docs/prog_regulations/081607_int_goals.pdf.* Accessed August 19, 2008.

U.S. Bureau of the Census. 1975. Historical Statistics of the United States: Colonial Times to 1970, Part 1. Washington, DC: U.S. Census Bureau.

U.S. Bureau of the Census. 2005. Table A1: Interim Projections for the Total Population for the United States: April 1, 2000 to July 1, 2030. Available online at *http://www.census.gov/population/projections/SummaryTabA1.pdf.* Accessed December 23, 2007.

U.S. Congress. 2007. Water Resources Development Act of 2007 Conference Report. Available online at *http://www.rules.house.gov/110/text/110_hr1495cr.pdf.* Accessed June 13, 2008.

USFWS (U.S. Fish and Wildlife Service). 2007. Everglade Snail Kite *Rostrhamus sociabilis plumbeus* 5-Year Review: Summary and Evaluation—5-year report on the kite. Vero Beach, FL: U.S. Fish and Wildlife Service South Florida Ecological Services Office.

Van de Koppel, J., D. v. d. Wal, J. P. Bakker, and P. M. J. Herman. 2004. Self-organization and vegetation collapse in salt marsh ecosystems. American Naturalist 165(1): E1–12.

Van der Valk, A., and F. H. Sklar. 2002. What we know and should know about tree islands. Pages 4499–4524 in F. H. Sklar and A. van der Valk, eds., Tree Islands of the Everglades. Boston, MA: Kluwer Academic.

Van Horn, S., C. Adorisio, C. Bedregal, J. Gomez, and J. Madden. 2008. Chapter 4: Phosphorus Source Controls for the Basins Tributary to the Everglades Protection Area. In 2008 South Florida Management Report. Available online at *https://my.sfwmd.gov/pls/portal/docs/PAGE/PG_GRP_SFWMD_SFER/PORTLET_SFER/TAB2236041/VOLUME1/chapters/v1_ch_4.pdf*. Accessed June 15, 2008.

Van Nes, E. H., and M. Scheffer. 2004. Large species shifts triggered by small forces. American Naturalist 164: 255–266.

Van Zee, R. 2007. Regional Simulation Model (RSM) Northern Everglades Features, Documentation and User Manual. West Palm Beach, FL: South Florida Water Management District.

Walters, C., and C. Holling. 1990. Large-scale management experiments and learning by doing. Ecology 71: 2060–2068.

Walters, J. R., S. R. Beissinger, J. W. Fitzpatrick, R. Greenberg, J. D. Nichols, H. R. Pulliam, and D. W. Winkler. 2000. The American Ornithologists Union Conservation Committee review of the biology, status, and management of Cape Sable seaside sparrows: Final report. Auk 117: 1093–1115.

Wang, D. L., M. W. Berry, and L. J. Gross. 2006. On parallelization of a spatially-explicit structured ecological model for integrated ecosystem simulation. International Journal of High Performance Computing Applications 20(4): 571–581.

Wanless, H. R., R. W. Parkinson, and L. P. Tedesco. 1994. Sea level control on stability of Everglades wetlands. Pages 199–222 in S. M. Davis, J. C. Ogden, and W. A. Park, eds., Everglades, the Ecosystem and Its Restoration. Fort Pierce, FL: St. Lucie Press.

Weinstein, M., J. Balletto, J. Teal, and D. Ludwig. 1997. Success criteria and adaptive management for large-scale wetland restoration projects. Wetlands Ecology and Management 4: 111–127

Welch, E. B., and G. D. Cooke, 1999. Effectiveness and longevity of phosphorus inactivation with alum. Lake and Reservoir Management 15(1): 5–27.

Willard, D. A., C. W. Holmes, M. S. Korvela, D. Mason, J. B. Murray, W. H. Orem, and D. T. Towles. 2002. Paleoecological insights on fixed tree island development in the Florida Everglades: I. environmental controls. Pages 117–151 in F. Sklar and A. van der Valk, eds., Tree Islands of the Everglades. Dordrecht, The Netherlands: Kluwer Academic Publishers.

Williams, B. 2008. Chapter 7A: Comprehensive Everglades Restoration Plan Annual Report. In 2008 South Florida Environmental Report. Available online at *https://my.sfwmd.gov/pls/portal/docs/PAGE/PG_GRP_SFWMD_SFER/PORTLET_SFER/TAB2236041/VOLUME1/chapters/v1_ch_7a.pdf*. Accessed June 26, 2008.

Wilson, C., L. Scotto, J. Scarpa, A. Volety, S. Larrimore, and D. Haunert. 2005. Survey of water quality, oysters reproduction and oyster health in the St. Lucie estuary. Journal of Shellfish Research 24: 157–165.

Working Group for Post-Hurricane Planning for the Louisiana Coast. 2006. A New Framework for Planning the Future of Coastal Louisiana after the Hurricanes of 2005. Available online at *http://www.umces.edu/la-restore/New%20Framework%20Final.pdf*. Accessed June 11, 2008.

Zaffke, M. 1983. Plant communities of water conservation Area 3A: Base-line documentation prior to the operation of S-339 and S-340. South Florida Water Management District. Technical Memorandum.

Zwick, P. D., and M. H. Carr. 2006. Florida 2060: A Population Distribution Scenario for the State of Florida. Available online at *http://www.1000friendsofflorida.org/PUBS/2060/Florida-2060-Report-Final.pdf*. Accessed January 11, 2008.

Acronyms

AMO	Atlantic Multidecadal Oscillation
ASR	aquifer storage and recovery
ATLSS	Across Trophic Level System Simulation
BACI	before-after control impact
BMP	best management practice
CERP	Comprehensive Everglades Restoration Plan
CFR	Code of Federal Regulations
CISRERP	Committee on Independent Scientific Review of Everglades Restoration Progress
CMMI	Capability Maturity Model Integrated
CROGEE	Committee on the Restoration of the Greater Everglades Ecosystem
C&SF	Central and Southern Florida
CSOP	Combined Structural and Operational Plan
CSSS	Cape Sable seaside sparrow
DAMP	Decomp Adaptive Management Plan
DASR	data access, storage, and retrieval
DEP	Department of Environmental Protection
DER	Department of Environmental Regulation
DOI	U.S. Department of the Interior
DPM	Decomp physical model
EAA	Everglades Agricultural Area
ELM	Everglades Landscape Model
ENP	Everglades National Park
ENPSM	Everglades National Park Seepage Management

ENSO	El Niño/Southern Oscillation
EPA	Everglades Protection Area
ESA	Endangered Species Act
FDACS	Florida Department of Agriculture and Consumer Services
FDEP	Florida Department of Environmental Protection
FDOT	Florida Department of Transportation
FWS	U.S. Fish and Wildlife Service
GAO	Government Accountability Office
GIS	geographic information system
GRR	General Reevaluation Report
HCP	habitat conservation plan
HSE	Hydrologic Simulation Engine
HSI	habitat suitability index
IAR	incremental adaptive restoration
IDM	information and data management
IDS	integrated delivery schedule
IMC	Interagency Modeling Center
IOP	Interim Operational Plan
IPCC	Intergovernmental Panel on Climate Change
IRL	Indian River Lagoon
IRL-S	Indian River Lagoon-South
ISOP	Interim Structural and Operational Plan
LILA	Loxahatchee Impoundment Landscape Assessment
LNWR	Loxahatchee National Wildlife Refuge
LOER	Lake Okeechobee and Estuary Recovery
LOPA	Lake Okeechobee Protection Act
LOPP	Lake Okeechobee Protection Plan
LOWCP-II	Lake Okeechobee Watershed Construction Project: Phase II Technical Plan
LRR	Limited Reevaluation Report
MAF	million acre-feet
MAP	monitoring and assessment plan
MGD	million gallons per day
MISP	Master Implementation Sequencing Plan

MSE	Management Simulation Engine
mt	metric tons
NAI	next added increment
NEEPP	Northern Everglades and Estuaries Protection Program
NERSM	Northern Everglades Regional Simulation Model
NGVD	National Geodetic Vertical Datum
NPDES	National Pollution Discharge Elimination System
NPS	National Park Service
NRC	National Research Council
NSM	Natural System Model
NSRSM	Natural System Regional Simulation Model
PDO	Pacific Decadal Oscillation
PIRs	program implementation reports
ppb	parts per billion
QASR	Quality Assurance Systems Requirements
RECOVER	Restoration, Coordination, and Verification
RLSA	Rural Land Stewardship Area
RSM	Regional Simulation Model
SAV	submerged aquatic vegetation
SEI	Sustainable Ecosystems Institute
SFEER	South Florida Everglades Ecosystem Restoration Program
SFERTF	South Florida Ecosystem Restoration Task Force
SFWMD	South Florida Water Management District
SFWMM	South Florida Water Management Model
SMA	square-mile area
SRS	Shark River Slough
SSR	System Status Report
STA	stormwater treatment area
SWIM	Surface Water Improvement and Management
TMDL	total maximum daily load
TP	total phosphorus
TRD	Transfer of Development Rights
USACE	U.S. Army Corps of Engineers

USGS	U.S. Geological Survey
WCA	Water Conservation Area
WMA	Wildlife Management Area
WRDA	Water Resources Development Act
WSE	Water Supply and Environmental
WSTB	Water Science and Technology Board

Glossary

8.5-square-mile area—The 8.5-square-mile area (SMA) is a low-lying, partially developed area near the northeast corner of Everglades National Park, west of the L-31 north canal. Flood protection was to have been provided under the original 1989 Mod Waters legislation, but years of subsequent study and negotiations with property owners resulted in a compromise in which a flood protection levee is to be built around approximately two-thirds of the 8.5 SMA while providing for purchase of approximately one-third of the private property and 12 homes in the western portion.

Acceler8—An expedited course of action for achieving Everglades restoration. Through Accler8, the State of Florida intends to implement 11 components of the CERP.

Across Trophic Level System Simulation (ATLSS)—A modeling system that uses topographic data to convert the 2 × 2 mile landscape of the regional hydrologic models to a 500 × 500 m landscape to which various ecological models are applied. These range from highly parameterized, mechanistic individual-based models (e.g., EVERKITE, SIMSPAR) to simpler, habitat-suitability models (Spatially-Explicit Species Index and Habitat Suitability Index). The objectives of the ATLSS project are to utilize the outputs of systems models to drive a variety of models that attempt to compare and contrast the relative impacts of alternative hydrologic scenarios on the biotic components of South Florida.

Aquifer storage and recovery (ASR)—A technology for storage of water in a suitable aquifer when excess water is available and recovery from the same aquifer when the water is needed to meet peak emergency or long-term water demands. Wells are used to pump water in and out of the aquifer.

Atlantic Multidecadal Oscillation (AMO)—An ongoing series of long-duration changes in the sea surface temperature of the North Atlantic Ocean, with cool and warm phases that may last for 20–40 years at a time and a difference of about 1°F between extremes. These changes are natural and have been occurring for at least the last 1,000 years. The AMO has affected air temperatures and rainfall over much of the Northern Hemisphere—particularly in North America and Europe—and is associated with changes in the frequency of North American droughts and of severe Atlantic hurricanes. Also, it alternately obscures and exaggerates the global increase in temperatures due to human activities.

Best management practices (BMPs)—Effective, practical methods that prevent or reduce the movement of sediment, nutrients, pesticides, and other pollutants resulting from agricultural, industrial, or other societal activities from the land to surface or groundwater or that optimize water use.

Central and Southern Florida (C&SF) Project for Flood Control and Other Purposes—A multipurpose project, first authorized by the U.S. Congress in 1948, to provide flood control, water supply protection, water quality protection, and natural resource protection.

Clean Water Act (CWA)—The Clean Water Act is the cornerstone of surface water quality protection in the United States. The statute employs a variety of regulatory and nonregulatory tools to reduce direct pollutant discharges into waterways, finance municipal wastewater treatment facilities, and manage polluted runoff. These tools help to achieve the broader goal of restoring and maintaining the chemical, physical, and biological integrity of the nation's waters so that they can support the protection and propagation of fish, shellfish, and wildlife and recreation in and on the water.

Comprehensive Everglades Restoration Plan (CERP)—The plan for the restoration of the South Florida ecosystem authorized by Congress in 2000.

Conceptual ecological models—Nonquantitative, verbal or diagrammatic hypotheses about the major anthropogenic and natural drivers and stressors on natural systems, the ecological effects of these stressors, and the biological attributes or indicators of these ecological responses. They are used as planning tools for research and adaptive management.

Critical Projects—Projects determined to be critical to the restoration of the South Florida ecosystem that were authorized in 1996 prior to the CERP. These projects are comparatively small and were undertaken by the U.S. Army Corps of Engineers and South Florida Water Management District. They are being implemented along with the CERP projects.

Decomp—Short title for Water Conservation Area 3 Decompartmentalization and Sheet Flow Enhancement—Part 1 project.

El Niño/Southern Oscillation—A coupled atmosphere-ocean phenomenon that occurs at timescales of 2 to about 7 years, collectively known as the El Niño/Southern Oscillation (ENSO). During an ENSO event, the prevailing trade winds in the South Pacific weaken, reducing upwelling of cold, deep water and altering ocean currents such that the sea surface temperatures warm. The warming further weakens the trade winds. The cold phase of ENSO is called La Niña. ENSO has global effects on climate, affecting agriculture, fisheries, and other human activities.

Endangered Species Act (ESA)—A United States law passed in 1973 to protect species listed by the federal government as threatened or endangered from extinction. It provides penalties for the taking of such species and requires any federal agency to consult with the U.S. Fish and Wildlife Service (or National Marine Fisheries Service for marine species) before undertaking or funding any action that could jeopardize the continued existence or recovery of listed species.

Estuary—The portion of the Earth's coastal zone where seawater, fresh water, and land, interact, typically arms of the sea where tide meets river currents.

Everglades—A mosaic of wetlands, uplands, and coastal areas that extends from the Kissimmee River basin to Florida Bay.

Everglades Agricultural Area (EAA)—Land in the northern Everglades south of Lake Okeechobee that was drained for agricultural use.

Everglades Construction Project—Twelve interrelated construction projects located between Lake Okeechobee and the Everglades. Six storm-water treatment areas (STAs, or constructed wetlands) totaling more than 47,000 acres are the cornerstone of the project. The STAs rely on physical and biological processes to reduce the level of total phosphorous entering the Everglades to an interim goal of 50 parts per billion.

Everglades Depth Estimation Network (EDEN)—A U.S. Geological Survey surface-water hydrologic monitoring network in support of the monitoring and assessment plan (MAP) projects that is intended to provide the hydrologic data necessary to integrate hydrologic and biological responses to the CERP during MAP performance measurement assessment and evaluation for the Greater Everglades module.

Everglades Landscape Model (ELM)—Model used to predict the landscape response to different water management scenarios. ELM consists of a set of integrated modules to understand ecosystem dynamics at a regional scale and simulates the biogeochemical processes associated with hydrology, nutrients, soil formation, and vegetation succession. Its main components include hydrology, water quality, soils, periphyton, and vegetation.

Everglades Protection Area—As defined in the Everglades Forever Act, the Everglades Protection Area comprises water conservation areas 1 (also known as the Arthur R. Marshall Loxahatchee National Wildlife Refuge), 2A, 2B, 3A, 3B; the Arthur R. Marshall Loxahatchee National Wildlife Refuge; and the Everglades National Park.

Exotic species—An introduced species not native to the place where it is found. Usually used for species introduced from outside a country's borders.

Extirpated species—A species that has become extinct in a given area.

Flow—The volume of water that passes a given point per unit of time, including in-stream flow requirements, minimum flow, and peak flow. "Flow" is used generically within the text to mean the movement of volumes of water across the landscape, and it incorporates the concepts of volumetric flow rate (e.g., cubic feet per second), velocity, and direction. Volumetric flow rate may be estimated for large averaging times, such as acre-feet per year, as in the South Florida Water Management Model and the Natural Systems Model, and also on a short-term ("instantaneous") basis by other models, as discussed in Chapter 4.

Flux—The rate of transfer of fluid, particles, or energy across a given surface.

Footprint—The area of productive land and aquatic ecosystems required to produce the resources used and to assimilate the wastes produced by a defined population at a specified material standard of living, wherever that land might be located.

Foundation projects—Non-CERP Everglades restoration activities, many of which are essential (the foundation) for completion of the CERP.

Geographic information system (GIS)—A map-based data storage and retrieval system.

Guidance memorandum—A document of prescribed format that officially captures decisions of the program managers and promulgates their guidance regarding implementation of the CERP. The guidance memoranda address an array of subjects including definitions, direction and procedures for reporting, Web management, financial management, and program controls.

Habitat Conservation Plan—A plan required by Section 10(a)(2)(A) of the ESA for an applicant for an incidental take permit. The plan is required to include, among other things, the impacts that are likely to result from the taking and the measures the permit applicant will undertake to minimize and mitigate such impacts. Habitat conservation plans reduce conflicts among listed species and economic use or development activities, allowing for the development of "creative partnerships" between the public and private sectors, designed to make the process work for both landowners and species.

Habitat Suitability Index (HSI)—Tool used to define, in relative terms, the quality of the habitat for various plant and animal species. HSIs can be used as the first approximation toward quantifying the relationships identified in various conceptual ecological models.

Herbert Hoover Dike—A dike system surrounding Lake Okeechobee that provides flood and storm damage reduction and other water control benefits in Central and South Florida. It consists of 143 miles of levees with 19 culverts, hurricane gates, and other water control structures.

Hydroperiod—Annual temporal pattern of water levels.

Incremental Adaptive Restoration (IAR)—An alternative framework called for in NRC (2007) for advancing natural resource restoration in the Everglades. The aim of IAR is to resolve decision-critical scientific uncertainties and to address project sequencing constraints to improve the pace of restoration. As conceived, the IAR approach makes investments in restoration project increments that are large enough to secure significant environmental benefits, while simultaneously testing hypotheses selected to resolve important scientific uncertainties about the response of the system to management interventions. Such steps would likely be smaller than the CERP projects because the purpose of IAR is to take actions that help address some sources of delay in the pace of restoration progress as well as to promote learning that can guide the remainder of project design. As an application of adaptive management, IAR would require rigorous monitoring and assessment to test hypotheses, yielding valuable information that can expedite future decision making and improve future project design.

Interagency Modeling Center (IMC)—An equal partnership between the U.S. Army Corps of Engineers Jacksonville District and the South Florida Water Management District that serves as the modeling services single point of responsibility for the CERP. It provides, coordinates, and oversees the modeling needs and efforts of each project delivery team and the Restoration. Coordination, and Verification Program, or RECOVER.

Interim goal—A means by which the restoration success of the CERP may be evaluated throughout the implementation process.

Interim target—A means by which the success of the CERP in providing for water-related needs of the region, including water supply and flood protection, may be evaluated throughout the implementation process.

Invasive species—Species of plants or animals, both native and exotic, that aggressively invade habitats and cause multiple ecological changes.

Marl—A type of wetland soil high in clay and carbonates. Hydroperiod is a critical determinant of marl formation.

Master Implementation Sequencing Plan (MISP)—Specifies the sequence in which CERP projects are planned, designed, and constructed.

National Pollution Discharge Elimination System (NPDES)—As authorized by the Clean Water Act, this permit program controls water pollution by regulating point sources that discharge pollutants into the waters of the United States.

Natural system—According to the Water Resources Development Act of 2000 (WRDA 2000), all land and water managed by the federal government or the state within the South Florida ecosystem, including water conservation areas, sovereign submerged land, Everglades National Park, Biscayne National Park, Big

Cypress National Preserve, other federal or state (including a political subdivision of a state) land that is designated and managed for conservation purposes, and any tribal land that is designated and managed for conservation purposes, as approved by the tribe.

Natural System Model (NSM)—Model that simulates hydropatterns before canals, levees, dikes, and pumps were built. The NSM mimics frequency, duration, depth, and spatial extent of water inundation under pre-management (i.e., natural) hydrologic conditions. In many cases, those pre-management water levels are used as a target for hydrologic restoration assuming that restoration of the hydrologic response that existed prior to drainage of the system would lead to restoration of natural habitats and biota.

Natural System Regional Simulation Model (NSRSM)—Application of the updated Regional Systems Model to simulate the natural system hydrology of South Florida. The use of refined input parameters, in combination with the model's improved hydrologic simulation engine, results in simulations that reasonably represent pre-drainage (mid-1800) hydrology within an estimated range of performance.

Original Everglades—The pre-drainage Everglades, or that which existed prior to the construction of drainage canals beginning in the late 1800s.

Part per billion (ppb)—A measure of concentration equivalent to one microgram of solute per liter of solution.

Part per million (ppm)—A measure of concentration equivalent to one milligram of solute per liter of solution.

Passive adaptive management—Adaptive management by which a preferred course of action is selected based on existing information and understanding. Outcomes are monitored and evaluated, and subsequent decisions (e.g., adjustments in design or operations, the design of subsequent projects, etc.) are adjusted based on improved understanding. It is distinguished from active adaptive management, which involves designing management actions as experimental activities, to enhance the learning process.

Performance measure—A quantifiable indicator of ecosystem response to changes in environmental conditions.

Periphyton—A biological community of algae, bacteria, fungi, protists, and other microorganisms. In the Everglades, periphyton grows on top of the soil surface—attached to the stems of rooted vegetation—and in the water column or at the water surface, sometimes in association with other floating vegetation.

Programmatic Regulations—Procedural framework and specific requirements called for in Section 601(h)(3) of WRDA 2000. The programmatic regulations are intended to guide implementation of the CERP and to ensure that the goals and purposes of the CERP are achieved. The final rule for the Programmatic Regulations (33 CFR § 385) was issued in November 2003.

Project delivery team (PDT)—An interdisciplinary group that includes representatives from the implementing agencies. PDTs develop the products necessary to deliver the project.

Project implementation report (PIR)—A decision document that bridges the gap between the conceptual design contained in the comprehensive plan and the detailed design necessary to proceed to construction.

Project management plan (PMP)—A document that establishes the project's scope, schedule, costs, funding requirements, and technical performance requirements (including the various functional area's performance and quality criteria) and that will be used to produce and deliver the products that comprise the project.

RECOVER—The Restoration, Coordination, and Verification Program (RECOVER) is an arm of the CERP responsible for linking science and the tools of science to a set of system-wide planning, evaluation, and assessment tasks. RECOVER's objectives are to evaluate and assess CERP performance; refine and improve the CERP during the implementation period; and ensure that a system-wide perspective is maintained throughout the restoration program. RECOVER conducts scientific and technical evaluations and assessments for improving CERP's ability to restore, preserve, and protect the South Florida ecosystem while providing for the region's other water-related needs. RECOVER communicates and coordinates the results of these evaluations and assessments.

Regional Simulation Model (RSM)—A regional finite-volume-based hydrologic model developed principally for application in South Florida that simulates the coupled movement and distribution of groundwater and surface water throughout the model domain using a Hydrologic Simulation Engine to simulate the natural hydrology and a Management Simulation Engine to simulate water control operations.

Ridge—Elevated areas of sawgrass habitat that rise above the foot-and-a-half deeper sloughs. A ridge may be submerged or above the water surface.

Savings Clause—Provision of WRDA 2000 that is designed to ensure that an existing legal source of water (e.g., agricultural or urban water supply, water supply for Everglades National Park, water supply for fish and wildlife) is not eliminated or transferred until a replacement source of water of comparable quantity and quality—as was available on the date of enactment of WRDA 2000—is available and that existing levels of flood protection are not reduced.

Sawgrass plain—An unbroken expanse of dense, tall (up to 10 feet) sawgrass that originally covered most of the northern Everglades. Agricultural crops, mainly sugar cane, have replaced most of the sawgrass plain area, but some tall sawgrass remains in the water conservation areas.

Sheet flow—Water movement as a broad front with shallow, uniform depth.

Slough—A depression associated with swamps and marshlands as part of a bayou, inlet, or backwater; contains areas of slightly deeper water and a slow current and can be thought of as the broad, shallow rivers of the Everglades.

South Florida ecosystem—An area consisting of the lands and waters within the boundary of the South Florida Water Management District, including the built environment, the Everglades, the Florida Keys, and the contiguous near-shore coastal waters of South Florida (also known as the Greater Everglades ecosystem).

South Florida Ecosystem Restoration Task Force (SFERTF or Task Force)—The Task Force was established by the WRDA of 1996 to coordinate policies, programs, and science activities among the many restoration partners in South Florida. Its 14 members include the secretaries of Interior (chair), Commerce, Army, Agriculture, and Transportation; the Attorney General; and the Administrator of the Environmental Protection Agency; or their designees. The Secretary of the Interior appoints one member each from the Seminole Tribe of Florida and the Miccosukee Tribe of Indians of Florida. The Secretary of the Interior also appoints, based on recommendations of the governor of Florida, two representatives of the state of Florida, one representative of the South Florida Water Management District, and two representatives of local Florida governments.

South Florida Water Management Model (SFWMM)—A model that simulates hydrology and water systems. It is widely accepted as the best available tool for analyzing structural and/or operational changes to the complex water management system in South Florida at the regional scale.

Storm-water Treatment Area (STA)—A human constructed wetland area to treat urban and agricultural runoff water before it is discharged to the natural areas.

Submerged aquatic vegetation (SAV)—Plants that grow completely below the water surface.

Total maximum daily load (TMDL)—A calculation of the maximum amount of a pollutant that a body of water can receive and still safely meet water quality standards.

Total phosphorus (TP)—Sum of phosphorus in dissolved and particulate forms.

Tree island—Patch of forest in the Everglades marsh occurring in the central peatlands and the peripheral marl prairies of the southern and southeastern Everglades and on higher ground than ridges. Sizes range from as small as one-hundredth of an acre to hundreds of acres.

Water conservation areas (WCAs)—Everglades marshland areas that were modified for use as storage to prevent flooding, to irrigate agriculture land and recharge well fields, to supply water for Everglades National Park, and for general water conservation. WCA-1, WCA-2A, WCA-2B, WCA-3A, and WCA-3B comprise five surface-water management basins in the Everglades; bounded by the Everglades Agricultural Area on the north and the Everglades National Park basin on the south, the WCAs are confined by levees and water control structures that regulate the inflows and outflows to each one of them. Restoration of more natural water levels and flows to the WCAs is a main objective of the CERP.

Water reservations—According to WRDA 2000, the state shall, under state law, make sufficient reservations of water provided by each CERP project for the natural system in accordance with the project implementation report for that project and consistent with the plan before water made available by a project is permitted for a consumptive use or otherwise made unavailable.

Water Resources Development Act (WRDA) of 2000—Legislation that authorized the Comprehensive Everglades Restoration Plan as a framework for modifying the Central and South Florida Project to increase future water supplies, with the appropriate quality, timing, and distribution, for environmental purposes so as to achieve a restored Everglades natural system as much as possible, while at the same time meeting other water-related needs of the ecosystem. WRDAs are passed periodically, the most recent one having been enacted in 2007; they provide the mechanism for authorizing CERP activities.

Water year—Time convention used as a basis for processing stream flow and other hydrologic data. In the Northern Hemisphere, the water year begins October 1 and ends September 30; in the Southern Hemisphere, it begins July 1 and ends June 30. The water year is designated by the calendar year in which it ends.

Wetlands—Areas that are inundated or saturated by surface water or groundwater at a frequency and duration sufficient to support a prevalence of vegetative or aquatic life that requires saturated or seasonally saturated soil conditions for growth and reproduction.

Yellow Book—This is the common name for the *Central and Southern Florida Comprehensive Review Study Final Integrated Feasibility Report and Programmatic Environmental Impact Statement* (USACE and SFWMD, 1999), which laid out the Comprehensive Everglades Restoration Plan.

Appendixes

Appendix A

National Research Council Everglades Reports

Progress Toward Restoring the Everglades: The First Biennial Review, 2006 **(2007)**

This report is the first in a congressionally mandated series of biennial evaluations of the progress being made by the Comprehensive Everglades Restoration Plan (CERP), a multibillion-dollar effort to restore historical water flows to the Everglades and return the ecosystem closer to its natural state, before it was transformed by drainage and by urban and agricultural development. The report finds that progress has been made in developing the scientific basis and management structures needed to support a massive effort to restore the Florida Everglades ecosystem. However, some important projects have been delayed due to several factors including budgetary restrictions and a project planning process that that can be stalled by unresolved scientific uncertainties. The report outlines an alternative approach that can help the initiative move forward even as it resolves remaining scientific uncertainties. The report calls for a boost in the rate of federal spending if the restoration of Everglades National Park and other projects are to be completed on schedule.

Re-engineering Water Storage in the Everglades: Risks and Opportunities **(2005)**

A Comprehensive Everglades Restoration Plan (CERP) was formulated in 1999 with the goal of restoring the original hydrologic conditions of the remaining Everglades. A major feature of this plan is providing enough storage capacity to meet human and ecological needs. This report reviews and evaluates not only storage options included in the plan, but also other options not considered in the plan. Along with providing hydrologic and ecological analyses of the size, location, and functioning of water storage components, the report also discusses and makes recommendations on related critical factors, such as timing of land acquisition, intermediate states of restoration, and trade-offs among competing goals and ecosystem objectives.

The CERP imposes some constraints on sequencing of its components. The report concludes that two criteria are most important in deciding how to sequence components of such a restoration project: (1) protecting against additional habitat loss by acquiring or protecting critical lands in and around the Everglades, and (2) providing ecological benefits as early as possible.

There is a considerable range in the degree to which various proposed storage components involve complex design and construction measures, rely on active controls and frequent equipment maintenance, and require fossil fuels or other energy sources for operation. The report recommends that, to the extent possible, the CERP should develop storage components that have fewer of those requirements, and are thus less vulnerable to failure and more likely to be sustainable in the long term.

Further, as new information becomes available and as the effectiveness and feasibility of various restoration components become clearer, some of the earlier adaptation and compromises might need to be revisited. The report recommends that methods be developed to allow trade-offs to be assessed over broad spatial and long temporal scales, especially for the entire ecosystem, and gives an example of what an overall performance indicator for the Everglades system might look like.

Adaptive Monitoring and Assessment for the Comprehensive Everglades Restoration Plan (2003)

A key premise of the CERP is that restoring the historical hydrologic regime in the remaining wetlands will reverse declines in many native species and biological communities. Given the uncertainties that will attend future responses of Everglades ecosystems to restored water regimes, a research, monitoring, and adaptive management program is planned. This report assessed the extent to which the restoration effort's "monitoring and assessment plan" included the following elements crucial to any adaptive management scheme: (1) clear restoration goals and targets, (2) a sound baseline description and conceptualization of the system, (3) an effective process for learning from management actions, and (4) feedback mechanisms for improving management based on the learning process.

The report concludes that monitoring needs must be prioritized, because many goals and targets that have been agreed to may not be achievable or internally consistent. Priorities could be established based on the degree of flexibility or reversibility of a component and its potential impact on future management decisions. Such a prioritization should be used for scheduling and sequencing of projects, for example. Monitoring that meets multiple objectives (e.g., adaptive management, regulatory compliance, and a "report card") should be given priority.

Ecosystem-level, system-wide indicators should be developed, such as land-cover and land-use measures, an index of biotic integrity, and diversity measures. Region-wide monitoring of human and environmental drivers of the ecosystem, especially population growth, land-use change, water demand, and sea-level rise are recommended. Monitoring, modeling, and research should be well integrated, especially with respect to defining the restoration reference state and using "active" adaptive management.

Does Water Flow Influence Everglades Landscape Patterns? (2003)

A commonly stated goal of the CERP is to "get the water right." This has largely meant restoring the timing and duration of water levels and the water quality in the Everglades. Water flow (speed, discharge, direction) has been considered mainly in the coastal and estuarine system, but not elsewhere. Should the restoration plan be setting targets for flows in other parts of the Everglades as well?

There are legitimate reasons why flow velocities and discharges have thus far not received greater emphasis in the plan. These include a relative lack of field information and poor resolution of numerical models for flows. There are, however, compelling reasons to believe that flow has important influences in the central Everglades ecosystem. The most important reason is the existence of major, ecologically important landforms—parallel ridges, sloughs, and "tree islands"—are aligned with present and inferred past flow directions. There are difficulties in interpreting this evidence, however, as it is essentially circumstantial and not quantitative.

Alternative mechanisms by which flow may influence this landscape can to some extent be evaluated from short-term research on underlying bedrock topography, detailed surface topographic mapping, and accumulation rates of suspended organic matter. Nonetheless, more extensive and long-term research will also be necessary, beginning with the development of alternative conceptual models of the formation and maintenance of the landscape to guide a research program. Research on maintenance rather than evolution of the landscape should have higher priority because of its direct impact on restoration. Monitoring should be designed for the full range of flow conditions, including extreme events.

Overall, flows approximating historical discharges, velocities, timing, and distribution should be considered in restoration design, but quantitative flow-related performance measures are not appropriate until there is a better scientific understanding of the underlying science. At present, neither a minimum nor a maximum flow to preserve the landscape can be established.

Florida Bay Research Programs and Their Relation to the Comprehensive Everglades Restoration Plan (2002)

This report of the Committee on Restoration of the Greater Everglades Ecosystem (CROGEE) evaluated Florida Bay studies and restoration activities that potentially affect the success of the CERP. Florida Bay is a large, shallow marine system immediately south of the Everglades, bounded by the Florida Keys and the Gulf of Mexico. Some of the water draining from the Everglades flows directly into northeast Florida Bay. Other freshwater drainage reaches the bay indirectly from the northwest.

For several decades until the late 1980s, clear water and dense seagrass meadows characterized most of Florida Bay. However, beginning around 1987, the seagrass beds began dying in the western and central bay. It is often assumed that increased flows to restore freshwater Everglades habitats will also help restoration of Florida Bay. However, the CERP may actually result in higher salinities in central Florida Bay than exist presently, and thus exacerbate the ecological problems. Further, some percentage of the proposed increase in fresh surface water flow discharging northwest of the bay will eventually reach the central bay, where its dissolved organic nitrogen may lead to algal blooms. Complicating the analysis of such issues is the lack of an operational bay circulation model.

The report notes the importance of additional research in the following areas: estimates of groundwater discharge to the bay; full characterization and quantification of surface runoff in major basins; transport and total loads of nitrogen and phosphorous from freshwater sources, especially in their organic forms; effects on nutrient fluxes of decreasing freshwater flows into the northeastern bay, and of increasing flows northwest of the bay; and the development of an operational Florida Bay circulation model to support a bay water quality model and facilitate analysis of CERP effects on the bay.

Science and the Greater Everglades Ecosystem Restoration: An Assessment of the Critical Ecosystems Study Initiative (2003)

The path to restoration will not be easy, but sound scientific information will increase the reliability of the restoration, help enable solutions for unanticipated problems, and potentially reduce long-term costs. The investment in scientific research relevant to restoration, however, decreased substantially within some agencies, including one major Department of the Interior (DOI) science program, the Critical Ecosystem Studies Initiative (CESI). In response to concerns regarding declining levels of funding for scientific research and the adequacy of science-based support for restoration decision making, the U.S. Congress instructed the DOI to commission the National Academy of Sciences to review the scientific component of the CESI and provide recommendations for program management, strategic planning, and information dissemination.

Although improvements should be made, this report notes that the CESI has contributed useful science in support of the DOI's resource stewardship interests and restoration responsibilities in South Florida. It recommends that the fundamental objectives of the CESI research program remain intact, with continued commitment to ecosystem research. Several improvements in CESI management are suggested, including broadening the distribution of requests for proposals and improving review standards for proposals and research products. The report asserts that funding for CESI science has been inconsistent and as of 2002 was less than that needed to support the DOI's interests in and responsibilities for restoration. The development of a mechanism for comprehensive restoration-wide science coordination and synthesis is recommended to enable improved integration of scientific findings into restoration planning.

Regional Issues in Aquifer Storage and Recovery for Everglades Restoration: A Review of the ASR Regional Study Project Management Plan of the Comprehensive Everglades Restoration Plan (2002)

The report reviews a comprehensive research plan on Everglades restoration drafted by federal and Florida officials that assesses a central feature of the restoration: a proposal to drill more than 300 wells funneling up to 1.7 billion gallons of water a day into underground aquifers, where it would be stored and then pumped back to the surface to replenish the Everglades during dry periods. The report says that the research plan goes a long way to providing information needed to settle remaining technical questions and clearly responds to suggestions offered by scientists in Florida and in a previous report by the NRC.

Aquifer Storage and Recovery in the Comprehensive Everglades Restoration Plan: A Critique of the Pilot Projects and Related Plans for ASR in the Lake Okeechobee and Western Hillsboro Areas (2001)

Aquifer storage and recovery (ASR) is a major component in the CERP, which was developed by the USACE and the SFWMD. The plan would use the upper

Floridan aquifer to store large quantities of surface water and shallow groundwater during wet periods for recovery during droughts.

ASR may limit evaporation losses and permit recovery of large volumes of water during multiyear droughts. However, the proposed scale is unprecedented and little subsurface information has been compiled. Key unknowns include impacts on existing aquifer uses, suitability of source waters for recharge, and environmental and/or human health impacts due to water quality changes during subsurface storage.

To address these issues, the USACE and the SFWMD proposed aquifer storage recharge pilot projects in two key areas. The CROGEE charge was to examine a draft of their plans from a perspective of adaptive management. The report concludes that regional hydrogeologic assessment should include development of a regional-scale groundwater flow model, extensive well drilling and water quality sampling, and a multi-objective approach to ASR facility siting. It also recommends that water quality studies include laboratory and field bioassays and ecotoxicological studies, studies to characterize organic carbon of the source water and anticipate its effects on subsurface biogeochemical processes, and laboratory studies. Finally, it recommends that pilot projects be part of adaptive assessment.

Appendix B

Summary from *Progress Toward Restoring the Everglades: The First Biennial Review - 2006*

Florida's Everglades have been transformed in the past century by urban and agricultural development. Once encompassing 3 million acres, they are now about half that size, and their waters are polluted with phosphorus, nitrogen, mercury, and pesticides. Associated drainage and flood-control structures have diverted large quantities of water to the ocean, reducing the freshwater inflows that defined the original ecosystem. The altered hydrologic system has contributed to dramatic declines in populations of wading birds, a 67 percent decline in the area of tree islands, and manifold changes in the ecosystem of Florida Bay. Invasive exotic species occupy much of the Everglades watershed, cattail has replaced vast areas of native sawgrass marsh, and 68 plant and animal species in South Florida are listed as federally threatened or endangered. Restoration of what remains of the Everglades ecosystem became the focus of activities that began in the 1990s and continue today, representing one of the most ambitious ecosystem restoration projects ever conceived.

The Comprehensive Everglades Restoration Plan (CERP) was unveiled in 1999 by the U.S. Army Corps of Engineers (USACE) and the South Florida Water Management District (SFWMD). The CERP aims to achieve ecological restoration by reestablishing hydrologic characteristics as close as possible to their pre-drainage conditions in what remains of the Everglades ecosystem, recognizing that irreversible changes to the landscape make restoration to full pre-drainage conditions impossible. The CERP includes more than 40 major projects and 68 project components to be constructed at an estimated cost of $10.9 billion in 2004 dollars. The projects embodied in the CERP are expected to take more than three decades to complete.

The Committee on Independent Scientific Review of Everglades Restoration Progress was established in 2004 in response to a request from the USACE, with support from the SFWMD and the U.S. Department of the Interior, based on Congress's mandate in the Water Resources Development Act of 2000 (WRDA 2000). The committee is charged to submit biennial reports that review the

BOX S-1
Statement of Task

This congressionally mandated activity will review the progress toward achieving the restoration goals of the Comprehensive Everglades Restoration Plan (CERP). The committee will meet approximately four times annually to receive briefings on the current status of the CERP and scientific issues involved in implementing the Plan. It will publish a report every other year providing:

1. an assessment of progress in restoring the natural system, which is defined by section 601(a) of WRDA 2000 as all the land and water managed by the federal government and state within the South Florida ecosystem;
2. discussion of significant accomplishments of the restoration;
3. discussion and evaluation of specific scientific and engineering issues that may impact progress in achieving the natural system restoration goals of the Plan; and
4. independent review of monitoring and assessment protocols to be used for evaluation of CERP progress (e.g., CERP performance measures, annual assessment reports, assessment strategies).

CERP's progress in restoring the natural system (see Box S-1). This is the committee's first report in a series of biennial evaluations that are scheduled to last the lifetime of the CERP.

The committee concludes that much good science has been developed to support the restoration efforts and that progress has been made in CERP program support, particularly in the monitoring and assessment program. However, no CERP projects have been completed to date, and anticipated restoration progress in the Water Conservation Areas (WCAs) and Everglades National Park appears to be lagging behind the production of natural system restoration benefits in other portions of the South Florida ecosystem. Additionally there have been some troubling delays in some projects that are important to the restoration of the Everglades ecosystem. These delays have resulted from several factors, including budgetary restrictions and a project planning process that that can be stalled by unresolved scientific uncertainties. Restoration benefits from early water storage projects remain uncertain because decisions have not yet been made regarding water allocations for the natural system.

SOUTH FLORIDA ECOSYSTEM RESTORATION

The South Florida Ecosystem Restoration Task Force (Task Force), an intergovernmental body established to facilitate coordination in the restoration effort,

has three broad strategic goals for the South Florida ecosystem:[1] (1) "get the water right;" (2) "restore, preserve, and protect natural habitats and species;" and (3) "foster compatibility of the built and natural systems." These goals encompass, but are not limited to, the CERP.

The goal of the CERP, as stated in WRDA 2000, is "restoration, preservation, and protection of the South Florida Ecosystem while providing for other water-related needs of the region, including water supply and flood protection." The Programmatic Regulations that guide implementation of the CERP further clarify this goal by defining restoration as "the recovery and protection of the South Florida ecosystem so that it once again achieves and sustains the essential hydrological and biological characteristics that defined the undisturbed South Florida ecosystem." These defining characteristics include a large areal extent of interconnected wetlands, extremely low concentrations of nutrients in freshwater wetlands, sheet flow, healthy and productive estuaries, resilient plant communities, and an abundance of native wetland animals. At the same time, the CERP is charged to maintain current levels of flood protection and to provide for other water-related needs, including water supply, for a rapidly growing human population in South Florida. Although the CERP contributes to each of the Task Force goals, it focuses primarily on restoring the hydrologic features of the undeveloped wetlands remaining in the South Florida ecosystem, on the assumption that improvements in ecological conditions should follow.

Both political and scientific issues contribute to the difficulty of specifying restoration goals. The goals, therefore, cannot be viewed as fixed endpoints but are instead approximations of the objectives that should be developed by careful analyses and reevaluated as new knowledge emerges. Even with clearly articulated restoration goals, disparate expectations for restoration may exist among stakeholders, including both its geographic extent and its functional characteristics. The Everglades restoration efforts are thus occurring in a challenging environment.

Restoration Activities

Several restoration programs, including the CERP—the largest of the initiatives—are now under way. The CERP, led by the USACE and the SFWMD, consists primarily of projects to increase storage capacity (e.g., conventional surface-water reservoirs, aquifer storage and recovery, in-ground reservoirs), improve water quality (e.g., stormwater treatment areas [STAs]), reduce loss of water from the system (e.g., seepage management, water reuse, and conserva-

[1]See Box 1-1 for definitions of geographic terms used in this report.

tion), and reestablish pre-drainage hydrologic patterns wherever possible (e.g., removing barriers to sheet flow, rainfall-driven water management). The largest portion of the budget is devoted to water storage and conservation and to acquiring the lands needed for those projects.

The CERP builds upon other activities of the state and federal government aimed at restoration (hereafter, non-CERP activities), many of which are essential to the success of the CERP. These include Modified Water Deliveries to Everglades National Park (Mod Waters) and modification of the C-111 canal—projects that will alter hydrologic patterns to more closely resemble pre-drainage conditions. Several non-CERP projects address water quality issues, including the Everglades Construction Project (construction of over 44,000 acres of STAs), restoration of the Kissimmee River, and restoration of Lake Okeechobee and its estuaries. In addition, research on and management of invasive species is important to the overall restoration program. Finally, the state of Florida's Acceler8 initiative is a mix of accelerated CERP project components and some non-CERP components.

What Natural System Restoration Requires

Although "getting the water right" is the oft-stated and immediate practical goal, the ultimate restoration goal is to reestablish the distinctive characteristics of the historical Everglades to what remains of the undeveloped South Florida ecosystem. Getting the water right is a means to an end, not the end in itself. **Natural system restoration will be best served by moving the system as quickly as possible toward physical, chemical, and biological conditions that previously molded and maintained the historical Everglades.** Toward this end, this committee judges five components of the Everglades restoration to be critical:

1. enough water-storage capacity combined with operations that provide appropriate volumes of water to support healthy estuaries and the return of sheet flow through the Everglades ecosystem while meeting other demands for water;
2. mechanisms for delivering and distributing the water to the natural system in a way that resembles historical flow patterns, affecting volume, depth, velocity, direction, distribution, and timing of flows;
3. barriers to eastward seepage of water so that higher water levels can be maintained in parts of the Everglades ecosystem without compromising the current levels of flood protection of developed areas as required by the CERP;
4. methods for securing water quality conditions compatible with restoration goals for a natural system that was inherently extremely nutrient poor, particularly with respect to phosphorus; and

5. retention, improvement, and expansion of the full range of habitats by preventing further losses of critical wetland and estuarine habitats and by protecting lands that could usefully be part of the restored ecosystem.

If these five critical components of restoration are achieved and the difficult problem of invasive species can be managed, then the basic physical, chemical, and biological processes that created the historical Everglades can once again create a functional mosaic of biotic communities that resemble what was distinctive about the historical Everglades. However, **the remaining Everglades landscape will continue to move away from conditions that support the defining ecosystem processes until greater progress is made in implementing CERP and non-CERP projects.**

Rapid population growth, with its attendant demands on land and water resources for development, water supply, flood protection, and recreation, only heightens the challenges facing the restoration efforts. Yet, despite new challenges and complexities, some positive examples of restoration progress offer hope that restoration is within reach given continued state and federal support.

PROMISING EXAMPLES OF RESTORATION PROGRESS

Restoring the Everglades is still in its early stages. **It is too early to evaluate the response of the ecosystem to the current restoration program, because no CERP projects have been constructed.** It is also too soon to fully assess the effects of non-CERP activities that are already under way, because the ecosystem is only beginning to respond to changes that these projects are designed to effect. However, several non-CERP activities are positive harbingers of future CERP programs.

For example, **the Kissimmee River Restoration Project has shown demonstrable ecological improvements and benefits to the natural system.** Improvements in the restored portions of the formerly channelized river include increases in river dissolved oxygen, increased density of wading birds, and colonization of the filled canal with wetland vegetation. Among several lessons learned from this project is that natural system restoration can be performed while continuing to maintain the flood-control function of the original channelization project. These achievements should be cause for cautious optimism that the CERP can achieve positive results as well.

Stormwater treatment areas and best management practices, implemented as part of non-CERP initiatives started in the 1990s, have proven remarkably effective at reducing phosphorus levels found in agricultural runoff. While falling short of the goal of 10 parts per billion (ppb) total phosphorus in the ambient waters, flow-weighted effluent concentrations from the STAs averaging 41 ppb are much reduced from influent concentrations that average 147 ppb. Because water quality is such a critical aspect of ecosystem restoration, additional research is needed to evaluate the need for additional acreage of STAs, to enhance removal of phosphorus and other constituents within these treatment wetlands, and to investigate their long-term sustainability.

The Mod Waters and C-111 projects have suffered long delays but are now moving forward, although Mod Waters should be completed without further delay. The Mod Waters and C-111 projects are non-CERP foundation projects that are necessary prerequisites to the CERP. Mod Waters represents a first major step toward restoration of the WCAs and Everglades National Park and a valuable opportunity to learn about the response of the natural system to restoration of sheet flow. Since the Mod Waters project is an assumed precursor for the WCA 3 Decompartmentalization and Sheet Flow Enhancement—Part 1 (Decomp) project, further delays in the project's completion may ultimately delay funding appropriations for Decomp. Additionally, limitations in its scope, such as in the extent of levee removal, may compromise the ultimate effectiveness of Decomp and restoration of flow to Northeast Shark River Slough.

CERP PROGRAM IMPLEMENTATION

During the first 6 years after WRDA 2000 was authorized, significant progress has been made in program support efforts, particularly in the monitoring and assessment program and the development of an adaptive management strategy, which represents the pathway by which science is used in support of decision making. Yet progress in CERP project implementation has been uneven, and many projects have been significantly delayed. Current barriers to project planning and implementation, highlighted below, threaten the timely delivery of restoration benefits.

Progress in the Use of Science in Decision Making

The committee reviewed three major science program documents that collectively provide a foundation for ensuring that scientific information needed to support restoration planning will be available in a timely way. The committee also examined the extensive set of models that have been developed to support restoration planning and adaptive management.

The Monitoring and Assessment Plan (MAP) documents reviewed describe a well-designed, statistically defensible monitoring program and an ambitious

assessment strategy. The plan provides for a continuous cycle of monitoring and experimentation, as well as regular and frequent assessment of the findings. In combination, the MAP provides an approach to reduce uncertainty associated with the conceptual ecological models that are the foundation of the monitoring plan and to create new knowledge for understanding old and emerging problems. The MAP should also help identify information gaps to support adaptive management.

Implementation of the monitoring plan is occurring more slowly than planned. The effectiveness of the MAP as a component of the adaptive management strategy can be determined only by implementation. Each of the components of the MAP needs to be in place and tested to enable integration of scientific information into the decision-making process. A spatially and temporally robust baseline of monitoring data is essential for a rigorous assessment of restoration progress, and a well-planned information management system is required to facilitate effective information sharing. Additional key staff and staff-support positions devoted to information management and implementation of the monitoring activities are needed to facilitate more rapid implementation of the MAP. Continuing to winnow the number of performance measures from 83 to an even smaller subset that includes a limited number of whole-system performance measures would help ensure that the MAP is sustainable over the lifetime of the CERP.

The CERP Adaptive Management Strategy provides a sound organizational model for the execution of a passive adaptive management program. The strategy should be implemented soon to test and refine the approach. The CERP Adaptive Management Strategy proposes a process for addressing uncertainty and supporting collaborative decision making. Although the objectives, mechanisms, and responsibilities are well specified in the Adaptive Management Strategy, the all-critical linkages among the planning, assessment, integration, and update activities require further development. The committee also judges that incorporating active adaptive management practices whenever possible will reduce the likelihood of making management mistakes and reduce the overall cost of the restoration. Regardless of which adaptive management approach is used, it remains to be seen how willing decision makers will be to make significant alterations to project design and sequencing, as opposed to limiting adaptive management to making modest adjustments in the operation of CERP projects after their construction.

A coordinated, multidisciplinary approach is required to improve modeling tools and focus modeling efforts toward direct support of the CERP adaptive management process. Models are used to forecast the short- and long-term responses of the South Florida ecosystem to CERP projects and, thus, are the critical starting point for adaptive management. An impressive variety of models has been developed to support the CERP, but better linkages between models, especially between hydrologic and ecological models, are needed to better integrate scientific knowledge and to extrapolate new information to the spatial scales at which decisions are made. In addition, hydrologic models suffer from the lack of high-resolution input data describing the basic terrain, so that their predictions are sometimes in error, and their connections to other more high-resolution ecosystem models is difficult. The development of quantitative ecological models is lagging behind the development of hydrologic models. Because models themselves must be improved through comparison with actual outcomes, coordination between modeling and monitoring efforts, within the adaptive management framework of iterative improvement, should be a high priority.

Status of CERP Planning and Coordination

The large size of the South Florida ecosystem as well as the cost, complexity, and number of years required to complete the CERP necessitates that the restoration effort be carefully planned and coordinated. Therefore, the committee reviewed several important planning, financing, and coordination issues that influence the progress being made on natural system restoration.

Although progress has been made in the planning, coordination, and program management functions required to implement the CERP, there have been significant delays in the expected completion dates of several construction projects that contribute to natural system restoration. Between 2000 and 2004 the USACE and SFWMD largely focused on developing a complex coordinating structure for planning and implementing CERP projects. However, while the management structures were being refined, all 10 of the CERP components that were scheduled for completion by 2005 were delayed. Additionally, six pilot projects originally scheduled for completion by 2004 are expected to be delayed on average by 8 years. The project implementation delays seem to be the result of a number of factors, including budgetary and manpower restrictions, the need to negotiate resolutions to major concerns or agency disagreements in the planning process, and a project planning process that can be stalled by unresolved scientific uncertainties, especially for complex or contentious projects. The observed project delays are of concern because they have affected projects on which substantial benefits to the natural ecosystem depend.

The Decomp project has been significantly delayed, although recent plans to implement an active adaptive management approach may move the project forward. Progress in implementing Decomp has been slowed by conflicts among stakeholders and inherent constraints in project planning in the face of scientific uncertainties. The committee is also concerned that project planning procedures may favor project alternatives that are limited in scope over project designs with less certain outcomes that have the potential to offer greater restoration benefits. Both the Decomp Physical Model and the Loxahatchee Impoundment Landscape Assessment experiments should help resolve some of the uncertainties that are constraining the project planning process. These are impressive adaptive management activities that should improve the likelihood of restoration success. Progress could be enhanced further if these experiments pave the way for additional experiments, some at even larger scales, that could be incorporated into an incremental approach to restoration.

Production of natural system restoration benefits within the Water Conservation Areas and Everglades National Park is lagging behind production of natural system restoration benefits in other portions of the South Florida ecosystem. The eight Acceler8 projects should provide ecological benefits primarily to the Lake Okeechobee region, the northern estuaries, the Ten Thousand Islands National Wildlife Refuge, and Biscayne Bay. Expected restoration benefits to the WCAs and Everglades National Park largely come from one project—the WCA 3A/B Seepage Management. The Acceler8 program may also provide momentum to the remaining restoration projects by hastening early construction efforts. Because determinations to allocate the water captured by the Acceler8 storage projects have not yet been finalized, future projections of benefits to the South Florida ecosystem remain unclear.

Federal funding will need to be significantly increased if the original CERP commitments are to be met on schedule. Inflation, project scope changes, and program coordination expenses have increased the original cost estimate of the CERP from $8.2 billion (in 1999 dollars) to $10.9 billion (in 2004 dollars). Further delays will add to this increase, particularly because of the escalating cost of real estate in South Florida. Despite these cost increases, current planned federal expenditures for fiscal year (FY) 2005 to FY 2009 fall far short of even those envisioned in the original CERP implementation plan. Although the CERP is intended to be a 50/50 cost-sharing arrangement between the federal and nonfederal (state and local) governments, federal expenditures from 2005 to 2009 are expected to be only 21 percent of the total. If federal funding for the CERP does not increase, major restoration projects directed toward the federal government's primary interests (e.g., Everglades National Park) may not be completed in a timely way.

The active land acquisition efforts should be continued, accompanied by monitoring and regular reporting on land conversion patterns in the South Florida ecosystem. Land management for a successful CERP depends on acquiring particular sites within the project area and protecting more general areas within the South

Florida ecosystem that could help meet the broad restoration goals. The committee commends the state of Florida for its aggressive and effective financial support for acquiring important parcels. Rapidly rising land costs imply that land within the project area should be acquired as soon as possible. Given the importance of wetland development and land-use conversion to the restoration potential of the CERP, the state should closely monitor and regularly report land conversion patterns within the South Florida ecosystem to stakeholders.

A significant challenge for the CERP is to implement the plan in a timely fashion while maintaining the federal and state partnership and the coalition of CERP stakeholders. The restoration of the Everglades rests on a fragile coalition of 66 signatory partners who agree in principle on the overarching goals of the CERP. Beyond the venerable notion of "getting the water right," virtually every signatory may find some part of the CERP with which to disagree and may have different views on the trade-offs that will need to be made as plan implementation begins. One particular concern expressed by stakeholders is whether the water supply goals of the CERP are being unduly emphasized in the current CERP implementation plan at the expense of the natural system restoration goals. Of the many partnerships, the most important is that between the state of Florida and the USACE. The state's Acceler8 initiative has raised concerns about disproportionate funding and control by the state over the implementation of the program. In the end, success will require cooperation among a disparate group of organizations with differing missions as the broad goal of getting the water right is more precisely defined.

AN ALTERNATIVE APPROACH TO ADVANCING NATURAL SYSTEM RESTORATION

To help address some sources of delay in the pace of restoration progress, including resolving conflicts over scientific uncertainty and addressing project sequencing constraints, the committee proposes an alternative framework for initiating and evaluating restoration actions, here called Incremental Adaptive Restoration (IAR).

To accelerate restoration of the natural system and overcome current constraints on restoration progress, many future investments in the South Florida ecosystem could profitably use an IAR approach. An IAR approach makes investments in restoration that are significant enough to secure environmental benefits while also resolving important scientific uncertainties about how the natural system will respond to management interventions. An IAR approach is not simply a reshuffling of priorities in the project implementation schedule. Instead it reflects an incremental approach using steps that are large enough to provide some restoration benefits and address critical scientific uncertainties, but generally smaller than the CERP projects or project components themselves, since the purpose of the IAR is to take actions that promote learning and that can guide the remainder of the project design. The improved understanding that results from an IAR approach will provide the foundation for more rapidly advancing restoration benefits. Without appropriate application of an IAR approach, valuable opportunities for learning would be lost, and subsequent actions would likely achieve fewer or smaller environmental benefits than they would if they had built upon previous knowledge. IAR is likely to be of particular value in devising management strategies for dealing with complex ecosystem restoration projects for which probable ecosystem responses are poorly known and, hence, difficult to predict (e.g., the role of flows in establishing and maintaining tree islands and ridge-and-slough vegetation). An IAR approach would also help address current constraints on restoration progress, including Savings Clause requirements (assurance that existing water supply and flood-control obligations will be met during CERP implementation; see Box 2-1), water reservation obligations, water quality considerations, and stakeholder disagreements.

An IAR approach would support the innovative adaptive management program now being developed for the CERP. IAR can be used in combination with a rigorous monitoring and assessment program to test hypotheses, thereby yielding valuable information that can expedite future decision making. A significant advantage of IAR over the present CERP adaptive management approach is that there may be early restoration benefits, as major restoration projects proceed incrementally in ways that enhance learning, improve efficiency of future actions, and potentially reduce long-term costs.

The existing authorization and budgeting process can be modified to accommodate the IAR process. To facilitate the IAR process and better support an adaptive management approach to the restoration effort, a modified programmatic authorization process would be needed that allows for the continuing reformulation and automatic authorization of added investment increments. This budgeting authority would still require securing individual appropriations for each new investment increment. This would constitute a variant of the current CERP programmatic authorization of groups of projects, where a project implementation report is required before the final authorization of a project is secured and funding can be requested.

OVERALL EVALUATION OF PROGRESS AND CHALLENGES

No CERP projects have been completed at this writing. Nonetheless, some conclusions are reasonably clear. First, the scientific program accompanying the restoration efforts has been of high quality and comprehensive. Important issues concerning scientific understanding, scientific coordination, and the incorporation of science into program planning and management remain, but the committee judges that no significant scientific uncertainty should stand in the way of restoration progress. Second, there have been some significant restoration achievements by non-CERP activities, most notably in reducing phosphorus inputs and loads and in restoring the Kissimmee River. Although those projects are not complete and the scientific and engineering challenges have not been entirely conquered, the achievements should be cause for cautious optimism that other elements of the program can achieve positive results as well.

Natural system restoration will be best served by moving the ecosystem as quickly as possible toward biological and physical conditions that previously molded and maintained the Everglades. However, restoration progress has been uneven and beset by delays. The state of Florida's Acceler8 and Lake Okeechobee and Estuary Recovery programs are providing a valuable surge in the pace of project implementation, especially in the northern portions of the ecosystem and its estuaries, although the expected ecosystem benefits from early water storage projects remain uncertain. Other important projects, including the work to reestablish sheet flow in the WCAs and Everglades National Park, are far behind the original schedule. Some of the sources of delay, such as the expansion of the aquifer storage and recovery pilot projects to address important uncertainties, are in the best interest of overall restoration success. Other sources of delay, including budgetary restrictions and a project planning and authorization process that can be stalled by unresolved scientific uncertainties, merit additional attention from senior managers and policy makers. Escalating land and other prices affect the restoration's budget, and federal funding has also fallen behind its original commitments. If federal funding for the CERP does not increase, restoration efforts focused on Everglades National Park and other federal interests may not be completed in a timely way. To help address the project planning concerns, the committee proposes an incremental adaptive-management-based approach, termed IAR, which can help resolve scientific uncertainties while enabling progress toward restoration goals. Finally, perhaps the largest challenge is maintaining the continued support of the coalition of stakeholders through the restoration process.

Appendix C

Status of Key Non-CERP Projects

KISSIMMEE RIVER RESTORATION

Status: This project will backfill a total of 22 miles of C-38 and reestablish approximately 40 miles of meandering river channel. Two of the four phases of the Kissimmee River Restoration Project to backfill the initial 10 miles of C-38 are complete, restoring an 18-mile section of the original river channel. All 102,061 acres of land needed for the restoration have been acquired. There are two remaining construction phases will backfill the final 12 miles of C-38.

Observed Benefits: About 6,300 acres of formerly drained portions of the river's floodplain are now experiencing enhanced inundation and converting back to wetland habitat. A comprehensive evaluation program for tracking environmental responses to the restoration is gauging the success of the project in meeting its goal of ecological integrity for the river and the floodplain. Densities of long-legged wading birds on the restored floodplain exceeded restoration expectations each year since 2002.

Start Date IFP (Integrated Financial Plan): 1994
Current Estimated Completion Date: 2013
Original Estimated Cost (WRDA 1992): $427 million (M)
2007 IFP Estimated Cost: $583M

EVERGLADES CONSTRUCTION PROJECT

Status: Stormwater treatment area (STA)-2 Cell 4, STA-5 Flow way 3, and STA-6 Section 2 became flow-capable December 2006. The completion of STA-6 Section 2 marks the completion of construction of the original Everglades Construction Project (ECP). Major efforts to rehabilitate STA-1W Cells 2, 4, 1B, and 5 were under way during water year (WY) 2007. Rehabilitation efforts included removal of the phosphorus-rich accrued layer that includes highly flocculent material, removal of tussock material, ground leveling to reduce flow constriction or short-circuiting, and rice planting to stabilize the soils. In early WY2008, a major submerged aquatic vegetation (SAV) inoculation effort was conducted using SAV harvested from STA-2 Cell 3 and deposited into STA-1W Cells 2B and 3.

Observed Benefits: Since 1994, the ECP STAs have retained over 900 mt of total phosphorus (TP) that would have otherwise entered into the Everglades Protection Area, reducing TP loads by 70 percent and phosphorus concentrations from an overall annual flow-weighted mean (FWM) TP of 145 parts per billion (ppb) down to 45 ppb.

Start Date: Authorized in 1994, Everglades Forever Act
Current Estimated Completion Date: Not available
Original Estimated Cost: $825M
Current Estimated Cost: $836.2M

MODIFICATIONS TO C-111 (SOUTH DADE)

Status: Currently, two interim pump stations and one permanent pump station have been completed, along with construction of three detention areas, replacement of the Taylor Slough Bridge, and removal of 4.75 miles of spoil mounds along lower C-111.

A land exchange of approximately 1,000 acres between Everglades National Park and the SFWMD was approved by Congress and executed in 2005. The project management plan (PMP) is being updated to detail and refinements in the design and accompanying costs and schedule. A supplemental Project Cooperation Agreement (PCA) to address the 50/50 cost share is forthcoming.

Construction of the earthwork for the retention/detention area is scheduled to be complete by September 2008. Construction on the S-331 command and control facility (cost shared with the Mod Waters project) is scheduled to be complete in March 2009. A construction contract to extend the S-332B north detention area and contain discharges from the 8.5 Square Mile Area STA component of the Mod Waters project is expected in 2010.

Observed Benefits: Not yet fully implemented. Distribution of flows has improved downstream of the Taylor Slough bridge replacement and C-111 spoil mounds removal areas.

Start Date IFP: 1994
Current Estimated Completion Date: 2014 (subject to appropriations)
Original Estimated Cost: $121M (1994)
2007 IFP Estimated Cost: $370M

MODIFIED WATER DELIVERIES TO EVERGLADES NATIONAL PARK

Status: Construction features completed:
1. Spillway structures S-355A and B in the L-29 Levee
2. S-333 modifications
3. Tigertail Camp elevation
4. Pump Station S-356 between L-31N Canal and L-29 Canal
5. Levees and a seepage collector canal to provide flood mitigation for the East Everglades residential area (8.5 square mile area)

Work in progress:
1. Degradation of the L-67 Extension Canal and Levee (4 of 9 miles degraded)
2. S-331 Command and Control (cost shared with the C-111 [South Dade])

Future work:
3. Structures S-345 A, B, and C through the L-67A and C Levees
4. Structures S-349 A, B, and C in the L-67A Borrow Canal
5. Osceola Camp elevation design and construction
6. L-29 weirs

The USACE has completed design of the bridges and road raising for the Tamiami Trail modifications feature. A final integrated Limited Reevaluation Report and Environmental Assessment is scheduled to be submitted to Congress in 2008.

Observed Benefits: Not yet implemented.

Start Date IFP: 1990
Current Estimated Completion Date: 2018
Original Estimated Cost: $98M (1989)
2007 IFP Estimated Cost: $398M

NORTHERN EVERGLADES AND ESTUARIES PROTECTION PROGRAM

Status: In February 2008, the South Florida Water Management District, in coordination with the Florida Department of Environmental Protection and the Florida Department of Agriculture and Consumer Services, issued the *Lake Okeechobee Watershed Construction Project Phase II Technical Plan* in response to the authorizing 2007 Florida state legislation. The plan identifies construction projects, along with on-site measures that prevent or reduce pollution at its source—such as agricultural and urban best management practices—and are needed to achieve water quality targets for the lake. In addition, it includes other projects for increasing water storage north of Lake Okeechobee to achieve healthier lake levels and reduce harmful discharges to the Caloosahatchee and St. Lucie estuaries. The plan consolidates the numerous initiatives currently under way through Florida's Lake Okeechobee Protection Plan and Lake Okeechobee and Estuary Recovery Plan (LOER). This plan is subject to ratification by the Florida legislature. Several LOER components are under construction, including 3 STAs, one STA expansion, one reservoir in association with the Taylor Creek STA, and the rerouting of flows to the Lakeside Ranch STA.

Observed Benefits: Not yet implemented.

Start Date: 2007
Current Estimated Completion Date: To be determined
Current Estimated Cost: Non-CERP features: \$260–\$320M; CERP features: \$1–\$1.4 billion

INVASIVE SPECIES MANAGEMENT

Status: Progress is being made through several programmatic initiatives. An interagency group, the Everglades Cooperative Invasive Species Management Area, has been assembled to support and enhance a weed management database (WEEDAR). Biocontrol agents have been successfully developed and introduced for *Melaleuca*; efforts to develop agents for Lygodium are continuing; and conventional controls (physical removal, herbicide applications) and airborne surveys are carried out regularly. Funding comes from specific projects under CERP (Melaleuca Eradication and Other Exotic Plants project, funded in 2002) and a variety of state-based projects. Surveys of invasive species are conducted by a variety of agencies (FLDEP, SFWMD, NPS). Shortages of funds for monitoring and assessment, and development of biocontrol agent hampers further progress. Management of exotic animal species lags well behind efforts for invasive exotic plants.

Observed Benefits: *Melaleuca* is thought to be under control, with most populations subject to maintenance control. Biocontrol agents are being introduced for *Lygodium* and *Schinus*; Lygodium is considered a major threat to ecosystem integrity.

Information costs and timelines not found

EVERGLADES AND SOUTH FLORIDA (E&SF) RESTORATION: CRITICAL PROJECTS

East Coast Canal Structures (C-4)

Status: Construction of a gated water control structure (S-380) in the C-4 basin in Dade County southeast of the Pennsuco wetlands is complete.

Observed Benefits: Raised surface and ground water levels to help preserve wetlands, increased aquifer recharge, and reduced seepage.

Start Date IFP: 1999
Completion Date IFP: 2003
2007 IFP Cost: \$3.7M

Tamiami Trail Culverts

Status: Original plans included Phase 1 placement of 77 culverts along Tamiami Trail (62 culverts west of SR 92 in the Picayune Strand area, plus 15 culverts east of SR92 near the Big Cypress Preserve area), and Phase 2 resurfacing of Tamiami Trail related to these efforts. Construction of the Western Tamiami Trail Culverts between SR92 and SR29 in Collier County was completed in May 2006. The remainder of Phase 1 and Phase 2 work is "on hold" pending funding. The western portion of Phase I of this project has now been included as a component of the Picayune Strand Restoration project and will be cost shared under that project instead of the initial Critical Projects authority.

Observed Benefits: Installation of Phase 1 culverts under Tamiami Trail established more natural hydropatterns north and south of the highway, which is expected to enhance biological restoration in the area.

Start Date IFP: 1998
Estimated Completion Date: 2011
2007 IFP Cost: \$8.9M

Florida Keys Carrying Capacity Study

Status: This project has been completed.

Observed Benefits: The South Florida Regional Planning Council has agreed to steward and maintain the Carrying Capacity Impact Assessment Model as a decision making tool. The Florida Marine Research Institute has also agreed to steward and maintain the databases.

Start Date IFP: 1997
Completion Date IFP: 2003
2007 IFP Cost: $6M

Western C-11 Water Quality Treatment

Status: Construction is complete for this project to improve the quality and timing of stormwater discharges to the Everglades Protection Area from the Western C-11 Basin located in south central Broward County. The structures have been turned over from the U.S. Army Corps of Engineers to the South Florida Water Management District for operation and maintenance.

Observed Benefits: The S-381 structure in the C-11 canal separates clean seepage flows from untreated agricultural and urban stormwater runoff. The S-9A pump station pumps clean flows into WCA-3A.

Start Date IFP: 1997
Completion Date IFP: 2006
2007 IFP Cost: $18.1 M

Seminole Tribe Big Cypress Reservation Water Conservation Plan

Status: Construction of the conveyance canal system is complete. Construction is under way on water control and treatment facilities in the western portion of Big Cypress Reservation.

Phase II of this project has been divided into four basins. The USACE awarded a contract for construction of the largest basin, Basin 1, in November 2006 expected to be completed in 2008. Basin 4 construction is scheduled to be awarded in 2008 with completion in 2009. The two remaining construction features, Basin 2 and Basin 3, are scheduled for construction award in 2009 and completion in 2010.

Projected Benefits: Should improve the quality of agricultural water runoff within the reservation, restore storage capacity, and return native vegetation.

Start Date IFP: 1997
Current Estimated Completion Date: 2010
Original Estimated Cost: \$75.3M (1996)
2007 IFP Estimated Cost: \$52.2M

Southern CREW Project Additions & Imperial River Flow Way

Status: This project aims to reestablish more natural flow patterns to 4,670 acres in the Southern Corkscrew Regional Ecosystem Watershed (CREW) and to restore Imperial River's natural flow way to Estero Bay and reduce river nutrient loads. Land acquisition, restoration construction, and exotics control for the project is ongoing. Because of escalating land costs, and difficulty in restoring the hydrology in the areas south of Kehl Canal, the project team is considering changes to the project footprint. The SFWMD may be able to partner with Lee County Conservation 20/20 to advance acquisition of remaining project lands. The SFWMD continues to acquire land and construct the project.

Observed Benefits: Several hundred acres of exotic species, primarily *Melaleuca*, have been treated. Exotic removal has taken place over approximately 2,560 acres. A number of canals have been plugged and berms breached and dirt roads removed to restore sheet flow in areas of the project footprint, restoring hydropatterns on approximately 640 acres of wetlands.

Start Date IFP: 1995
Current Estimated Completion Date: 2015
Yellow Book Original Estimated Cost: \$33.5M (\$3.4M Construction and \$30.1M Real Estate)
2007 IFP Estimated Cost: \$29.4M

Lake Okeechobee Water Retention & Phosphorous Removal

Status: Construction of two new stormwater treatment areas within the Taylor Creek/Nubbin Slough basin is physically complete. The interim construction and testing phases are in progress thru October 2008 and September 2009, respectively.

Projected Benefits: To improve the quality of water flowing into Lake Okeechobee.

Start Date IFP: 1997
Completion Date IFP: 2006 (for construction; testing continues to 2009)
2007 IFP Cost: $21.9M

Ten Mile Creek Water Preserve Area

Status: Construction of an aboveground reservoir, pump station, and gated water level control structure was complete in 2006. Since that time, interim operations, testing and monitoring have been under way by the SFWMD and the USACE in accordance with the water quality permit and project cooperation agreement. During the process to transfer the project to the SFWMD for full operations, the USACE and the SFWMD immediately began identifying all concerns and planning a course of action toward remediation. The additional project needs that have been identified have significant associated costs. Due to limitations on funding, reauthorization will likely be required to proceed.

Projected Benefits: Will provides 6,000 acre-feet of seasonal or temporary storage of storm water from the Ten Mile Creek basin on 526 acres of land, which will moderate high-volume freshwater flows and salinity fluctuations in the St. Lucie Estuary and reduce sediment and nutrient loads to benefit 2,740 acres of estuarine habitat.

Start Date IFP: 1997
Completion Date IFP: 2006 (construction only; monitoring continues)
2007 IFP Cost: $40.7M

Lake Trafford Restoration

Status: Construction and muck removal should have been completed by December 2007, but dredging was delayed due to dry weather and low water. There was insufficient funding to award a contract when plans and specs were first completed. The SFWMD has assumed 100 percent of the cost of revamping the design and the construction with the intent of receiving credit and/or reimbursement upon completion and approval by the USACE. The containment facility and much of the dredging have been completed.

Observed and Projected Benefits: Approximately 3 million cubic yards of organic sediments that blanketed the bottom of the lake were removed. Expectations include improving water quality, reestablishing native vegetation, and improving subsequent flows to Corkscrew Swamp Sanctuary and the Florida Panther National Wildlife Refuge.

Start Date IFP: 1999
Current Estimated Completion Date: 2011
Yellow Book Original Estimated Cost: $15.4M
2007 IFP Estimated Cost: $35.3M

SOURCES: SFERTF (2006; 2007a); SFWMD (2007); USACE (2007c); Williams (2008); *http://www.saj.usace.army.mil/projects/ index.html,* G. Landers, USACE, personal communication, 2008.

Appendix D

Primary Purposes and Reported Natural System Benefits of Project Components Scheduled for Completion in MISP Band 1 (2005–2010)

Band 1 Project Components	Primary Purpose	Reported Potential Natural System Benefits
Caloosahatchee (C-43) River ASR Pilot	Improved design and reduction of uncertainty	Minimal.
Hillsboro ASR Pilot Project	Improved design and reduction of uncertainty	Minimal.
Melaleuca Eradication and Other Exotic Plants (PIR)	Habitat restoration	Enhance efforts to control the spread of *Melaleuca* and other exotic plants that are flourishing throughout the greater Everglades ecosystem.
Winsberg Farm Wetlands Restoration	Habitat restoration	Created wetlands in developed area of Palm Beach County will provide habitat for wildlife and native plants.
L-30N Seepage Management Pilot	Improved design and reduction of uncertainty	Minimal; construction will reduce seepage loss to east and save some water for Everglades National Park.
Lake Okeechobee ASR Pilot	Improved design and reduction of uncertainty	Minimal.
Biscayne Bay Coastal Wetlands (Phase 1)	Habitat restoration	Restore freshwater sheet flow towards Biscayne Bay thereby improving its freshwater and tidal wetlands, near-shore bay habitat, marine nursery habitat, oysters and the oyster reef community.
Picayune Strand Hydrologic Restoration	Habitat restoration	Freshwater habitat restoration and estuarine salinity stabilization.
Indian River LagoonSouth (IRL-S): C-44 Reservoir	Water storage	Moderate damaging freshwater discharges to Indian River Lagoon, thereby improving the ecology of the lagoon.

Band 1 Project Components	Primary Purpose	Reported Potential Natural System Benefits
IRL-S: Natural Areas Real Estate Acquisition (Phase 1)	Habitat restoration	Preserve natural habitat.
Broward County Water Preserve Area: C-9 Impoundment	Water storage	Divert urban runoff into impoundments.
Broward County Water Preservation Area (WPA): C-11 Impoundment	Water storage	Divert urban runoff into impoundments.
Broward County WPA: WCA 3A-3B Seepage Management	Seepage management	Reduce water seepage losses from WCA 3A/3B.
Acme Basin B Discharge	Water storage	Provide water and water quality treatment for Arthur R. Marshall Loxahatchee National Wildlife Refuge.
Site 1 Impoundment	Water storage	Reduce water demands on Lake Okeechobee and Arthur R. Marshall Loxahatchee National Wildlife Refuge.
C-111 Spreader Canal	Habitat restoration	Reestablish sheet flow in South Dade County.
North Palm Beach County: C-51 and L-8 Basin Reservoir	Water storage	Improve timing and volume of discharges to Loxahatchee Slough and Lake Worth Lagoon and improve hydropattern in wildlife management area.
Everglades Agricultural Area Storage Reservoir, Part 1, Phase 1	Water storage	Improve timing of deliveries to WCA 2A and 3A and moderate high stages in Lake Okeechobee as well as water discharges to the estuaries from the lake.
Lake Okeechobee Watershed: Lake Istokpoga Regulation Schedule	Habitat restoration	Enhance fish and wildlife habitat in Lake Istokpoga littoral zone.
Modify Rotenberger Wildlife Management Area Operation Plan	Habitat restoration	Enhance plant and animal habitat.
Lakes Park Restoration	Habitat restoration	Reduce exotic species and enhance watershed biodiversity in Hendry Creek.
C-43 Basin Storage Reservoir	Water storage	Improve timing and water quality of freshwater discharges to Caloosahatchee Estuary.

NOTE: Reported natural system benefits were obtained from the project descriptions and supporting project materials found at *www.evergladesplan.org/ pm/projects/project_list.cfm*. The primary project purpose represents the committee's judgment based on the same materials. Among the primary purposes, water storage could provide benefits to both the natural system and to the human environment, depending on the water reservations ultimately determined. Gray shading indicates those projects being constructed by the South Florida Water Management District.

Appendix E

GAO Report Appendix II: Project Status and Cost by CERP, CERP-Related, and Non-CERP Categories

This table is a direct excerpt from the Government Accountability Office's 2007 report, *South Florida Ecosystem: Restoration Is Moving Forward but Is Facing Significant Delays, Implementation Challenges, and Rising Costs.*

Table 6: 222 Restoration Projects, Sponsor, Primary Purpose, Completion Date, and Project Cost

Project name	Sponsor(s)	Primary purpose	Completion date	Cost[a] (in millions)
60 CERP projects				
Acme Basin B Discharge	Corps/SFWMD	Habitat acquisition and improvement	2008[b]	$26.5
Aquifer Storage and Recovery Regional Study	Corps/SFWMD	Study	2010	73.4
Big Cypress / L-28 Interceptor Modifications	Corps/SFWMD	Water quality	2022	51.4
Biscayne Bay Coastal Wetlands	Corps/SFWMD	Habitat acquisition and improvement	2011[b]	386.9
Broward County Secondary Canal System	Corps/SFWMD	Water storage and flow	2014	15.5
Broward County Water Preserve Areas	Corps/SFWMD	Water quality	2009[b]	408.3
C-4 Structure	Corps/SFWMD	Water storage and flow	2013	2.8
C-43 Basin Storage Reservoir - Part 1	Corps/SFWMD	Water storage and flow	2011[b]	530.6
C-43 Basin Aquifer Storage and Recovery - Part 2	Corps/SFWMD	Water storage and flow	2019	[c]
C-111 Spreader Canal	Corps/SFWMD	Water quality	2015[b]	117.6
Caloosahatchee Backpumping with Stormwater Treatment	Corps/SFWMD	Water quality	2018	99.7

Project name	Sponsor(s)	Primary purpose	Completion date	Cost[a] (in millions)
Caloosahatchee River (C-43) Aquifer Storage and Recovery Pilot	Corps/SFWMD	Water storage and flow (pilot)	2009	7.9
Central Lake Belt Storage	Corps/SFWMD	Water storage and flow	2035	155.4
Change Coastal Wellfield Operations	Corps/SFWMD	Water supply	To be decided	[d]
Comprehensive Integrated Water Quality Feasibility Study	Corps/FDEP	Study	2014	9.3
Environmental Water Supply Deliveries to St. Lucie Estuary	Corps/SFWMD	Habitat acquisition and improvement	To be decided	[d]
Environmental Water Supply Deliveries to the Caloosahatchee Estuary	Corps/SFWMD	Habitat acquisition and improvement	To be decided	[d]
Everglades Agricultural Storage Reservoir	Corps/SFWMD	Water storage and flow	2015[b]	542.2
Everglades National Park Seepage Management	Corps/SFWMD	Water storage and flow	2015	390.9
Everglades Rain Driven Operations	Corps/SFWMD	Water storage and flow	To be decided	[d]
Florida Bay and the Florida Keys Feasibility Study	Corps/SFWMD	Study	2012	6.3
Florida Keys Tidal Restoration	Corps/SFWMD	Water storage and flow	2010	1.5
Flow to Northwest and Central Water Conservation Area 3A	Corps/SFWMD	Water storage and flow	2018	36.3
Flows to Eastern Water Conservation Area	Corps/SFWMD	Water storage and flow	2017	8.0
Henderson Creek / Belle Meade Restoration	Corps/FDEP	Water quality	2011	5.8
Hillsboro Aquifer Storage and Recovery - Phase 2	Corps/SFWMD	Water storage and flow	2020	[c]
Hillsboro Aquifer Storage and Recovery Pilot	Corps/SFWMD	Water storage and flow (pilot)	2009	9.4
Indian River Lagoon-South	Corps/SFWMD	Water storage and flow	2022[b,e]	1,309.7
L-31N (L-30) Seepage Management Pilot	Corps/SFWMD	Water storage and flow (pilot)	2010	11.3
Lake Belt In-Ground Reservoir Technology Pilot	Corps/SFWMD	Water storage and flow (pilot)	2026	26.5
Lake Okeechobee Aquifer Storage and Recovery	Corps/SFWMD	Water storage and flow	2027	1,223.4

Project name	Sponsor(s)	Primary purpose	Completion date	Cost[a] (in millions)
Lake Okeechobee Aquifer Storage and Recovery Pilot	Corps/SFWMD	Water storage and flow (pilot)	2009	32.3
Lake Okeechobee Regulation Schedule	Corps/SFWMD	Water storage and flow	2007	1.1
Lake Okeechobee Watershed	Corps/SFWMD	Water storage and flow	2014	575.5
Lakes Park Restoration	Corps/Lee County	Habitat acquisition and improvement	2009	6.0
Lower East Coast Utility Water Conservation	Corps/SFWMD	Water supply	To be decided	[d]
Loxahatchee National Wildlife Refuge Internal Canal Structures	Corps/SFWMD	Water storage and flow	2015	9.1
Melaleuca Eradication and Other Exotic Plants	Corps/SFWMD	Invasive species control	2025	6.6
Miccosukee Water Management Plan	Corps/ Miccosukee	Water quality	2016	29.0
Modify Holey Land Wildlife Management Area Operation Plan	Corps/SFWMD	Water storage and flow	2011	[d]
Modify Rotenberger Wildlife Management Area Operation Plan	Corps/SFWMD	Water storage and flow	2009	[d]
North Lake Belt Storage Area	Corps/SFWMD	Water storage and flow	2035	308.2
North Palm Beach County - Part 1	Corps/SFWMD	Water quality	2015[f]	533.2
North Palm Beach County - Part 2	Corps/SFWMD	Water storage and flow	2019	203.9
Operational Modification to Southern Portion of L-31N and C-111	Corps/SFWMD	Water storage and flow	To be decided	[d]
Palm Beach County Agriculture Reserve Reservoir - Part 1	Corps/SFWMD	Water storage and flow	2016	154.4
Palm Beach County Agriculture Reserve Aquifer Storage and Recovery - Part 2	Corps/SFWMD	Water storage and flow	2018	[c]
Picayune Strand Restoration	Corps/SFWMD	Habitat acquisition and improvement	2009[b,g]	362.6
Restoration of Pineland and Hardwood Hammocks in C-111 Basin	Corps/Miami-Dade County	Habitat acquisition and improvement	2021	0.7
Seminole Tribe Big Cypress Reservation Water Conservation Plan	Corps/Seminole	Water quality	2021	89.5
Site 1 Impoundment	Corps/SFWMD	Water storage and flow	2009[b,h]	153.7

Project name	Sponsor(s)	Primary purpose	Completion date	Cost[a] (in millions)
South Miami-Dade Reuse	Corps/Miami-Dade County	Water supply	2022	430.6
Southwest Florida Feasibility Study	Corps/SFWMD	Study	2009	12.0
Strazzulla Wetlands	Corps/SFWMD	Habitat acquisition and improvement	2010	70.4
Wastewater Reuse Technology Pilot	Corps/SFWMD	Water supply (pilot)	2021	35.4
Water Conservation Area 2B Flows to Everglades National Park	Corps/SFWMD	Water storage and flow	2021	539.4
Water Conservation Area 3 Decompartmentaliza-tion and Sheet Flow Enhancement (Decomp)	Corps/SFWMD	Water storage and flow	2020	253.4
Water Preserve Area Conveyance	Corps/SFWMD	Water storage and flow	2016	331.7
West Miami-Dade Reuse	Corps/Miami-Dade County	Water supply	2022	518.1
Winsberg Farm Wetlands Restoration	Corps/Palm Beach County	Habitat acquisition and improvement	2008	17.1

28 CERP-related projects				
C-111 (South Dade)	Corps/SFWMD	Water storage and flow	2012	287.6
Chapter 298 Districts / Lease 3420 Improvements	SFWMD	Water quality	2005	24.1
Critical Project: Additional Water Conveyance Structures Under Tamiami Trail	Corps/SFWMD	Water storage and flow	To be decided[i]	16.5
Critical Project: East Coast Canal Structures (C-4)	Corps/SFWMD	Water storage and flow	2003	3.7
Critical Project: Keys Carrying Capacity Study	Corps/FDCA	Study	2003	6.0
Critical Project: Lake Okeechobee Water Retention / Phosphorus Removal	Corps/SFWMD	Water quality	2006	21.9
Critical Project: Lake Trafford	Corps/SFWMD	Water quality	2007	30.0
Critical Project: Seminole Big Cypress Reservation Water Conservation Plan	Corps/Seminole	Water storage and flow	2010	52.2

Project name	Sponsor(s)	Primary purpose	Completion date	Cost[a] (in millions)
Critical Project: Southern CREW	Corps/SFWMD	Water storage and flow	To be decided	33.3
Critical Project: Ten Mile Creek	Corps/SFWMD	Water storage and flow	2006	40.7
Critical Project: Western C-11 Water Quality Treatment	Corps/SFWMD	Water quality	2006	18.1
East Water Conservation Area 3A Hydropattern Restoration	SFWMD	Water storage and flow	2012	5.3
Everglades Agricultural Area (EAA) Storm-water Treatment Areas Expansion	SFWMD	Water quality	2010	226.7
Indian River Lagoon Restoration Feasibility Study	Corps/SFWMD	Study	2002	7.9
Kissimmee River Restoration	Corps/SFWMD	Water storage and flow	2016[j]	575.4
Manatee Pass Gates	Corps/SFWMD	Habitat acquisition and improvement	2010	13.8
Melaleuca Quarantine Facility	USDA (ARS)	Invasive species control	2004	8.0
Modified Water Deliveries to Everglades National Park (Mod Waters)	NPS/Corps	Water storage and flow	2009	398.4
Rotenberger Restoration	SFWMD	Water storage and flow	2005	3.6
Storm-water Treatment Area 1 Inflow and Distribution Works	SFWMD	Water quality	2005	12.7
Storm-water Treatment Area 1 West Works and Outflow Pump Station (G-310)	SFWMD	Water quality	2000	82.1
Storm-water Treatment Area 2 Works and Outflow Pump Station (G-335)	SFWMD	Water quality	2000	100.4
Storm-water Treatment Area 3/4 Works	SFWMD	Water quality	2005	170.4
Storm-water Treatment Area 5 Works	SFWMD	Water quality	2005	36.2
Storm-water Treatment Area 6 (includes Sections 1 and 2)	SFWMD	Water quality	2006	14.6
Water Conservation Area 2A Hydropattern Restoration	SFWMD	Water storage and flow	2012	4.9
West Palm Beach Canal (C-51) and Storm-water Treatment Area 1E	Corps/SFWMD	Water quality	2008	288.6

Project name	Sponsor(s)	Primary purpose	Completion date	Cost[a] (in millions)
West Water Conservation Area 3A Hydropattern Restoration	SFWMD	Water storage and flow	2012	7.4

NOTE: Ten projects had primary purposes—such as recreation or soil monitoring—that fell outside of our established categories. These project purposes are designated "Other" in this table.

[a] Project cost shown is reported cost for completed projects and estimated cost for all other projects.
[b] SFWMD is expediting the design and construction of this project with its own funds in advance of congressional authorization, which may result in earlier project completion.
[c] The estimated cost of this aquifer storage and recovery (ASR) project is included in the cost estimate for the project's initial part or phase. Specifically, the estimated cost of the C-43 Basin ASR is included in the cost estimate for the C-43 Basin Storage Reservoir; the estimated cost of the Hillsboro ASR is included in the cost estimate for the Site 1 Impoundment; and the estimated cost of the Palm Beach County Agriculture Reserve ASR is included in the cost estimate for the Palm Beach County Agriculture Reserve Reservoir.
[d] We did not receive cost information for this project.
[e] A project implementation report was submitted to the U.S. Congress in 2005 for this project, but it has not yet received authorization.
[f] The South Florida Water Management District is expediting a portion of this project with its own funds in advance of congressional authorization. It is constructing a water storage reservoir that it expects to finish by 2008.
[g] This project is currently being reviewed by the Office of Management and Budget before its project implementation report is submitted to the Congress for authorization.
[h] This project is currently being reviewed by the assistant secretary of the U.S. Army before its project implementation report is submitted to the Congress for authorization.
[i] Phase 1 of this project has been completed; phase 2 is on hold pending additional funding.
[j] This date encompasses construction completion and several years of post-construction monitoring.

SOURCE: Excerpted from GAO (2007).

Appendix F

Performance Measures

Performance Measures	Used for Assessment (A) and/or Evaluation (E)?	Required for Compliance Purposes (e.g., ESA, NEPA, etc)?	Has an Interim Goal?
Lake Okeechobee Performance Measures			
• Lake Okeechobee Stage	A, E		yes
• Lake Okeechobee Water Quality	A	TMDL	
• Lake Okeechobee Diatom-Cyanobacteria Ratio	A		
• Lake Okeechobee Vegetation Mosaic	A		
• Lake Okeechobee Fish Population	A		
• Lake Okeechobee Macroinvertebrates	A		
Northern Estuaries Performance Measures			
• Northern Estuaries Salinity	A , E		
• Northern Estuaries Water Quality	A		
• Northern Estuaries Oyster Habitat	A		yes
• Northern Estuaries Benthic Macroinvertebrates	A		yes
• Northern Estuaries Submerged Aquatic Vegetation	A		
• Northern Estuaries Fish Communities	A		
Greater Everglades Wetlands Performance Measures			
• Sheet flow in the Everglades Ridge and Slough Landscape (under review)	E		yes
• Wet priarie (under review)	E		
• Number and Duration of Dry Events for Shark River Slough	E		
• Inundation Pattern in Greater Everglades Wetlands	E		yes
• Extreme High and Low Water Levels in Greater Everglades Wetlands	E		
• Greater Everglades Wetlands TP Concentrations in Surface Water	A, E planned		yes
• Greater Everglades Wetlands Basin-wide TP Loading and Flow-weighted Mean Concentration in Inflows	A, E planned		
• Greater Everglades Wetlands Nutrient TN Concentrations in Surface Water	A, E planned		

Performance Measures	Used for Assessment (A) and/or Evaluation (E)?	Required for Compliance Purposes (e.g., ESA, NEPA, etc)?	Has an Interim Goal?
• TN Loads/Flow-weighted Mean Concentration in Inflows to Greater Everglades Wetlands	A, E planned		
• TP Concentrations in Soil	A, E planned		
• Greater Everglades Tracer of Storm-water Treatment Area Bypass Flows	A		
• Greater Everglades Wetlands Sulfate Concentrations in Surface Water	A		
• Greater Everglades Wetlands Conductivity in Surface Water	A		
• Greater Everglades Wetlands Coastal Salinity Gradients	A		
• Wetland Landscape Patterns - Freshwater and Estuarine Vegetation Mosaics	A		
• Wetland Landscape Patterns - Marl Prairie Cape Sable Sparrow Habitat	A		
• Wetland Landscape Patterns - Ridge and Slough Community Sustainability	A planned		yes
• Wetland Landscape Patterns - Tidal Creek Sustainability	A		
• Wetland Trophic Relationships - Periphyton	A		
• Wetland Trophic Relationships - Mangrove Forest Production/Soil Accretion	A		
• Wetland Trophic Relationships - Regional Populations of Fishes, Crayfish, Grass Shrimp and Amphibians	A		yes
• Wetland Trophic Relationships - Wading Bird Foraging Patterns	A		
• Wetland Trophic Relationships - Wading Bird Nesting Patterns	A		yes
• Roseate Spoonbill Nesting Patterns	A	ESA	
• Wetland Trophic Relationships - American Alligator Distribution, Size, Nesting, and Condition	A		yes
• American Crocodile – Juvenile Growth and Survival	A	ESA	yes
Southern Estuaries Performance Measures			
• Southern Estuaries Salinity	A, E		yes
• Water Level at Regionally Significant Gauge Stations in Everglades National Park	A, E		
• Southern Estuaries Submerged Aquatic Vegetation	A		yes
• Southern Estuaries Juvenile Pink Shrimp and Associated Epifauna	A		yes
• Southern Estuaries Fish Community	A		
• Southern Estuaries Water Quality	A		
Water Supply and Flood Protection Performance Measures			
• Frequency of Water Restrictions for Lake Okeechobee Service Area	A, E		
• Frequency of Water Restrictions for Lower East Coast Service Area	A, E		

Performance Measures	Used for Assessment (A) and/or Evaluation (E)?	Required for Compliance Purposes (e.g., ESA, NEPA, etc)?	Has an Interim Goal?
• Potential for High Water Levels in South Miami-Dade Agricultural Area	A, E		
• Prevent Saltwater Intrusion of Biscayne Aquifer - Meet Minimum Flows and Levels Criteria for Biscayne Aquifer	A, E		
• Prevent Saltwater Intrusion of Biscayne Aquifer in South Miami-Dade County	A, E		
• Comparison of Stage Differences of Water Levels in South Miami-Dade Agricultural Area	A, E		
Total System Performance Measures			
• Snail Kite Foraging Conditions	A		
• White Tail Deer Breeding Potential	A		
• Mercury Bioaccumulation	A		

SOURCE: RECOVER (2007b).

Appendix G

Interim Restoration Goals for the CERP

1. Northern Estuaries Region

1.1 American Oysters in Northern Estuaries

- *Increase the areal coverage of American oysters in the Caloosahatchee, St. Lucie and Loxahatchee Estuaries, and the Lake Worth Lagoon*

1.2 Submerged Aquatic Vegetation in Northern Estuaries

- *Increase the areal coverage and improve the functionality of submerged aquatic vegetation in the northern estuaries*

1.3 Flows to the Northern Estuaries

- *Reduce high-volume flows (monthly average flows in excess of 2,800 cfs) to the Caloosahatchee Estuary and low-volume flows (monthly flows from October to July below 450 cfs) as measured at the S-79 structure*
- *Reduce high-volume flows (monthly average flows in excess of 2,000 cfs) to the St. Lucie Estuary and low-volume flows (monthly flows below 350 cfs) as measured using the combined flows from the S-80, S-49 and S-97 structures*
- *Reduce high-volume flows (flows in excess of 500 cfs daily over a 7-day moving average) to the Lake Worth Lagoon as measured at the S-155 structure*

2. Lake Okeechobee Region

2.1 Lake Okeechobee Phosphorus

- *Reduce phosphorus concentrations in Lake Okeechobee*

2.2 Water Levels in Lake Okeechobee

- *Reduce the frequency of both harmful high water stages above 17 feet and harmful stages above 15 feet occurring for longer than 12 consecutive months*

- *Reduce the frequency of harmful low stages below 11 feet*
- *Increase the frequency of natural spring recession events, i.e., stage decline from approximately 15.5 feet in January to approximately 12.5 feet in June*

2.3 Lake Okeechobee Algal Blooms
- *Reduce the frequency of harmful algal blooms in Lake Okeechobee*

2.4 Lake Okeechobee Aquatic Vegetation
- *Increase the areal coverage of desirable native vegetation in Lake Okeechobee*

3. Everglades Region

3.1 Water Volume
- *Distribute water across the ecosystem in a manner that reflects natural conditions while providing for other water-related needs of the region*

3.2 Sheet Flow
- *Establish more historic magnitudes and directions of sheet flow in the natural areas of the Everglades*

3.3 Hydropattern
- *Restore the natural timing and pattern of inundation throughout the ecological communities of South Florida, including sawgrass plains, ridge and slough and marl marshes*

3.4 System-Wide Spatial Extent of Natural Habitat
- *Increase spatial extent of natural habitat*

3.5 Everglades Total Phosphorus
- *Achieve water column phosphorus concentrations of 10 micrograms per liter in the Everglades*

3.6 Periphvton Mat Cover, Structure and Composition
- *Restore periphyton mat cover, structure and composition that were characteristic of the spatially distinct hydroperiods (short and long hydroperiods) and low nutrient conditions in the greater Everglades wetland communities*

3.7 Ridge and Slough Pattern
- *Restore the historical ridge and slough landscape directionality and pattern*

3.8 Everglades Tree Islands

- *Improve tree island health and maintain healthy tree islands*

3.9 Aquatic Fauna Regional Populations in Everglades Wetlands

- *Increase the abundance of fish to levels that approximate those predicted for pre-drainage conditions*

3.10 American Alligator

- *Restore more natural numbers and distribution patterns for alligators across South Florida's major freshwater and estuarine landscapes*

3.11 System-wide Wading Bird Nesting Pattern

- *Increase the total number of nesting pairs in the Everglades*
- *Increase the percentage of wading bird pairs nesting in estuarine locations*
- *Increase the frequency of super colony events*
- *Establish conditions that encourage wood storks to initiate nesting earlier in winter*

3.12 Snail Kite

- *Increase the areal extent of suitable foraging habitat for snail kites*

3.13 Flows to Northern Boundaries of the Water Conservation Areas

- *Provide more natural surface water flows to the northern boundaries of the water conservation areas*

3.14 Flows to Everglades National Park

- *Provide more natural surface water flows to Everglades National Park*

4. Southern Estuaries Region

4.1 Salinity Patterns in Florida Bay and Biscayne Bay

- *Reduce the intensity, duration, frequency and spatial extent of high salinity events, reestablish low salinity conditions in mainland nearshore areas, and reduce the frequency of and rapidity of salinity fluctuations resulting from pulse releases of fresh water from canals*

4.2 Submerged Aquatic Vegetation in Southern Estuaries

- *Reestablish a diverse seagrass community with moderate plant densities and more natural seasonality, and increase the percentage of Florida Bay having suitable habitat for seagrass growth*

4.3 Juvenile Shrimp Densities in Florida Bay and Biscayne Bay
- *Increase densities of juvenile shrimp within the various basins of Florida Bay and Biscayne Bay*

4.4 American Crocodile
- *Increase the frequency of salinities less than 20 parts per thousand in Florida Bay to foster optimal growth and survival of juvenile crocodiles*

4.5 Florida Bay Algal Blooms
- *Minimize the magnitude, duration and spatial extent of algal blooms in Florida Bay*

4.6 Freshwater Flows to Florida Bay
- *Increase freshwater flows to Florida Bay*

4.7 Freshwater Flows to Biscayne Bay
- *Increase freshwater flows to Biscayne Bay*

5. System-wide Water Volume

5.1 Quantity of Freshwater Lost to Tide
- *Reduce the quantity of fresh water lost to tide*

6. Predicted Plan Capability

The following provides a qualitative interim goal statement for surface water storage capacity, stormwater treatment areas and wetland/natural areas restoration, along with incremental performance predictions for the end of 2010 and 2015 based on the best available information regarding: (1) project costs; (2) future federal and state budgets for the Plan; and (3) schedules for completion of Project Implementation Reports, obtaining project authorizations, and executing Project Cooperation Agreements.

- Provide increased surface water storage capacity
  - 2010—60,000 acre-feet
  - 2015—430,000 acre-feet

- Provide increased water quality treatment capacity through additional STAs
  - 2010—5,000 acres of additional STA
  - 2015—10,000 acres of additional STA

- Increase spatial extent of wetlands/natural areas
  - 2010—15,000 acres of restored or enhanced wetlands/natural areas
  - 2015—70,000 acres of restored or enhanced wetlands/natural areas

SOURCE: USACE et al. (2007).

Appendix H

Standard Content of a Performance Measure Specification

Component	Description
Performance measure title	Clear description of the performance measure content, and description of the module or region to which it applies
Last revised date	To be modified as needed, to clearly indicate the most recent version
Acceptance dates	Dates on which the performance measure is approved for evaluation and for assessment
Desired restoration condition	This section specifies (a) a narrative description of the desired restoration goal, i.e., that state of the performance measure variables that would be found in a healthy and sustainable ecosystem; (b) a quantitative projection, based on simulation modeling or other methods, of the target state for evaluation; (c) specification of the parameters and target values within the performance measure that will serve as the basis for assessment of response to CERP actions, including specification of the variables to be monitored. This latter portion may be a quantitative value for measured variables, a minimum acceptable level, or a trend of change.
Justification	This section provides a narrative description of the pre-drainage state of the indicator, a description of its current state, and a description of how CERP implementation will affect the indicator.
Scientific and/or regulatory basis	This section relates the performance measure to relevant conceptual ecological models, and establishes through cited literature its application to the model(s) as an attribute or a stressor. The CEM model diagram, and the specific Adaptive Assessment Hypothesis Clusters that involve this performance measure are cited. For performance measures developed with respect to regulatory goals, the statute requiring the performance measure is given.
Evaluation application	Information needed to use the performance measure to evaluate program alternatives includes (a) measurement protocols and the protocol used to conduct the evaluation; (b) normalized performance output, in which the projected performance measure values are scaled to the expected value (1.0) at full restoration, and a value of 0 indicating fully degraded condition; (c) examples of output from models or other predictive tools are given to provide an example of the interpretation of the output; (d) a discussion of the uncertainty associated with the performance measure, and the use of error statistics to describe that uncertainty.

Component	Description
Monitoring & assessment application	Information needed to use the performance measure for monitoring and assessment, including (a) the appropriate modules and components as specified in the CERP Monitoring and Assessment Plan Part I; (b) interim goals and interim targets, as applicable, and (c) the appropriate assessment procedures, as specified in CERP Monitoring and Assessment Plan Part II.
Future tool development	This section describes any additional tools need to use the performance measure effectively for both evaluation and assessment. This section justifies the procurement of resources to develop such tools.
Notes	Any additional relevant information
Working group members	A list of the working group members, with agency affiliations, that were responsible for developing the performance measure and drafting the documentation
References	All documents cited in the performance measure description are cited; they are updated with each iteration of the performance measure documentation.
Resources for completing the documentation sheet	CEMs—*www.evergladesplan.org/pm/recover/cems.aspx* Adaptive hypothesis clusters—(*www.evergladesplan.org/pm/recover/perf_systemwide.aspx*), plus links therein Other links to hypothesis clusters—*http://www.evergladesplan.org/pm/recover/recover_docs/et/101707*

SOURCE: RECOVER (2007b).

Appendix I

Water Science and Technology Board and Board on Environmental Studies and Toxicology

WATER SCIENCE AND TECHNOLOGY BOARD

CLAIRE WELTY, *Chair*, University of Maryland, Baltimore County
JOAN G. EHRENFELD, Rutgers University, New Brunswick, New Jersey
GERALD E. GALLOWAY, University of Maryland, College Park
SIMON GONZALEZ, National Autonomous University of Mexico, Mexico City
CHARLES N. HAAS, Drexel University, Philadelphia, Pennsylvania
KENNETH R. HERD, Southwest Florida Water Management District, Brooksville, FL
JAMES M. HUGHES, Emory University, Atlanta, Georgia
THEODORE L. HULLAR, Cornell University, Ithaca, New York
KIMBERLY L. JONES, Howard University, Washington, DC
G. TRACY MEHAN, The Cadmus Group, Inc., Arlington, Virginia
DAVID H. MOREAU, University of North Carolina, Chapel Hill
THOMAS O'ROURKE, Cornell University, Ithaca, New York
DONALD I. SIEGEL, Syracuse University, Syracuse, New York
SOROOSH SOROOSHIAN, University of California, Irvine
HAME M. WATT, Independent Consultant, Washington, DC
JAMES L. WESCOAT, JR., Massachusetts Institute of Technology, Cambridge

Staff

STEPHEN D. PARKER, Director
LAUREN E. ALEXANDER, Senior Program Officer
LAURA J. EHLERS, Senior Program Officer
JEFFREY W. JACOBS, Scholar
STEPHANIE E. JOHNSON, Senior Program Officer
WILLIAM S. LOGAN, Senior Program Officer
LAURA E. HELSABECK, Associate Program Officer

M. JEANNE AQUILINO, Financial and Administrative Associate
ELLEN A. DE GUZMAN, Research Associate
ANITA A. HALL, Senior Program Associate
DOROTHY K. WEIR, Research Associate
MICHAEL J. STOEVER, Senior Program Assistant
STEPHEN T. RUSSELL, Program Assistant

BOARD ON ENVIRONMENTAL STUDIES AND TOXICOLOGY

JONATHAN M. SAMET *Chair*, Johns Hopkins University, Baltimore, Maryland
RAMÓN ALVAREZ, Environmental Defense Fund, Austin, Texas
JOHN M. BALBUS, Environmental Defense Fund, Washington, DC
DALLAS BURTRAW, Resources for the Future, Washington, DC
JAMES S. BUS, Dow Chemical Company, Midland, Michigan
RUTH DEFRIES, University of Maryland, College Park
COSTEL D. DENSON, University of Delaware, Newark
E. DONALD ELLIOTT, Willkie Farr & Gallagher LLP, Washington, DC
MARY R. ENGLISH, University of Tennessee, Knoxville
J. PAUL GILMAN, Covanta Energy Corporation, Fairfield, New Jersey
SHERRI W. GOODMAN, Center for Naval Analyses, Alexandria, Virginia
JUDITH A. GRAHAM (Retired), Pittsboro, North Carolina
WILLIAM P. HORN, Birch, Horton, Bittner and Cherot, Washington, DC
WILLIAM M. LEWIS, JR., University of Colorado, Boulder
JUDITH L. MEYER, University of Georgia, Athens
DENNIS D. MURPHY, University of Nevada, Reno
PATRICK Y. O'BRIEN, ChevronTexaco Energy Technology Company, Richmond, California
DOROTHY E. PATTON (Retired), U.S. Environmental Protection Agency, Chicago, Illinois
DANNY D. REIBLE, University of Texas, Austin
JOSEPH V. RODRICKS, ENVIRON International Corporation, Arlington, Virginia
ARMISTEAD G. RUSSELL, Georgia Institute of Technology, Atlanta
ROBERT F. SAWYER, University of California, Berkeley
KIMBERLY M. THOMPSON, Massachusetts Institute of Technology, Cambridge
MONICA G. TURNER, University of Wisconsin, Madison
MARK J. UTELL, University of Rochester Medical Center, Rochester, New York
CHRIS G. WHIPPLE, ENVIRON International Corporation, Emeryville, California
LAUREN ZEISE, California Environmental Protection Agency, Oakland

Senior Staff

JAMES J. REISA, Director
DAVID J. POLICANSKY, Scholar
RAYMOND A. WASSEL, Senior Program Officer for Environmental Studies
EILEEN N. ABT, Senior Program Officer for Risk Analysis
SUSAN N.J. MARTEL, Senior Program Officer for Toxicology
ELLEN K. MANTUS, Senior Program Officer
RUTH E. CROSSGROVE, Senior Editor

Appendix J

Biographical Sketches of Committee Members and Staff

William L. Graf, *Chair,* is Foundation University Professor and professor and chair of the Department of Geography at the University of South Carolina. His expertise is in fluvial geomorphology and hydrology, as well as policy for public land and water. Dr. Graf's research and teaching have focused on river-channel change, human impacts on river processes, morphology, and ecology, along with contaminant transport and storage in river systems. His present work emphasizes the downstream effects of dams on rivers. In the arena of public policy, he has emphasized the interaction of science and decision making, and the resolution of conflicts among economic development, historical preservation, and environmental restoration for rivers. Dr. Graf has served as member of the NRC's Water Science and Technology Board and Board on Earth Sciences and Resources; he served on the NRC's Panel to Review the Critical Ecosystem Studies Initiative, the Committee on Restoration of the Greater Everglades Ecosystem, and the first Committee on Independent Scientific Review of Everglades Restoration Progress. He is also a National Associate of the National Academies. Dr. Graf earned a Ph.D. from the University of Wisconsin, Madison, in 1974.

Steven R. Beissinger holds the A. Starker Leopold Chair in Wildlife Biology at the University of California at Berkeley in the Department of Environmental Science, Policy, and Management where he previously served as department chair. He teaches undergraduate and graduate courses in conservation biology and population ecology. Dr. Beissinger's research has been conducted primarily with birds but has also included work with plants, mammals, aquatic invertebrates, and herps. His current work focuses on (1) ecology of endangered or exploited species; (2) modeling population viability; and (3) parental care strategies. Dr. Beissinger has studied wetland birds in the Everglades, California, and internationally. He has worked extensively with federal and state agencies as a member of recovery teams, to develop regional monitoring plans, and as a training instructor. Dr. Beissinger is a Fellow of the London Zoological Society and the American Ornithologists' Union and a member of the Board of Directors of the National Audubon Society. He has served on several NRC committees including the U.S. National Committee of the International Union of Biological Sciences, the U.S. National Committee for DIVERSITAS, and the Committee on Scientific Basis for Recovering the Mariana Crow. Dr. Beissinger earned a B.S. and M.S. in zoology at Miami University and a Ph.D. in natural resource ecology at the University of Michigan.

Linda K. Blum is research associate professor in the Department of Environmental Sciences at the University of Virginia. Her current research projects include study of mechanisms controlling bacterial community abundance, productivity, and structure in tidal marsh creeks; impacts of microbial processes on water quality; organic matter accretion in salt marsh sediments; and rhizosphere effects on organic matter decay in anaerobic sediments. Dr. Blum was previously the chair of the NRC's Panel to Review the Critical Ecosystem Studies Initiative and a member of the Committee on Restoration of the Greater Everglades Ecosystem and the first Committee on Independent Scientific Review of Everglades Restoration Progress. She earned a B.S. and M.S. in forestry from Michigan Technological University and a Ph.D. in soil science from Cornell University.

Donald F. Boesch is a professor of marine science and president of the University of Maryland Center for Environmental Science. Dr. Boesch is a biological oceanographer who has conducted research in coastal and continental shelf environments along the Atlantic Coast and in the Gulf of Mexico, eastern Australia, and the East China Sea. He has served as science adviser to many state and federal agencies and regional, national, and international programs. In 1980, Dr. Boesch was appointed as the first executive director of the Louisiana Universities Marine Consortium, where he was also a professor of marine science at Louisiana State University. Earlier he was a Fulbright Postdoctoral Fellow at the University of Queensland and subsequently served on the faculty of the Virginia Institute of Marine Science. Dr. Boesch is a member of the NRC's Ocean Studies Board and served on the Committee to Assess the U.S. Army Corps of Engineers Methods of Analysis and Peer Review for Water Resources Planning and the first Committee on Independent Scientific Review of Everglades Restoration Progress. He received his B.S. from Tulane University and Ph.D. from the College of William and Mary.

Frank W. Davis is professor at the Donald Bren School of Environmental Science & Management at the University of California at Santa Barbara. His research interests are in landscape ecology and conservation planning. Dr. Davis' ecological research has focused on the coupling of ecological pattern and process in California oak woodland and chaparral ecosystems. His research in conservation planning and reserve network design includes a long involvement with the U.S. Gap Analysis Program creating spatial databases to evaluate the conservation status of plant communities and wildlife species. Dr. Davis has worked in California on several statewide and regional conservation efforts, directing the California Gap Analysis, as a science team member on the Sierra Nevada Ecosystem Project, and as the research team leader for the California Legacy Project. A fellow in the Aldo Leopold Leadership Program and a Trustee of the Nature Conservancy of California, he increasingly works at the interface between environmental science and policy. Dr. Davis served on the NRC's Committee on the Restoration of the Greater Everglades Ecosystem. He earned his Ph.D. in geography from Johns Hopkins University in 1982.

Charles T. Driscoll is University Professor in the Department of Civil and Environmental Engineering at Syracuse University where he also serves as the director of the Center for Environmental Systems Engineering. His teaching and research interests are in the area of environmental chemistry, biogeochemistry, and environmental quality modeling. A principal research focus has been the response of forest, aquatic, and coastal ecosystems to disturbance, including air pollution, land use change, and elevated inputs of nutrients and mercury. Dr. Driscoll is currently the principal investigator of the National Science Foundation's Long Term Ecological Research Network's project at the Hubbard Brook Experimental Forest in New Hampshire. He was a member of the NRC's Panel on Process of Lake Acidification and the Committee on the Collaborative Large-scale Engineering Analysis Network for Environmental Research (CLEANER). Dr. Driscoll received his B.S. degree in civil engineering from the University of Maine and his M.S. and Ph.D. in environmental engineering from Cornell University.

Joan G. Ehrenfeld is a professor in the Department of Ecology, Evolution, and Natural Resources at Rutgers University and is also the director of the New Jersey Water Resources Research Institute, a federally funded program of water-related research and outreach. Her research is in the area of wetland ecology and ecosystems ecology and focuses on plant-soil interactions. Dr. Ehrenfeld's current research includes several studies of the interactions of exotic invasive plants and forest soils and studies of nitrogen cycling in forested wetlands affected by urbanization. Her work covers a wide variety of ecosystems in New Jersey, including the Pinelands, the hardwood forests of the northwestern hills, and the red maple swamps of the northeastern Piedmont province. Dr. Ehrenfeld is a member of the Water Science and Technology Board and was a member of the

Committee on Assessment of Water Resources Research. She received her B.A. in biology from Columbia University, her M.A. in biology from Harvard University, and her Ph.D. in biology from City University of New York.

Chris T. Hendrickson is the Duquesne Light Company Professor of Engineering and codirector of the Green Design Institute at Carnegie Mellon University. His research, teaching, and consulting are in the general area of engineering planning and management, including design for the environment, system performance, project management, finance, and computer applications. Dr. Hendrickson's current research projects include environmental life-cycle assessment methodology development, heavy metal material flow analysis, infrastructure requirements for alternative transportation fuels, and sustainable infrastructure. He has served on several NRC committees including the first Committee on Independent Scientific Review of Everglades Restoration Progress, the Committee on Assessing the Results of External Independent Reviews for U.S. Department of Energy Projects, and the Committee for Review of the Project Management Practices Employed on the Boston Central Artery ("Big Dig") Project. Dr. Hendrickson holds B.S. and M.S. degrees from Stanford University, a master of philosophy degree in economics from Oxford University, and a Ph.D. from the Massachusetts Institute of Technology.

William P. Horn is a partner in the law firm of Birch, Horton, Bittner and Cherot in Washington, DC. Prior to entering private practice, Mr. Horn served in a variety of congressional and executive posts including as Assistant Secretary of the Interior for Fish, Wildlife, and Parks, and as Deputy Under Secretary of the Interior with responsibilities for western water rights negotiations, international fishery negotiations, and Alaska programs. He specializes in natural resources law and has expertise in land acquisition and appraisal, wildlife law including the Endangered Species Act and the Migratory Bird Treaty Act, National Park concessions, Forest Service matters, recreational permits, and other public land and related regulatory matters. Mr. Horn is a member of the Bar of the District of Columbia, a recipient of the Interior Department's Outstanding Services Award, and the International Academy of Trial Lawyers Advocacy Award. He earned his J.D. in 1983 from American University.

Wayne C. Huber is professor in the Department of Civil, Construction, and Environmental Engineering at Oregon State University. Prior to moving to Oregon State in 1991, he served 23 years on the faculty of the Department of Environmental Engineering Sciences at the University of Florida where he engaged in several studies involving the hydrology and water quality of South Florida regions. Dr. Huber's technical interests are principally in the areas of surface hydrology, stormwater management, non-point-source pollution, and transport processes related to water quality. He is one of the original authors of the Environmental Protection Agency's Storm-Water Management Model. Dr. Huber is a former member of the Committee on Restoration of the Greater Everglades Ecosystem and served as chair of the first Committee on Independent Scientific Review of Everglades Restoration Progress. He holds a B.S. in engineering from the California Institute of Technology and an M.S. and Ph.D. in civil engineering from the Massachusetts Institute of Technology.

David H. Moreau is professor in the Departments of City and Regional Planning and Environmental Sciences and Engineering at the University of North Carolina at Chapel Hill where he teaches water resources planning and regional environmental planning. His research interests include analysis, planning, financing, and evaluation of water resource and related environmental programs. Dr. Moreau is engaged in water resources planning at the local, state, and national levels. He has chaired or served on several NRC committees, most recently as a member of the Committee on New Orleans Regional Hurricane Protection Projects and the Committee to Review the Lake Ontario–St. Lawrence River Studies. Dr. Moreau serves as chairman of the North Carolina Environmental Management Commission, the state's regulatory commission for water quality, air quality, and water allocation. He received his B.S. and M.S. degrees from Mississippi State University and North Carolina State University, respectively, and his Ph.D. degree from Harvard University.

Jean-Yves Parlange (NAE) is professor in the Department of Biological and Environmental Engineering at Cornell University where his research centers on the application of mathematics to agricultural and environmental problems. His current projects focus on water movement in porous media, solute transport in soils, surface and subsurface hydrology, erosion and sediment transport, and similarity solutions of the nonlinear diffusion equation. The goal of Dr. Parlange's work is to provide analytical descriptions of complex problems that are accurate and can be used for practical management purposes. He was elected to the National Academy of Engineering in 2006 for fundamental contributions to the formulation of water flow and solute transport in soils and groundwater. Dr. Parlange earned his Ph.D. from Brown University.

K. Ramesh Reddy is graduate research professor and chair of the Department of Soil and Water Science at the University of Florida. His research areas include soil quality, ecological indicators, wetlands, and aquatic systems. Dr. Reddy investigates biogeochemical cycling of nutrients (including redox-related processes) in natural ecosystems—including wetlands, shallow lakes, estuaries, and constructed wetlands—and he develops biogeochemical indicators to evaluate changes in ecosystem functions. He is a member of the U.S. National Committee for Soil Sciences in the National Academy's Policy and Global Affairs Division. Dr. Reddy earned his Ph.D. in agronomy and soil science from Louisiana State University and Agricultural and Mechanical College in 1976.

STAFF

Stephanie E. Johnson, study director, is a senior program officer with the Water Science and Technology Board. Since joining the NRC in 2002, she has served as study director for seven committees, including the Panel to Review the Critical Ecosystem Studies Initiative and the Committee on Advancing Desalination Technology Research. She has also worked on NRC studies on contaminant source remediation, the disposal of coal combustion wastes, and water security. Dr. Johnson received her B.A. from Vanderbilt University in chemistry and geology, and her M.S. and Ph.D. in environmental sciences from the University of Virginia on the subject of pesticide transport and microbial bioavailability in soils.

David J. Policansky is a scholar and director of the Program in Applied Ecology and Natural Resources in the Board on Environmental Studies and Toxicology. He earned a Ph.D. in biology from the University of Oregon. Dr. Policansky has directed approximately 35 National Research Council studies and his areas of expertise include genetics; evolution; ecology, including fishery biology; natural resource management; and the use of science in policy making.

Dorothy K. Weir is a research associate with the Water Science and Technology Board. She has worked on a number of studies including Water System Security Research and Colorado River Basin Water Management. Ms. Weir has also served as a study director for the Committee on Collaborative Large-scale Engineering Analysis Network for Environmental Research and the Committee on the Review of the Water and Environmental Research Systems Network. She received a B.S. in biology from Rhodes College in Memphis, Tennessee and an M.S. degree in environmental science and policy from John Hopkins University. She joined the NRC in 2003.